T0094327

Introduction to Quantum Cryptography

This book offers an accessible and engaging introduction to quantum cryptography, assuming no prior knowledge in quantum computing. Essential background theory and mathematical techniques are introduced and applied to the analysis and design of quantum cryptographic protocols. The title explores several important applications such as quantum key distribution, quantum money, and delegated quantum computation, while also serving as a self-contained introduction to the field of quantum computing. With frequent illustrations and simple examples relevant to quantum cryptography, this title focuses on building intuition and challenges readers to understand the basis of cryptographic security. Worked examples and mid-chapter exercises allow readers to extend their understanding, and in-text quizzes, end-of-chapter homework problems, and recommended further reading reinforce and broaden understanding. Online resources available to instructors include interactive computational problems in Julia, videos, lecture slides, and a fully worked solutions manual.

THOMAS VIDICK is Professor in the Department of Computer Science and Applied Mathematics at the Weizmann Institute of Science in Israel. He received his PhD from UC Berkeley in 2011, followed by two years of postdoctoral research at the Massachusetts Institute of Technology, where he worked on quantum complexity and cryptography. He joined the Computer Science Faculty at the California Institute of Technology in 2014, and became Professor in 2018. In 2022 he took leave from CalTech to start his current position at the Weizmann Institute. Professor Vidick is best known for his research on device independence, including the first security proof of device-independent quantum key distribution. In 2019 he received a Presidential Early Career Award, and in 2021 was named a Simons Investigator at CalTech. In 2023 he received the Michael and Sheila Held prize from the National Academy of Sciences for his work in quantum cryptography and quantum complexity.

STEPHANIE WEHNER is Antoni van Leeuwenhoek Professor at Delft University of Technology, QuTech, and Director of the Quantum Internet Alliance. A member of the Royal Netherlands Academy of Arts and Sciences, she is also a co-founder of QCRYPT, the largest international conference in quantum cryptography. In a former life, she worked as a professional hacker. Her research is focused on manipulating the laws of quantum mechanics to construct better information networks and computer systems. Together with the Quantum Internet Alliance she is working on realizing a large-scale quantum computer network.

"If you are intrigued by the prospects of quantum cryptography but not yet familiar with the formalism behind it, then this book is the perfect starting point for you. It playfully introduces the most important concepts in modern quantum cryptography and at the same time gently but purposefully helps you discover the mathematical framework required to make formal statements."
Marco Tomamichel, National University of Singapore

"Vidick and Wehner cover quantum cryptography in its full beauty and depth. Packed with enlightening examples and comprehensive exercises, this book will likely become an indispensable companion next time I hold lectures on the subject."
Renato Renner, ETH Zurich

"Thomas Vidick and Stephanie Wehner take readers on an insightful exploration of the full landscape of quantum cryptography, skillfully weaving together theory and applications and providing pedagogical quizzes and exercises. The mathematical formalism is rigorous yet approachable, making this book an excellent introduction to this captivating area."
Anne Broadbent, University of Ottawa

"I recommend exploring the intriguing world of quantum cryptography with this definitive guide. Stephanie and Thomas, with expert precision, clarify the conceptual and mathematical complexities of quantum mechanics and secure communication, bringing clarity where confusion often reigns. It is an essential resource for anyone involved in quantum information science, from rookies to veterans."
Artur Ekert, University of Oxford and National University of Singapore

Introduction to
Quantum Cryptography

Thomas Vidick
California Institute of Technology, USA
Weizmann Institute of Science, Israel

Stephanie Wehner
Delft University of Technology, The Netherlands

CAMBRIDGE
UNIVERSITY PRESS

Shaftesbury Road, Cambridge CB2 8EA, United Kingdom

One Liberty Plaza, 20th Floor, New York, NY 10006, USA

477 Williamstown Road, Port Melbourne, VIC 3207, Australia

314–321, 3rd Floor, Plot 3, Splendor Forum, Jasola District Centre, New Delhi – 110025, India

103 Penang Road, #05–06/07, Visioncrest Commercial, Singapore 238467

Cambridge University Press is part of Cambridge University Press & Assessment,
a department of the University of Cambridge.

We share the University's mission to contribute to society through the pursuit of
education, learning and research at the highest international levels of excellence.

www.cambridge.org
Information on this title: www.cambridge.org/highereducation/isbn/9781316515655
DOI: 10.1017/9781009026208

First published 2024

A catalogue record for this publication is available from the British Library

A Cataloging-in-Publication data record for this book is available from the Library of Congress

ISBN 978-1-316-51565-5 Hardback

Additional resources for this publication at www.cambridge.org/vidick-wehner

Contents

Preface

Welcome! We are excited to introduce you to one of our favorite topics: quantum cryptography. With this book, we would like to provide you with all the basics needed to understand and analyze fundamental quantum cryptographic protocols, and even motivate you to design your own. Most of all, we hope you have fun exploring the adventures – and occasional mishaps – of our chief protagonists, Alice and Bob, as they attempt to use quantum communication to solve cryptographic challenges.

This book is meant to provide a textbook introduction to the theory of quantum cryptography, presented in an engaging yet largely mathematically precise manner. Our writing is voluntarily playful and places a high emphasis on developing intuition, a style we find to be appropriate to the discussion of cryptographic tasks; we will frequently invite you to devise cryptographic schemes and break them. However, the formalism is introduced in mathematical detail, and proofs are included when they are useful to build understanding. Whenever we do provide a proof, we give one that requires minimal background knowledge and fosters physical intuition, rather than providing the most sharp statements obtained using advanced quantum information tools. End of chapter notes provide references to the literature, where more information can be found. We emphasize that the book is meant as an engaging introduction, not as a reference manual.

We intend the book primarily for undergraduate students, as well as graduate students with a strong background in mathematics, physics, or computer science, but not necessarily any prior knowledge of quantum computing or cryptography. We do assume, however, that students have taken undergraduate courses on linear algebra and statistics before embarking on reading this book. The book can be used to teach a one-semester course on quantum cryptography (see examples below), or be used for self-study.

For the purpose of this book, quantum cryptography can be defined as the study of those cryptographic tasks that can be implemented using quantum hardware, i.e. devices capable of manipulating quantum information. The most prominent such task, quantum key distribution, acts as a focal point throughout the book. Other tasks, such as quantum commitments, quantum money, and others, are discussed. The reader should be warned that this is not a book about post-quantum cryptography, and neither quantum algorithms nor classical cryptographic schemes that are resistant to them are meant to be discussed in any detail.

Organization

The book is organized in a progressive manner, starting from the basic formalism of quantum communication, building intuition about properties such as no-cloning and uncertainty principles, and applying these properties to the design and analysis of quantum cryptosystems. In more detail, the book is composed of 12 chapters, plus an introductory chapter, not counting this preface:

Chapter 1 forms a first introduction to quantum information, assuming a background in linear algebra and probability. It is not meant to replace a thorough introduction to quantum information, for which many good textbooks exist, but aims to provide sufficient knowledge to follow the rest of this book. In this chapter we give a first application of quantum information by studying a quantum random number generator. A reader already familiar with quantum information may skip this chapter, and it may be omitted in a course series where students have already taken an introduction to quantum information.

Chapter 2 introduces several quantum tools that are often not discussed in introductory courses in quantum information but are crucial for cryptography, such as density matrices and the partial trace. This chapter also introduces the notion of secure communication and contains a first quantum protocol, namely the quantum one-time pad. Chapter 2 assumes students are familiar with the material covered in Chapter 1.

Chapter 3 introduces one of the very first ideas in the development of quantum cryptography, quantum money. This chapter assumes the students are familiar with the material covered in Chapters 1 and 2.

Chapter 4 introduces the concept of quantum entanglement, as well as fundamental notions such as Bell nonlocality and the monogamy of entanglement. These notions play an important role in understanding quantum key distribution in later chapters. Chapter 4 assumes the students are familiar with the material covered in Chapters 1 and 2.

Chapter 5 discusses how to quantify information in quantum cryptography, providing a cornerstone for security definitions. It also introduces a tripartite uncertainty game that will play an important role in our security analysis of quantum key distribution. This chapter assumes the students are familiar with the material covered in Chapters 1, 2, and 4.

Chapter 6 introduces the concept of privacy amplification and its realization using randomness extractors. This chapter assumes familiarity with the material covered in Chapters 1 and 2.

Chapter 7 formally introduces the task of key distribution, and gives examples of key distribution protocols in simplified settings. This chapter assumes the students are familiar with the material covered in Chapters 1, 2, 5, and 6.

Chapter 8 introduces the BB'84 protocol and provides a sketch of its security analysis, using the purified protocol and uncertainty relations. Chapter 8 assumes the reader is familiar with the material covered in Chapters 1, 2, 4, 5, 6, and 7.

Chapter 9 discusses device-independent quantum key distribution. A partial security analysis, assuming independent rounds, is made using the notion of a guessing game. This chapter assumes the students know the material presented in Chapters 1, 2, 4, 5, 6, 7, and 8.

Chapter 10 examines the use of quantum communication for two-party cryptographic tasks such as coin flipping, bit commitment, and oblivious transfer. Chapter 10 assumes familiarity with Chapters 1, 2, 4, and 5.

Chapter 11 introduces the use of physical assumptions in combination with quantum communication to solve cryptographic challenges such as bit commitment and oblivious transfer. This chapter assumes the students have studied Chapters 1, 2, 5, 6, and 10.

Chapter 12 discusses security notions for quantum encryption. It assumes familiarity with the material covered in Chapters 1 and 2.

Chapter 13 introduces the problem of delegating quantum computations to a quantum server in the cloud. This chapter assumes familiarity with the material covered in Chapters 1, 2, and 5.

Suggestions for Teaching

This book can be used to teach a variety of classes for undergraduate and graduate students. The following are suggestions organized by duration of the course and background of the students. We provide approximate time estimates based on our experience in teaching the material. More, or significantly less, time could be appropriate depending on the level and prior knowledge of the students, as well as the level of detail with which one wishes to cover the material.

A First Introduction to Quantum Information and Cryptography
This course is aimed at undergraduate students as an introduction to quantum information with some applications to quantum cryptography. The course would cover Chapters 1, 2, 3, and possibly 4, introducing quantum applications such as the quantum random number generator, the quantum one-time pad, and quantum money, in roughly nine interactive lectures of two hours each. If more time is available, parts of Chapters 5, 7, and 8 could be used to introduce quantum key distribution, omitting many of the formal details and proofs.

Introduction to Quantum Information and Quantum Key Distribution
Such a course is aimed at advanced undergraduate students with no prior knowledge of quantum information. The course introduces the students to the basic notions of quantum information, and works its way up to quantum key distribution and its security analysis. This could cover Chapters 1, 2, 4, 5, 7, and 8, assuming roughly 12 interactive lectures of two hours each. Depending on the level of the students, one might add device-independent quantum key distribution in Chapter 9.

Quantum Key Distribution
This course is aimed at advanced undergraduate students who are already familiar with quantum information through, for example, a class focusing mainly on quantum computing. Such a class would typically cover the materials covered in Chapter 1, but often not more advanced notions such as density matrices or the partial trace

introduced in Chapter 2. The course could cover Chapters 2, 4, 5, 7, 8, and 9, assuming roughly 10 interactive lectures of two hours each. Depending on the level or interest of the students, one might add any of the further topics in this book from Chapters 3, 10, 11, or 13.

Quantum Cryptography beyond Quantum Key Distribution
This course is aimed at advanced undergraduate or graduate students who are already familiar with quantum information and who have at least a passing familiarity with quantum key distribution (QKD). Its focus would be on quantum cryptography beyond QKD. The course could cover Chapters 2, 4, and 5 as necessary, and continue with quantum protocols beyond QKD using Chapters 3, 11, 12, and 13.

Quantum Cryptography
A course for graduate students could cover the entirety of this book in a linear fashion, going faster over the first two chapters. Depending on the students, these could also be assigned as preparatory reading, with the students using the quiz questions to test their understanding. The course could proceed more slowly over the technically more advanced sections, such as the end of Chapters 6 and 9 and Chapter 13.

Resources

When the web icon shown here appears on the page, a supporting resource can be found on the website. Several resources accompany this book, and are available at www.cambridge.org/vidick-wehner. These include:

Julia sheets: Online interactive exercise sheets in the Julia language to enable you to explore and play with the material presented here.
Videos: Videos from our online MOOC that match the relevant portions of the text.
End of chapter notes: References to the literature mentioned in a chapter.
Quizzes: Throughout the book you are encouraged to test yourself using short quizzes. The answers to the quizzes may be found at the end of each chapter.
Exercises: Several longer exercises are provided throughout the text to challenge yourself. Additional homework problems may be found at the end of each chapter.
Solution manual: A solution manual for the problems given at the end of each chapter.
Slide materials: Materials are available that may be used for slides for a course using this book.

We also recommend two additional resources to explore some of the materials in this book, and expand your knowledge further. The Quantum Protocol Zoo (https://wiki.veriqloud.fr/index.php?title=Main_Page) provides an overview of many cryptographic protocols not included in this book. The Quantum Network Explorer (http://quantum-network.com) lets you program quantum cryptographic protocols, and also provides a graphical interface to help you understand the effects of hardware imperfections on some of the protocols we discuss in this book.

Origins and Acknowledgments

This book grew over the years out of our offline and online classes, chiefly our MOOC QuCryptoX "Quantum cryptography" initially offered on edX in Fall 2017. Lecture notes created for the MOOC were refined over multiple years of teaching the course in a hybrid format in Delft and in Caltech, and formed the basis for this now expanded book.

Since this book evolved from our lecture notes, a great many people have made contributions to the book over the years. Generations of teaching assistants for both offline and online classes have made diverse contributions to developing these notes, exercises, figures, and corrections. In roughly chronological order they are, from TU Delft, Nelly Ng, Jed Kaniewski, Corsin Pfister, Willem Hekman, Jeremy Ribeiro, Kenneth Goodenough, Filip Rozpędek, Jonas Helsen, Victoria Lipinska, Guus Avis, Francisco Horta Ferreira da Silva, Alvaro G. Iñesta, Scarlett Gauthier, Ravisankar Ashok Kumar Vattekkat, and Hana Jirovska. We also thank guest lecturer David Elkouss for his contributions to explaining information reconciliation. From Caltech, they are Chinmay Nirkhe, Jalex Stark, Andrea Colandangelo, and Tina Zhang. Last but not least, we would like to thank all the students who have taken our classes over the years and provided useful feedback that helped us to improve this book.

Chapter title images are from Hanson Lab, QuTech, Delft University of Technology.

1

Background Material

In this chapter, we give a gentle introduction to quantum information. We will introduce the basics of working with quantum bits – qubits – and examine how to write down simple measurements at the mathematical level. At the end, we will apply our newfound knowledge to see how we can use quantum information to design a deterministic machine that produces true randomness – a feat that is impossible classically. Such a machine is called a quantum random number generator. Quantum random number generators are one of the first commercially available applications of quantum information to cryptography, and it is exciting that we can describe their underlying principle in our first chapter! We conclude with a brief overview of the state of the art of quantum communication technologies needed to realize quantum crytographic protocols in the real world.

For this chapter, and throughout the book, we assume that you already know how to perform calculations involving complex numbers, and that you are familiar with basic notions from linear algebra such as finite-dimensional vector spaces, vectors, and matrices. For suggestions on how to pick up the necessary background, as well as additional resources to learn about how qubits can be realized physically, see the chapter notes at the end of the chapter.

1.1 Mathematical Notation

Let us start by recalling common notation that we use throughout this book. We will use $\mathbb{C}$ to denote the field of complex numbers, and write $i = \sqrt{-1} \in \mathbb{C}$ for the imaginary unit. Remember that any complex number c can be written as $c = a + ib \in \mathbb{C}$ for some real numbers $a, b \in \mathbb{R}$. In this context, we call a the *real part* of c, and b the *imaginary part* of c respectively. The *complex conjugate* of a complex number $c \in \mathbb{C}$ can be written as $c^* = a - ib$. For a complex number, we can define its *absolute value* (sometimes also called *modulus*) as follows.

> **Definition 1.1.1 (Absolute value of a complex number).** *Consider a complex number* $c \in \mathbb{C}$ *expressed as* $c = a + ib$, *where* $a, b \in \mathbb{R}$. *The* absolute value *of* c *is given by*
> $$|c| := \sqrt{c^* c} = \sqrt{a^2 + b^2}. \tag{1.1}$$

For example, the absolute value of $c = 1 + i2$ is $|c| = \sqrt{1^2 + 2^2} = \sqrt{5}$.

Remember that a vector space V over $\mathbb{C}$ is a collection of vectors with complex coefficients, such that V contains the all 0 vector and is stable under vector addition and multiplication by scalars (in this case, the complex numbers). In quantum information vectors are written in a special way known as the "bra-ket" or "Dirac" notation. While it may look a little cumbersome at first, it turns out to provide a convenient way of dealing with the many operations that we will perform with such vectors. To explain the Dirac notation, let us start with two examples. We write $|v\rangle \in \mathbb{C}^2$ to denote a vector in a 2-dimensional vector space $V = \mathbb{C}^2$. For example,

$$|v\rangle = \begin{pmatrix} 1+i \\ 0 \end{pmatrix}. \tag{1.2}$$

The vector $|v\rangle$ is called a "ket" vector. The "bra" of this vector is its conjugate transpose, which looks like

$$\langle v| := ((|v\rangle)^*)^T = \begin{pmatrix} (1+i)^* \\ 0^* \end{pmatrix}^T = (1-i \quad 0). \tag{1.3}$$

Here and throughout the book we use the notation ":=" to indicate a definition. The general definition of the "bra-ket" notation is as follows.

Definition 1.1.2 (bra-ket notation). *A* ket, *denoted* $|\cdot\rangle$, *represents a d-dimensional column vector in the complex vector space* $\mathbb{C}^d$. *(The dimension d is usually left implicit in the notation.) A* bra, *denoted* $\langle\cdot|$, *is a d-dimensional row vector equal to the complex conjugate of the corresponding ket, namely*

$$\langle\cdot| = (|\cdot\rangle^*)^T, \tag{1.4}$$

where * *denotes the entry-wise conjugate and T denotes the transpose.*

We will frequently use the "dagger" notation for the conjugate-transpose: for any vector $|u\rangle \in \mathbb{C}^d$,

$$|u\rangle^\dagger := (|u\rangle^*)^T = \langle u|.$$

This notation extends to matrices in the natural way, $A^\dagger := (A^*)^T$.

In quantum information we very often need to compute the inner product of two vectors. The "bra-ket" notation makes this operation very convenient.

Definition 1.1.3 (Inner product). *Given two d-dimensional vectors*

$$|v_1\rangle = \begin{pmatrix} a_1 \\ \vdots \\ a_d \end{pmatrix} \quad \text{and} \quad |v_2\rangle = \begin{pmatrix} b_1 \\ \vdots \\ b_d \end{pmatrix}, \tag{1.5}$$

their inner product *is given by* $\langle v_1|v_2\rangle := \langle v_1| \cdot |v_2\rangle = \sum_{i=1}^d a_i^* b_i$.

Note that the inner product of two vectors $|v_1\rangle, |v_2\rangle \in \mathbb{C}^d$ is in general a complex number. Later on, we will see that the modulus squared of the inner product $|\langle v_1|v_2\rangle|^2$

has a physical significance when it comes to measuring qubits. As an example, let us consider the inner product of the vector $|v\rangle$ given in (1.2) and

$$|w\rangle = \begin{pmatrix} 2i \\ 3 \end{pmatrix}. \tag{1.6}$$

We have

$$\langle v|w\rangle = (1-i \quad 0)\begin{pmatrix} 2i \\ 3 \end{pmatrix} = (1-i)\cdot 2i + 0 \cdot 3 = 2i - 2i^2 = 2 + 2i. \tag{1.7}$$

Exercise 1.1.1 Show that for any two vectors $|v_1\rangle$ and $|v_2\rangle$,

$$|\langle v_1|v_2\rangle|^2 = \langle v_1|v_2\rangle\langle v_2|v_1\rangle.$$

[Hint: first prove the relation $(\langle v_1|v_2\rangle)^* = \langle v_2|v_1\rangle.$]

It is convenient to have a notion of the "length" of a vector. For this we use the *Euclidean norm.*

Definition 1.1.4 (Norm of a ket vector). *Consider a ket vector*

$$|v\rangle = \begin{pmatrix} a_1 \\ \vdots \\ a_d \end{pmatrix}. \tag{1.8}$$

The length, *or norm, of* $|v\rangle$ *is given by*

$$\||v\rangle\|_2 := \sqrt{\langle v|v\rangle} = \sqrt{\sum_{i=1}^{d} a_i^* a_i} = \sqrt{\sum_{i=1}^{d} |a_i|^2}. \tag{1.9}$$

If $\||v\rangle\|_2 = 1$ *we say that* $|v\rangle$ *has norm* 1 *or simply that* $|v\rangle$ *is normalized.*

Example 1.1.1 Consider a ket $|v\rangle = \frac{1}{2}\begin{pmatrix} 1+i \\ 1-i \end{pmatrix} \in \mathbb{C}^2$. The corresponding bra is given by $\langle v| = \frac{1}{2}(1-i \quad 1+i)$, and the norm of $|v\rangle$ is

$$\sqrt{\langle v|v\rangle} = \sqrt{\frac{1}{4}\cdot 2 \cdot (1+i)(1-i)} = \sqrt{\frac{1}{2}(1+i-i-i^2)} = \sqrt{\frac{1}{2}\cdot 2} = 1. \tag{1.10}$$

∎

You should be familiar with the notion of an orthonormal basis for a vector space V from linear algebra. We often write such a basis as $\mathcal{B} = \{|b\rangle\}_b$, which is shorthand for $\{|0\rangle, |1\rangle, \ldots, |d-1\rangle\}$, where d is the dimension of the vector space V in which the kets live, and is often implicit.[1] The condition of being orthonormal can be expressed

---

1 By convention, in quantum information bases are usually indexed starting at 0, rather than 1. So the standard orthonormal basis of $\mathbb{C}^2$ will be written $\{|0\rangle, |1\rangle\}$.

succinctly as $\langle b|b'\rangle = \delta_{bb'}$ for all $b,b' \in \{0,\ldots,d-1\}$, where δ_{ab} is the *Kronecker symbol*, defined as $\delta_{ab} = 0$ if $a \neq b$ and $\delta_{ab} = 1$ for $a = b$. That is, the different vectors of the basis are orthogonal, and are each normalized to have length 1. Recall that if $\mathcal{B}$ is the basis for a vector space V, then any vector $|v\rangle \in V$ can be expressed as $|v\rangle = \sum_b c_b |b\rangle$, for some coefficients $c_0,\ldots,c_{d-1} \in \mathbb{C}$.

1.2   What Are Quantum Bits?

We are all familiar with the notion of a "bit" in classical computing: mathematically, a bit is a value $b \in \{0,1\}$ that represents some information that is stored and manipulated by an algorithmic procedure. Physically, classical bits can be realized in hardware in many different ways, as long as the two physical states corresponding to "0" and "1" can be distinguished sufficiently clearly. For example, when transmitting data over a fiber-optic cable, the presence of a light pulse can be used to represent a "1" and its absence a "0". Typically, computing and communication systems need more than a single bit to operate, and one talks about a *string of bits* $b = (b_1,\ldots,b_n) \in \{0,1\}^n$.

How do quantum bits differ from classical bits? To define a quantum bit, let us start by writing classical bits somewhat differently. Instead of writing them as "0" and "1", we associate a 2-dimensional vector to each of them as

$$0 \to |0\rangle = \begin{pmatrix} 1 \\ 0 \end{pmatrix} \quad \text{and} \quad 1 \to |1\rangle = \begin{pmatrix} 0 \\ 1 \end{pmatrix}. \tag{1.11}$$

The main difference between quantum bits and classical bits is that while a physical classical bit can be in only one of the two states $|0\rangle$ or $|1\rangle$, a qubit can be in any state of the form $\alpha|0\rangle + \beta|1\rangle$ with $\alpha,\beta \in \mathbb{C}$ such that $|\alpha|^2 + |\beta|^2 = 1$. We often say that the quantum bit is in a "superposition" of $|0\rangle$ and $|1\rangle$, with "amplitudes" α and β. As we will see later, such amplitudes are directly related to the probabilities of obtaining certain outcomes when measuring the qubit, and the demand that $|\alpha|^2 + |\beta|^2 = 1$ is needed to ensure that these probabilities add up to 1. Since "quantum bit" is somewhat long, researchers use the term "qubit" to refer to a quantum bit. To recap, a qubit is a normalized vector $|v\rangle \in \mathbb{C}^2$, and the vector space $\mathbb{C}^2$ is also known as the *state space* of the qubit.

Physically, qubits can be realized in many different ways. In the context of quantum communication, $|0\rangle$ and $|1\rangle$ can be realized, for example, by the presence and absence of a photon, in direct analogy to the example from classical communication given above. Amazingly, it is also possible to create a *superposition* between the presence and absence of a photon, and thus realize a qubit.

Definition 1.2.1 (Qubit). *A pure state of a qubit can be represented by a 2-dimensional ket vector* $|\psi\rangle \in \mathbb{C}^2$,

$$|\psi\rangle = \alpha|0\rangle + \beta|1\rangle, \quad \text{where} \quad \alpha,\beta \in \mathbb{C} \quad \text{and} \quad |\alpha|^2 + |\beta|^2 = 1. \tag{1.12}$$

Whenever the condition on α and β is satisfied we say that $|\psi\rangle$ is normalized. The complex numbers α and β are called amplitudes *of $|\psi\rangle$.*

You probably noticed the use of the word "pure" in the definition. This is because there is a more general notion of qubit, called a "mixed" state, which we introduce in the next chapter.

Example 1.2.1 Some examples of qubits that we will frequently encounter in quantum cryptography are

$$|+\rangle := \frac{1}{\sqrt{2}}(|0\rangle + |1\rangle) , \quad \text{and} \quad |-\rangle := \frac{1}{\sqrt{2}}(|0\rangle - |1\rangle) . \tag{1.13}$$

∎

QUIZ 1.2.1 *Is $|\psi\rangle = \frac{1}{4}|0\rangle + \frac{1}{8}|1\rangle$ a valid quantum state?*

(a) *Yes*
(b) *No*

QUIZ 1.2.2 *Is $|\psi\rangle = \begin{pmatrix} 1 \\ 0 \end{pmatrix}$ a valid quantum state?*

(a) *Yes*
(b) *No*

Exercise 1.2.1   Verify that for all real values of θ, $|\psi_\theta\rangle = \cos(\theta)|0\rangle + \sin(\theta)|1\rangle$ is a valid pure state of a qubit.

Throughout the book we mostly focus on encoding information in qubits. In general, quantum information can also be encoded in higher-dimensional systems. Indeed, one can define a qu*d*it as follows.

Definition 1.2.2 (Qudit). *A pure state of a* qudit *can be represented as a d-dimensional ket vector $|\psi\rangle \in \mathbb{C}^d$,*

$$|\psi\rangle = \sum_{i=0}^{d-1} \alpha_i |i\rangle , \quad \text{where} \quad \forall i, \; \alpha_i \in \mathbb{C} \text{ and } \sum_{i=0}^{d-1} |\alpha_i|^2 = 1. \tag{1.14}$$

In our definition of qubits we started from a way to write classical bits as vectors $|0\rangle$ and $|1\rangle$. Note that these two vectors are orthonormal, which in the quantum notation can be expressed as $\langle 1|0\rangle = 0$ and $\langle 1|1\rangle = \langle 0|0\rangle = 1$. These two vectors thus form a basis for $\mathbb{C}^2$, so that any vector $|v\rangle \in \mathbb{C}^2$ can be written as $|v\rangle = \alpha|0\rangle + \beta|1\rangle$ for some coefficients $\alpha, \beta \in \mathbb{C}$. This basis corresponding to "classical" bits is used so often that it carries a special name.

Definition 1.2.3 (Standard basis). *The* standard basis, *also known as the* computational basis, *of* $\mathbb{C}^2$ *is the orthonormal basis* $\mathcal{S} = \{|0\rangle, |1\rangle\}$ *where*

$$|0\rangle = \begin{pmatrix} 1 \\ 0 \end{pmatrix} \text{ and } |1\rangle = \begin{pmatrix} 0 \\ 1 \end{pmatrix}. \tag{1.15}$$

There are many other bases for $\mathbb{C}^2$. Another favorite basis is the Hadamard basis.

Definition 1.2.4 (Hadamard basis). *The* Hadamard basis *of* $\mathbb{C}^2$ *is the orthonormal basis* $\mathcal{H} = \{|+\rangle, |-\rangle\}$ *where*

$$|+\rangle = \frac{1}{\sqrt{2}}(|0\rangle + |1\rangle) = \frac{1}{\sqrt{2}}\begin{pmatrix} 1 \\ 1 \end{pmatrix} \quad \text{and} \quad |-\rangle = \frac{1}{\sqrt{2}}(|0\rangle - |1\rangle) = \frac{1}{\sqrt{2}}\begin{pmatrix} 1 \\ -1 \end{pmatrix}. \tag{1.16}$$

Let us verify that this is indeed an orthonormal basis using the "bra-ket" notation:

$$\langle+|+\rangle = \frac{1}{2}\begin{pmatrix} 1 & 1 \end{pmatrix}\begin{pmatrix} 1 \\ 1 \end{pmatrix} = \frac{1}{2} \cdot 2 = 1, \quad \implies \quad \sqrt{\langle+|+\rangle} = 1, \tag{1.17}$$

so $|+\rangle$ is normalized. A similar calculation gives that $|-\rangle$ is normalized as well. You may wish to verify that this normalization already follows from the more general Exercise 1.2.1, by observing that $|+\rangle = |\psi_{\pi/4}\rangle$ and $|-\rangle = |\psi_{3\pi/4}\rangle$ as defined there. Furthermore, the inner product

$$\langle+|-\rangle = \frac{1}{2}\begin{pmatrix} 1 & 1 \end{pmatrix}\begin{pmatrix} 1 \\ -1 \end{pmatrix} = 0, \tag{1.18}$$

so $|+\rangle$ and $|-\rangle$ are orthogonal to each other.

Exercise 1.2.2 Decompose the state $|1\rangle$ in the Hadamard basis. In other words, find coefficients α and β such that $|1\rangle = \alpha|+\rangle + \beta|-\rangle$. Verify that $|\alpha|^2 + |\beta|^2 = 1$. This reflects the fact that the formula for the length of a vector given in Definition 1.1.4 does not depend on the choice of the orthonormal basis.

1.3   Multiple Qubits

Classically we can write the state of two bits as a string "00", "01", and so forth. What is the state of two qubits? Proceeding as we did earlier, we can first associate a vector to each of the four possible strings of two classical bits $x_1, x_2 \in \{0, 1\}^2$. This gives us a mapping from two-bit strings to 4-dimensional vectors as

$$00 \to |00\rangle = \begin{pmatrix} 1 \\ 0 \\ 0 \\ 0 \end{pmatrix} \qquad\qquad 01 \to |01\rangle = \begin{pmatrix} 0 \\ 1 \\ 0 \\ 0 \end{pmatrix}$$

$$10 \to |10\rangle = \begin{pmatrix} 0 \\ 0 \\ 1 \\ 0 \end{pmatrix} \qquad\qquad 11 \to |11\rangle = \begin{pmatrix} 0 \\ 0 \\ 0 \\ 1 \end{pmatrix}$$

More generally, a pure state of two qubits can always be expressed as a normalized vector $|\psi\rangle \in \mathbb{C}^4$. Since the four vectors above form an orthonormal basis of $\mathbb{C}^4$, any such $|\psi\rangle$ has a decomposition as a linear combination of the four basis vectors:

$$|\psi\rangle = \alpha_{00}|00\rangle + \alpha_{01}|01\rangle + \alpha_{10}|10\rangle + \alpha_{11}|11\rangle .$$

In quantum-speak we say that $|\psi\rangle$ is a "superposition" of the four basis vectors, with "amplitudes" α_{00}, α_{01}, α_{10}, and α_{11}.

As a concrete example, let us consider a state $|\psi\rangle$ that is an equal superposition of all four standard basis vectors for the space of two qubits:

$$|\psi\rangle_{AB} = \frac{1}{2}|00\rangle + \frac{1}{2}|01\rangle + \frac{1}{2}|10\rangle + \frac{1}{2}|11\rangle$$

$$= \frac{1}{2}\begin{pmatrix} 1 \\ 0 \\ 0 \\ 0 \end{pmatrix} + \frac{1}{2}\begin{pmatrix} 0 \\ 1 \\ 0 \\ 0 \end{pmatrix} + \frac{1}{2}\begin{pmatrix} 0 \\ 0 \\ 1 \\ 0 \end{pmatrix} + \frac{1}{2}\begin{pmatrix} 0 \\ 0 \\ 0 \\ 1 \end{pmatrix}$$

$$= \frac{1}{2}\begin{pmatrix} 1 \\ 1 \\ 1 \\ 1 \end{pmatrix}. \tag{1.19}$$

The sum of four amplitudes $\frac{1}{2}$ squared is $4 \cdot \frac{1}{2^2} = 1$, therefore $|\psi\rangle$ is a valid two-qubit quantum state.

We can proceed analogously to define a pure state of n qubits, for $n = 1, 2, 3, \ldots$. To see how such a state can be represented we first look at the vector representation for multiple classical bits. There is a total of $d = 2^n$ strings of n bits. Each such string x can be associated to a basis vector $|x\rangle \in \mathbb{C}^d$, where x is 0 everywhere, except at the coordinate indexed by the integer $i \in \{0, \ldots, d-1\}$ of which x is the binary representation (specifically, $i = x_1 + 2x_2 + \cdots + 2^{n-1}x_n$). A general pure state of n qubits can then be expressed as

$$|\psi\rangle = \sum_{x \in \{0.1\}^n} \alpha_x |x\rangle , \tag{1.20}$$

with $\alpha_x \in \mathbb{C}$ and $\sum_x |\alpha_x|^2 = 1$. The numbers α_x are again called *amplitudes*. It is worth noticing that the dimension of the vector space $\mathbb{C}^{2^n}$ increases exponentially with the number n of bits. The space $\mathbb{C}^d$ with $d = 2^n$ is called the *state space of n qubits*. Analogously to the case of a single qubit, the basis given by the set of vectors $\{|x\rangle \mid x \in \{0,1\}^n\}$ is called the *standard* (or *computational*) basis.

Definition 1.3.1 (Standard basis for *n* qubits). *Consider the state space of n qubits* $\mathbb{C}^d$, *where* $d = 2^n$. *For each distinct string* $x \in \{0,1\}^n$, *associate with x the integer* $i \in \{0,1,2,\ldots d\}$ *of which it is the binary representation. The standard basis for* $\mathbb{C}^d$ *is the orthonormal basis* $\{|x\rangle\}_{x \in \{0,1\}^n}$, *where for* $x \in \{0,1\}^n$, $|x\rangle$ *is the d-dimensional vector*

$$|x\rangle = \begin{pmatrix} 0 \\ \vdots \\ 1 \\ \vdots \\ 0 \end{pmatrix} \longrightarrow i\text{-th position.} \tag{1.21}$$

An n-qubit pure state $|\psi\rangle \in \mathbb{C}^d$ *with* $d = 2^n$ *can be written as a superposition of standard basis vectors*

$$|\psi\rangle = \sum_{x \in \{0,1\}^n} \alpha_x |x\rangle, \qquad \text{where } \forall x, \alpha_x \in \mathbb{C} \text{ and } \sum_{x \in \{0,1\}^n} |\alpha_x|^2 = 1. \tag{1.22}$$

We look at two examples of two-qubit states. The first is so famous it carries a special name, and we will see it very frequently throughout the book.

Example 1.3.1 The two-qubit state known as the *EPR pair* is defined as:[2]

$$|\text{EPR}\rangle = \frac{1}{\sqrt{2}}(|00\rangle + |11\rangle) = \frac{1}{\sqrt{2}} \left(\begin{pmatrix} 1 \\ 0 \\ 0 \\ 0 \end{pmatrix} + \begin{pmatrix} 0 \\ 0 \\ 0 \\ 1 \end{pmatrix} \right) = \frac{1}{\sqrt{2}} \begin{pmatrix} 1 \\ 0 \\ 0 \\ 1 \end{pmatrix}, \tag{1.23}$$

which is an equal superposition between the vectors $|00\rangle$ and $|11\rangle$. ∎

It is a useful exercise to verify that the state $|\text{EPR}\rangle$ is normalized. For this we compute the inner product

$$\langle \text{EPR}|\text{EPR}\rangle = \frac{1}{\sqrt{2}}(\langle 00| + \langle 11|) \cdot \frac{1}{\sqrt{2}}(|00\rangle + |11\rangle)$$

$$= \frac{1}{2}(\underbrace{\langle 00|00\rangle}_{1} + \underbrace{\langle 00|11\rangle}_{0} + \underbrace{\langle 11|00\rangle}_{0} + \underbrace{\langle 11|11\rangle}_{1})$$

$$= \frac{1}{2} \cdot 2 = 1, \implies \sqrt{\langle \text{EPR}|\text{EPR}\rangle} = 1. \tag{1.24}$$

Example 1.3.2 Consider the two-qubit state

$$|\psi\rangle = \frac{1}{\sqrt{2}}(|01\rangle + |11\rangle) = \frac{1}{\sqrt{2}} \begin{pmatrix} 0 \\ 1 \\ 0 \\ 1 \end{pmatrix}. \tag{1.25}$$

---

2 The abbreviation EPR stands for Einstein, Podolsky, and Rosen. Later we will show that this state is "entangled."

For this state, the second qubit always corresponds to the bit 1. We will later see that this state is significantly different from $|\text{EPR}\rangle$. (Hint: it is not entangled!) ∎

QUIZ 1.3.1 *Let* $|\psi\rangle = \frac{1}{\sqrt{2}} \begin{pmatrix} 1 \\ 1 \\ 1 \\ 1 \end{pmatrix}$. *Is this a valid two-qubit state?*

(a) *Yes*
(b) *No*

1.4 Combining Qubits Using the Tensor Product

So far we have learned how to represent the state of one qubit, of two qubits, and more generally of any number n of qubits as a $d = 2^n$-dimensional vector. This normalized vector can be expressed as a linear combination of basis vectors associated with the n-bit strings.

Let us now imagine that we have two qubits, A and B, and that we can write the state of qubit A as $|\psi\rangle_A \in \mathbb{C}^2$ and the state of qubit B as $|\phi\rangle_B \in \mathbb{C}^2$ respectively. How can we find the vector that represents the state of both qubit A and qubit B at the same time? When talking about multiple qubits, we will often refer to A and B as "systems," or "registers"; these words are used interchangeably to designate abstract quantum systems A and B, which could represent physical quantum states situated in different physical locations. Later on, A and B might consist of more than one qubit, and correspond to quantum systems held by different participants such as Alice (A) and Bob (B). We will use AB to denote the joint quantum system, consisting of the qubit(s) of A and the qubit(s) of B. In general, we will use subscripts (here A and B) to denote this, e.g. vector $|\psi\rangle_A$ denotes the state of system A, and $|\phi\rangle_B$ the state of B. Note that from a mathematical standpoint there is no difference between $|0\rangle_A$ and $|0\rangle_B$: both are given by the same vector $|0\rangle = \begin{pmatrix} 1 \\ 0 \end{pmatrix}$.

Let us introduce a new piece of mathematics that allows us to write down the vector for the system AB using our knowledge of the vectors for A and B. The rule that will allow us to do this is known as the *tensor product* (sometimes also called the Kronecker product). For the example of two single-qubit states we know that it is always possible to express

$$|\psi\rangle_A = \alpha_A |0\rangle_A + \beta_A |1\rangle_A = \begin{pmatrix} \alpha_A \\ \beta_A \end{pmatrix}, \tag{1.26}$$

$$|\phi\rangle_B = \alpha_B |0\rangle_B + \beta_B |1\rangle_B = \begin{pmatrix} \alpha_B \\ \beta_B \end{pmatrix}. \tag{1.27}$$

The joint state $|\psi\rangle_{AB} \in \mathbb{C}^2 \otimes \mathbb{C}^2$ of both qubits is obtained as the tensor product of the individual vectors $|\psi\rangle_A$ and $|\phi\rangle_B$, which by definition evaluates to

$$|\psi\rangle_{AB} = |\psi\rangle_A \otimes |\phi\rangle_B = \begin{pmatrix} \alpha_A \\ \beta_A \end{pmatrix} \otimes |\psi\rangle_B = \begin{pmatrix} \alpha_A |\psi\rangle_B \\ \beta_A |\psi\rangle_B \end{pmatrix} = \begin{pmatrix} \alpha_A \alpha_B \\ \alpha_A \beta_B \\ \beta_A \alpha_B \\ \beta_A \beta_B \end{pmatrix}. \tag{1.28}$$

More generally, for quantum systems A and B that are larger than just one qubit, the definition of the tensor product is as follows.

Definition 1.4.1. *For vectors $|\psi_1\rangle \in \mathbb{C}^{d_1}$ and $|\psi_2\rangle \in \mathbb{C}^{d_2}$, their tensor product is the vector $|\psi_1\rangle \otimes |\psi_2\rangle \in \mathbb{C}^{d_1} \otimes \mathbb{C}^{d_2}$ given by*

$$|\psi_1\rangle \otimes |\psi_2\rangle = \begin{pmatrix} \alpha_1 \\ \vdots \\ \alpha_d \end{pmatrix} \otimes |\psi_2\rangle = \begin{pmatrix} \alpha_1 |\psi_2\rangle \\ \vdots \\ \alpha_d |\psi_2\rangle \end{pmatrix}. \tag{1.29}$$

The following simplified (also known as "lazy") notations are commonly used:

$$\text{Omitting the tensor product symbol: } |\psi\rangle_A \otimes |\psi\rangle_B = |\psi\rangle_A |\psi\rangle_B. \tag{1.30}$$
$$\text{Writing classical bits as a string: } |0\rangle_A \otimes |0\rangle_B = |0\rangle_A |0\rangle_B = |00\rangle_{AB}. \tag{1.31}$$
$$\text{Combining several identical states: } |\psi\rangle_1 \otimes |\psi\rangle_2 \cdots \otimes |\psi\rangle_n = |\psi\rangle^{\otimes n}. \tag{1.32}$$

The tensor product satisfies a few important properties, which we will use frequently throughout the book.

Proposition 1.4.1 *Properties of the tensor product:*

1. *Distributivity:* $|\psi_1\rangle \otimes (|\psi_2\rangle + |\psi_3\rangle) = |\psi_1\rangle \otimes |\psi_2\rangle + |\psi_1\rangle \otimes |\psi_3\rangle$. *Similarly,* $(|\psi_1\rangle + |\psi_2\rangle) \otimes |\psi_3\rangle = |\psi_1\rangle \otimes |\psi_3\rangle + |\psi_2\rangle \otimes |\psi_3\rangle$.

2. *Associativity:* $|\psi_1\rangle \otimes (|\psi_2\rangle \otimes |\psi_3\rangle) = (|\psi_1\rangle \otimes |\psi_2\rangle) \otimes |\psi_3\rangle$.

These relations hold not only for kets, but also for bras.

Be careful that the tensor product is NOT commutative: in general, $|\psi_1\rangle \otimes |\psi_2\rangle \neq |\psi_2\rangle \otimes |\psi_1\rangle$, unless of course $|\psi_1\rangle = |\psi_2\rangle$. You may convince yourself of this fact by computing the representation as 4-dimensional vectors, using the rule (1.28), of $|0\rangle \otimes |1\rangle$ and $|1\rangle \otimes |0\rangle$.

To practice with the definition of the tensor product, let us have a look at a few examples. The first shows how the tensor product can be applied to construct a basis for the space of n qubits from a basis for the space of a single qubit.

Example 1.4.1 Recall that the standard basis for two qubits A and B is given by

$$|00\rangle_{AB} = \begin{pmatrix} 1 \\ 0 \\ 0 \\ 0 \end{pmatrix}, \quad |01\rangle_{AB} = \begin{pmatrix} 0 \\ 1 \\ 0 \\ 0 \end{pmatrix}, \quad |10\rangle_{AB} = \begin{pmatrix} 0 \\ 0 \\ 1 \\ 0 \end{pmatrix}, \quad |11\rangle_{AB} = \begin{pmatrix} 0 \\ 0 \\ 0 \\ 1 \end{pmatrix}.$$

This basis can be obtained by taking the tensor product of standard basis elements for the individual qubits: $|0\rangle_A \otimes |0\rangle_B, |0\rangle_A \otimes |1\rangle_B, |1\rangle_A \otimes |0\rangle_B, |1\rangle_A \otimes |1\rangle_B$. For example, consider

$$|1\rangle_A \otimes |0\rangle_B = \begin{pmatrix} 0 \\ 1 \end{pmatrix} \otimes |0\rangle_B = \begin{pmatrix} 0\,|0\rangle_B \\ 1\,|0\rangle_B \end{pmatrix} = \begin{pmatrix} 0\cdot 1 \\ 0\cdot 0 \\ 1\cdot 1 \\ 1\cdot 0 \end{pmatrix} = \begin{pmatrix} 0 \\ 0 \\ 1 \\ 0 \end{pmatrix} = |10\rangle_{AB}. \tag{1.33}$$

■

We have seen a few examples of two-qubit states. Let us see whether we can recover them from individual qubit states by taking the tensor product.

Example 1.4.2 Consider the states $|+\rangle_A = \frac{1}{\sqrt{2}}(|0\rangle_A + |1\rangle_A)$ and $|1\rangle_B$. The joint state $|\psi\rangle_{AB}$ is given by

$$|\psi\rangle_{AB} = |+\rangle_A \otimes |1\rangle_B = \frac{1}{\sqrt{2}} \begin{pmatrix} 1 \\ 1 \end{pmatrix}_A \otimes |1\rangle_B = \frac{1}{\sqrt{2}} \begin{pmatrix} 1\cdot |1\rangle_B \\ 1\cdot |1\rangle_B \end{pmatrix} = \frac{1}{\sqrt{2}} \begin{pmatrix} 0 \\ 1 \\ 0 \\ 1 \end{pmatrix}. \tag{1.34}$$

One can also express the joint state in the standard basis by

$$|\psi\rangle_{AB} = \frac{1}{\sqrt{2}}(|0\rangle_A + |1\rangle_A) \otimes |1\rangle_B$$

$$= \frac{1}{\sqrt{2}}(|0\rangle_A \otimes |1\rangle_B + |1\rangle_A \otimes |1\rangle_B)$$

$$= \frac{1}{\sqrt{2}}(|01\rangle_{AB} + |11\rangle_{AB}).$$

This is the state from Example 1.3.2.   ■

Example 1.4.3 Consider the states $|+\rangle_A = \frac{1}{\sqrt{2}}(|0\rangle_A + |1\rangle_A)$ and $|+\rangle_B = \frac{1}{\sqrt{2}}(|0\rangle_B + |1\rangle_B)$. The joint state $|\psi\rangle_{AB}$ is

$$|\psi\rangle_{AB} = \frac{1}{\sqrt{2}}(|0\rangle_A + |1\rangle_A) \otimes \frac{1}{\sqrt{2}}(|0\rangle_B + |1\rangle_B)$$

$$= \frac{1}{2}(|00\rangle_{AB} + |01\rangle_{AB} + |10\rangle_{AB} + |11\rangle_{AB})$$

$$= \frac{1}{2} \begin{pmatrix} 1 \\ 1 \\ 1 \\ 1 \end{pmatrix}.$$

This is the state we have seen in (1.19), which is an equal superposition of all standard basis states for the two qubits.　∎

QUIZ 1.4.1 $|\psi\rangle \otimes |\phi\rangle = |\phi\rangle \otimes |\psi\rangle$ *for all* $|\psi\rangle$ *and* $|\phi\rangle$. *True or false?*

QUIZ 1.4.2 *Consider a two-qubit state* $|\psi\rangle = (\alpha_1 |0\rangle + \beta_1 |1\rangle) \otimes (\alpha_2 |0\rangle + \beta_2 |1\rangle)$. *How do you write this state in a vector form in the standard basis? In other words, compute* $|\psi\rangle = (\alpha_1 |0\rangle + \beta_1 |1\rangle) \otimes (\alpha_2 |0\rangle + \beta_2 |1\rangle)$.

(a) $|\psi\rangle = \begin{pmatrix} \alpha_1\beta_1 \\ \alpha_2\beta_2 \\ \alpha_1\beta_2 \\ \alpha_2\beta_1 \end{pmatrix}$

(b) $|\psi\rangle = \begin{pmatrix} \alpha_1\beta_2 \\ \beta_1\alpha_2 \\ 0 \\ 0 \end{pmatrix}$

(c) $|\psi\rangle = \begin{pmatrix} \alpha_1\alpha_2 \\ \alpha_1\beta_2 \\ \beta_1\alpha_2 \\ \beta_1\beta_2 \end{pmatrix}$

QUIZ 1.4.3 *Consider the following state* $|\psi\rangle$ *of two qubits:* $|\psi\rangle = |-\rangle \otimes |-\rangle$, *where* $|-\rangle = \frac{1}{\sqrt{2}}(|0\rangle - |1\rangle)$. *Written in the standard basis, this state can be expressed as* $|\psi\rangle = \frac{1}{2}(|00\rangle - |01\rangle - |10\rangle - |11\rangle)$. *True or false?*

Looking at these examples, one may wonder whether *any* state $|\psi\rangle_{AB}$ of two qubits may be expressed as the tensor product of two states $|\psi\rangle_A$ and $|\psi\rangle_B$. It turns out that this is *not* the case! Later, we will see that such states have special properties (in Chapter 4 we will learn that they are *entangled*), and without them much of quantum cryptography would not be possible. Let's see an example of such a state, for which it is impossible to find any such $|\psi\rangle_A$ and $|\psi\rangle_B$. To avoid any confusion, our example also illustrates that the state $|\psi\rangle_{AB}$ can of course still be expressed as a linear combination of the standard basis for two qubits.

Example 1.4.4 Consider the state of two qubits

$$|\psi\rangle_{AB} = \frac{1}{\sqrt{2}}\left(|+\rangle_A|+\rangle_B + |-\rangle_A|-\rangle_B\right). \qquad (1.35)$$

Let us express this state in terms of the standard basis, by expanding the terms

$$|+\rangle_A |+\rangle_B = \frac{1}{2}(|0\rangle_A + |1\rangle_A)(|0\rangle_B + |1\rangle_B)$$

$$= \frac{1}{2}(|00\rangle_{AB} + |10\rangle_{AB} + |01\rangle_{AB} + |11\rangle_{AB}),$$

$$|-\rangle_A |-\rangle_B = \frac{1}{2}(|0\rangle_A - |1\rangle_A)(|0\rangle_B - |1\rangle_B)$$

$$= \frac{1}{2}(|00\rangle_{AB} - |10\rangle_{AB} - |01\rangle_{AB} + |11\rangle_{AB}).$$

Substituting this into Eq. (1.35) gives

$$|\psi\rangle_{AB} = \frac{1}{\sqrt{2}}(|+\rangle_A |+\rangle_B + |-\rangle_A |-\rangle_B)$$

$$= \frac{1}{2\sqrt{2}}(|00\rangle_{AB} + |10\rangle_{AB} + |01\rangle_{AB} + |11\rangle_{AB} + |00\rangle_{AB} - |10\rangle_{AB} - |01\rangle_{AB} + |11\rangle_{AB})$$

$$= \frac{1}{\sqrt{2}}(|00\rangle_{AB} + |11\rangle_{AB}) = |\text{EPR}\rangle_{AB}, \tag{1.36}$$

where $|\text{EPR}\rangle_{AB}$ is the state we have seen previously in Example 1.3.1. We see that the coefficients of $|\text{EPR}\rangle_{AB}$ are the same whether we write it in the Hadamard basis or the standard basis. As you will show in Problem 1.2, this state *cannot* be written as $|\psi\rangle_{AB} = |\psi\rangle_A \otimes |\phi\rangle_B$, for any choice of single-qubit states $|\psi\rangle_A$ and $|\phi\rangle_B$. Nevertheless, it can still be decomposed as a linear combination of multiple such states, in more than one way, such as (1.35) and (1.36). ∎

1.5   Simple Measurements

Let us now examine how we can mathematically describe the simplest possible measurements on our qubit(s). Thinking back to the example encoding of classical bits "0" and "1" by the absence and presence of a light pulse, a way to physically measure them immediately presents itself: we could simply "look" whether light is present or not. When we see light, we record a "1" and when we see no light, we write down a "0".

How about measuring a qubit? In our example of using the absence of a photon to represent a $|0\rangle$ and the presence a $|1\rangle$, an idea for a measurement immediately presents itself: we could install a detector capable of measuring a single photon and record a "0" when no photon is detected and a "1" otherwise.

1.5.1   Measurement in the Standard Basis

It turns out that this idea of measuring is known mathematically as measuring the qubit in the standard (or computational) basis. Remember that a state of a qubit can be represented by a normalized vector, $|\psi\rangle \in \mathbb{C}^2$, that can always be expressed as a superposition of the basis vectors $|0\rangle$ and $|1\rangle$, with amplitudes $\alpha, \beta \in \mathbb{C}$ such that $|\alpha|^2 + |\beta|^2 = 1$. A good way to think about a quantum measurement is as a question that can be asked about such a state. The measurement rule then provides a way to answer the question. For example, by analogy with the classical setting in the example above we are asking the question: "Is $|\psi\rangle$ in state $|0\rangle$ (no photon) or in state $|1\rangle$ (photon)?" Given that $|\psi\rangle$ is in general neither of these – it is in a *superposition* of the two basis states – how do we answer such a question? The measurement rule gives a way to do this. Quantum measurements are special in two significant ways: first, in general they result in probabilistic outcomes; second, they perturb the quantum state on which they are performed.

For our example the probability of each possible outcome, for example the outcome "0", can be computed by, roughly speaking, "looking at how much '$|0\rangle$' is present in the state of the qubit." The way this is quantified is by taking the squared inner product between $|\psi\rangle$ and $|0\rangle$. Concretely, if $|\psi\rangle = \alpha |0\rangle + \beta |1\rangle$, then the measurement associated with the question "Is $|\psi\rangle$ in state $|0\rangle$ or in state $|1\rangle$?" returns the outcome "$|0\rangle$" with probability p_0, and "$|1\rangle$" with probability p_1, where

$$
\begin{aligned}
p_0 &= |\langle \psi | 0 \rangle|^2 = \left| (\alpha^* \quad \beta^*) \begin{pmatrix} 1 \\ 0 \end{pmatrix} \right|^2 = |\alpha|^2, \\
p_1 &= |\langle \psi | 1 \rangle|^2 = \left| (\alpha^* \quad \beta^*) \begin{pmatrix} 0 \\ 1 \end{pmatrix} \right|^2 = |\beta|^2.
\end{aligned}
\tag{1.37}
$$

We now see a good reason for the condition $|\alpha|^2 + |\beta|^2 = 1$: it means that $p_0 + p_1 = 1$; that is, the probabilities of observing "$|0\rangle$" and "$|1\rangle$" add up to 1.

QUIZ 1.5.1 *The state $|1\rangle$ is measured in the standard basis. What is the probability of obtaining the outcome 0?*

(a) 0
(b) $\frac{1}{2}$
(c) $\frac{3}{4}$
(d) $\frac{\sqrt{3}}{2}$

What happens after the measurement? Measuring our qubit in the standard basis destroys the superposition. Thinking back to our physical example, once we have detected a photon, we are in the state $|1\rangle$. If we do not detect a photon, $|0\rangle$. There is no way for us to recreate the superposition, and we will say that the state has *collapsed*.

In quantum information we label the outcomes "0" for "$|0\rangle$" and "1" for "$|1\rangle$",[3] while in physics people often use "$+1$" for "$|0\rangle$" and "-1" for "$|1\rangle$". In this book we will mostly use the first convention, though we may sometimes use the second when convenient; which one will always be clear in context.

---

3 And more generally, x for outcome "$|x\rangle$", for a bit string x.

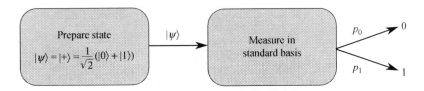

Fig. 1.1   Generation of true randomness from the deterministic preparation of a qubit in superposition.

1.5.2   Application: Randomness from a Deterministic Process

Can we do anything interesting with what we have learned so far? It turns out that the answer is yes: by preparing just single qubits and measuring them in the standard basis we can achieve a task that is impossible classically. Namely, we can build a deterministic machine – i.e. a machine that uses no randomness itself – that nevertheless produces true random numbers.

Consider the following process illustrated in Figure 1.1: first, prepare a qubit in the state $|+\rangle = \frac{1}{\sqrt{2}}(|0\rangle + |1\rangle)$. Next, measure this state in the standard basis. The probability of obtaining each outcome can be calculated by evaluating the inner products, using the recipe given in (1.37):

$$p_0 = |\langle +|0\rangle|^2 = \left| \frac{1}{\sqrt{2}}(\langle 0| + \langle 1|)|0\rangle \right|^2 = \left| \frac{1}{\sqrt{2}}(\underbrace{\langle 0|0\rangle}_{1} + \underbrace{\langle 1|0\rangle}_{0}) \right|^2 = \frac{1}{(\sqrt{2})^2} = \frac{1}{2},$$

$$p_1 = |\langle +|1\rangle|^2 = \left| \frac{1}{\sqrt{2}}(\langle 0| + \langle 1|)|1\rangle \right|^2 = \left| \frac{1}{\sqrt{2}}(\underbrace{\langle 0|1\rangle}_{0} + \underbrace{\langle 1|1\rangle}_{1}) \right|^2 = \frac{1}{(\sqrt{2})^2} = \frac{1}{2}.$$

This simple example tells us something about the power of quantum information: it is in principle possible to build a machine that deterministically prepares the qubit $|+\rangle$ and subsequently measures it in the standard basis. Since $p_0 = p_1 = 1/2$, this machine obtains an outcome that is perfectly uniformly distributed between "0" and "1".

Even though the machine is perfectly deterministic (it always does exactly the same thing), each time the process is executed the outcome is unpredictable. This intrinsic randomness is a consequence of the rules of quantum mechanics as we have presented them, and is an integral part of the power of quantum information for cryptography: as we will see throughout the book, uncertainty, or ignorance, is the key to security. Moreover, machines exploiting such ideas have already been built, see e.g. Figure 1.2.

We have described the measurement rule for the case of a single qubit, measured in the standard basis. The rule generalizes directly to a measurement of an n-qubit state in the standard basis. Indeed, consider an arbitrary n-qubit quantum state expressed as a superposition in the standard basis:

$$|\psi\rangle = \sum_{x \in \{0,1\}^n} \alpha_x |x\rangle . \tag{1.38}$$

When $|\psi\rangle$ is measured in the standard basis $\{|x\rangle\}_x$, the probability of obtaining the outcome x is naturally given by $p_x = |\langle x|\psi\rangle|^2 = |\alpha_x|^2$. Once again, the normalization condition on the vector $|\psi\rangle$ shows that these probabilities sum to 1, as expected.

Fig. 1.2   Chip generating quantum random numbers (Photo: QuSide Technologies).

BOX 1.1   Expectation Values

Physicists (but also computer scientists!) like to compute expectation values of measurement outcomes, as they provide an indication of average behavior, if one performs a measurement many times. To see what this means, suppose that we measure a qubit $|\psi\rangle$ in the standard basis $\{|0\rangle, |1\rangle\}$. For this discussion we adopt the physics convention of labeling the two possible outcomes as $+1$ and -1 respectively: "$+1$" for "$|0\rangle$", and "-1" for "$|1\rangle$". Then the expectation value of the outcome obtained when measuring $|\psi\rangle$ is by definition $E = 1 \cdot |\langle 0|\psi\rangle|^2 - 1 \cdot |\langle 1|\psi\rangle|^2$. Since $|\langle 0|\psi\rangle|^2 = \langle\psi|0\rangle\langle 0|\psi\rangle$, we have $E = \langle\psi|\left(|0\rangle\langle 0| - |1\rangle\langle 1|\right)|\psi\rangle = \langle\psi|Z|\psi\rangle$, where $Z = |0\rangle\langle 0| - |1\rangle\langle 1| = \begin{pmatrix} 1 & 0 \\ 0 & -1 \end{pmatrix}$ is a 2×2 matrix. We will encounter Z frequently in the future; it is called the "Pauli Z observable."

1.5.3   Measuring a Qubit in an Arbitrary Basis

 What other kinds of observations, or measurements, are allowed in quantum mechanics? As it turns out, any orthonormal basis for the state space of one (or multiple) qubit can be used to define a valid measurement. Indeed, abstractly speaking there is nothing special about the standard basis: it is "a" basis of the state space $\mathbb{C}^d$, but many other bases exist.

To find out how to analyze this more general setting, let us first take a step back and consider how we found the probabilities in the case of measurements in the standard basis. To obtain them, we first expressed an arbitrary quantum state as a superposition over elements of the standard basis, and then took the square of the amplitudes to obtain the outcome probabilities.

When measuring a qubit in a different orthonormal basis, given by vectors $\{|v_0\rangle, |v_1\rangle\}$, we proceed in a similar way: first, we expand the quantum state as a superposition over vectors from the new basis, i.e. find amplitudes $\hat{\alpha}$ and $\hat{\beta}$ such that

$$|\psi\rangle = \alpha|0\rangle + \beta|1\rangle = \hat{\alpha}|v_0\rangle + \hat{\beta}|v_1\rangle \, . \tag{1.39}$$

Due to the assumption that $\{|v_0\rangle, |v_1\rangle\}$ is a basis, the complex numbers $\hat{\alpha}$ and $\hat{\beta}$ are uniquely defined and can be found by simple linear algebra. Second, take the modulus squared of the associated amplitudes $\hat{\alpha}$ and $\hat{\beta}$ to obtain the probability of each outcome: here, the outcome is "$|v_0\rangle$" with probability $|\hat{\alpha}|^2$ and "$|v_1\rangle$" with probability $|\hat{\beta}|^2$.

Example 1.5.1 Consider the qubit $|+\rangle = \frac{1}{\sqrt{2}}(|0\rangle + |1\rangle)$. Instead of measuring it in the standard basis, let us now measure it in the basis $\{|+\rangle, |-\rangle\}$ given by the two orthonormal vectors of the Hadamard basis, $|+\rangle$ and $|-\rangle = \frac{1}{\sqrt{2}}(|0\rangle - |1\rangle)$. Clearly, we can write the qubit as $1 \cdot |+\rangle + 0 \cdot |-\rangle$. Thus, in this case the probability of obtaining measurement outcome "$|+\rangle$" is $|1|^2 = 1$, and the probability of outcome "$|-\rangle$" is 0. The probabilities of measurement outcomes depend dramatically on the basis in which we measure: for this measurement, there is no randomness in the outcomes! ∎

Example 1.5.2 Consider measuring an arbitrary qubit $|\psi\rangle = \alpha|0\rangle + \beta|1\rangle$ in the basis $\{|+\rangle, |-\rangle\}$. To find out how to express the qubit in this other basis, it is convenient to determine what the basis elements $|0\rangle$ and $|1\rangle$ look like in that basis. We find that

$$|0\rangle = \frac{1}{2}\big((|0\rangle + |1\rangle) + (|0\rangle - |1\rangle)\big) = \frac{1}{\sqrt{2}}(|+\rangle + |-\rangle) , \tag{1.40}$$

$$|1\rangle = \frac{1}{2}\big((|0\rangle + |1\rangle) - (|0\rangle - |1\rangle)\big) = \frac{1}{\sqrt{2}}(|+\rangle - |-\rangle) . \tag{1.41}$$

Substituting in the definition of $|\psi\rangle$, we get

$$\alpha|0\rangle + \beta|1\rangle = \frac{1}{\sqrt{2}}\big(\alpha(|+\rangle + |-\rangle) + \beta(|+\rangle - |-\rangle)\big) \tag{1.42}$$

$$= \frac{\alpha + \beta}{\sqrt{2}}|+\rangle + \frac{\alpha - \beta}{\sqrt{2}}|-\rangle . \tag{1.43}$$

This means that upon measuring the qubit $|\psi\rangle$ in the basis $\{|+\rangle, |-\rangle\}$ the outcome "$|+\rangle$" is obtained with probability $|\alpha + \beta|^2/2$ and the outcome "$|-\rangle$" is obtained with probability $|\alpha - \beta|^2/2$. In particular, you can check that this calculation recovers the one performed in the previous example as a special case. ∎

Quite often we do not care about the entire probability distribution, but just the probability of one specific outcome. Is there a more efficient way to find this probability than to rewrite the entire state $|\psi\rangle$ in another basis? To investigate this, let us consider a single qubit

$$|\psi\rangle = \alpha|0\rangle + \beta|1\rangle . \tag{1.44}$$

Remember that the elements of the standard basis are orthonormal. As a result, we could have found the desired probabilities by simply computing the inner product between two vectors, as described above. Specifically, when given the qubit $|\psi\rangle = \alpha|0\rangle + \beta|1\rangle$ we obtain outcomes "$|0\rangle$" and "$|1\rangle$" with probabilities

$$p_0 = |\langle 0|\psi\rangle|^2 = \left|(1\ 0)\begin{pmatrix}\alpha\\\beta\end{pmatrix}\right|^2 = |\alpha|^2, \tag{1.45}$$

$$p_1 = |\langle 1|\psi\rangle|^2 = \left|(0\ 1)\begin{pmatrix}\alpha\\\beta\end{pmatrix}\right|^2 = |\beta|^2. \tag{1.46}$$

Example 1.5.3 Suppose we measure $|0\rangle$ in the Hadamard basis. The probabilities of observing outcomes "$|+\rangle$" and "$|-\rangle$" are given by

$$p_+ = |\langle +|0\rangle|^2 = \left|\left(\frac{1}{\sqrt{2}}\ \frac{1}{\sqrt{2}}\right)\begin{pmatrix}1\\0\end{pmatrix}\right|^2 = \frac{1}{2}, \tag{1.47}$$

$$p_- = |\langle -|0\rangle|^2 = \left|\left(\frac{1}{\sqrt{2}}\ \frac{-1}{\sqrt{2}}\right)\begin{pmatrix}1\\0\end{pmatrix}\right|^2 = \frac{1}{2}. \tag{1.48}$$

$\blacksquare$

For qudits, the rule for finding probabilities is analogous.

Definition 1.5.1. *Suppose that $|\psi\rangle \in \mathbb{C}^d$ is a pure quantum state in dimension d. Suppose that $|\psi\rangle$ is measured in the orthonormal basis $\{|b_j\rangle\}_{j=1}^d$ of $\mathbb{C}^d$. Then the probability of obtaining the outcome "$|b_j\rangle$" is*

$$p_j = |\langle b_j|\psi\rangle|^2. \tag{1.49}$$

Let us now consider some examples to gain intuition on measuring quantum states in different bases. First, let us have a look at another single-qubit example.

Example 1.5.4 Consider the single-qubit state $|\psi\rangle = \frac{1}{\sqrt{2}}(|0\rangle + i|1\rangle)$. Measure the qubit in the basis $\{|+\rangle, |-\rangle\}$. The probabilities of obtaining outcomes "$+$" and "$-$" can be evaluated as follows:

$$p_+ = |\langle +|\psi\rangle|^2 = \left|\frac{1}{2}((\langle 0| + \langle 1|)(|0\rangle + i|1\rangle))\right|^2$$

$$= \frac{1}{4}\left|\langle 0|0\rangle + \langle 1|0\rangle + i\langle 1|0\rangle + i\langle 1|1\rangle\right|^2$$

$$= \frac{1}{4}|1 + i|^2$$

$$= \frac{1}{4}(1 - i)(1 + i) = \frac{1}{2},$$

$$p_- = |\langle -|\psi\rangle|^2 = \left|\frac{1}{2}((\langle 0| - \langle 1|)(|0\rangle + i|1\rangle))\right|^2$$

$$= \frac{1}{4}\left|\langle 0|0\rangle - \langle 1|0\rangle + i\langle 0|1\rangle - i\langle 1|1\rangle\right|^2$$

$$= \frac{1}{4}|1 - i|^2$$

$$= \frac{1}{4}(1 + i)(1 - i) = \frac{1}{2}. \qquad \blacksquare$$

QUIZ 1.5.2 *The state* $|1\rangle$ *is measured in the basis* $\{|0'\rangle, |1'\rangle\}$, *where* $|0'\rangle = \frac{1}{2}(|0\rangle + \sqrt{3}|1\rangle)$ *and* $|1'\rangle = \frac{1}{2}(\sqrt{3}|0\rangle - |1\rangle)$. *What is the probability of obtaining the outcome* $0'$?

(a) 0

(b) $\frac{1}{2}$

(c) $\frac{3}{4}$

(d) $\frac{\sqrt{3}}{2}$

While we will generally talk about states of qubits, we may occasionally consider quantum states in d dimensions, where d is not necessarily a power of 2.

Example 1.5.5 Consider a *qutrit* $|\psi\rangle \in \mathbb{C}^3$, which is a 3-dimensional quantum system, represented by the vector

$$|\psi\rangle = \frac{1}{\sqrt{2}}\begin{pmatrix}1\\0\\0\end{pmatrix} + \frac{1}{2}\begin{pmatrix}0\\1\\0\end{pmatrix} + \frac{1}{2}\begin{pmatrix}0\\0\\1\end{pmatrix}. \qquad (1.50)$$

Suppose that $|\psi\rangle$ is measured in the orthonormal basis $\{|b_1\rangle, |b_2\rangle, |b_3\rangle\}$, where

$$|b_1\rangle = \begin{pmatrix}1\\0\\0\end{pmatrix}, \qquad |b_2\rangle = \frac{1}{\sqrt{2}}\begin{pmatrix}0\\1\\1\end{pmatrix}, \qquad |b_3\rangle = \frac{1}{\sqrt{2}}\begin{pmatrix}0\\1\\-1\end{pmatrix}. \qquad (1.51)$$

The probabilities of obtaining each outcome can be calculated as follows:

$$p_{b_1} = |\langle b_1|\psi\rangle|^2 = \frac{1}{2}, \qquad (1.52)$$

$$p_{b_2} = |\langle b_2|\psi\rangle|^2 = \langle b_2|v\rangle\langle v|b_2\rangle = \frac{1}{2\sqrt{2}}(1+1) \cdot \frac{1}{2\sqrt{2}}(1+1) = \frac{1}{2}, \qquad (1.53)$$

$$p_{b_3} = |\langle b_3|\psi\rangle|^2 = \langle b_3|v\rangle\langle v|b_3\rangle = \frac{1}{2\sqrt{2}}(1-1) \cdot \frac{1}{2\sqrt{2}}(1-1) = 0. \qquad (1.54)$$

$\blacksquare$

1.5.4 Measuring Multiple Qubits

Since we can always consider a state of n qubits as a single quantum state of dimension $d = 2^n$, the rule for describing measurements of arbitrary-dimensional states given in the previous section can be applied to the case of an n-qubit state. Nevertheless, it is often more convenient not to forget the qubit structure of the state. Let us see explicitly

---

BOX 1.2   Distinguishing Quantum States

Suppose we are given a quantum state $|\psi\rangle \in \mathbb{C}^d$. We do not know $|\psi\rangle$ exactly, but we do know that $|\psi\rangle$ is one of two possible quantum states, $|\psi_0\rangle$ or $|\psi_1\rangle$. How is it possible to find out which of the two states $|\psi_0\rangle$ and $|\psi_1\rangle$ we have? If $|\psi_0\rangle$ and $|\psi_1\rangle$ are orthogonal then there is always an orthonormal basis $\{|b_0\rangle, |b_1\rangle, \ldots, |b_{d-1}\rangle\}$ such that $|b_0\rangle = |\psi_0\rangle$ and $|b_1\rangle = |\psi_1\rangle$. A measurement of $|\psi\rangle$ in this basis will yield the outcome b_0 if and only if $|\psi\rangle = |\psi_0\rangle$ and the outcome b_1 if and only if $|\psi\rangle = |\psi_1\rangle$, hence the two states can be perfectly distinguished. If, however, $|\psi_0\rangle$ and $|\psi_1\rangle$ are not orthogonal then it is impossible to distinguish them with certainty using a quantum measurement. We return to this question in Chapter 5.

---

what happens when such a state is measured. Let us do it in general: consider a 2-qudit state in the space $|\psi\rangle_{AB} \in \mathbb{C}_A^{d_A} \otimes \mathbb{C}_B^{d_B}$, for arbitrary dimension $d_A, d_B \geq 1$. First, remember how a basis for this space can be obtained from bases for the individual state spaces $\mathbb{C}_A^{d_A}$ and $\mathbb{C}_B^{d_B}$: if $\{|b_j^A\rangle\}_j$ is a basis for $\mathbb{C}_A^{d_A}$ and $\{|b_k^B\rangle\}_k$ is a basis for the state space $\mathbb{C}_B^{d_B}$, then the set of vectors $\{\{|b_j^A\rangle \otimes |b_k^B\rangle\}_{j=0}^{d_A}\}_{k=0}^{d_B}$ gives a basis for $\mathbb{C}_A^{d_A} \otimes \mathbb{C}_B^{d_B}$.

Example 1.5.6 Consider the basis $\{|0\rangle_A, |1\rangle_A\}$ for qubit A, and the basis $\{|+\rangle_B, |-\rangle_B\}$ for qubit B. A basis for the joint system AB is given by

$$\{|0\rangle_A|+\rangle_B, |0\rangle_A|-\rangle_B, |1\rangle_A|+\rangle_B, |1\rangle_A|-\rangle_B\} .$$

■

Suppose now that we would like to measure qudit A in the basis $\{|b_j^A\rangle\}_j$, and qudit B in the basis $\{|b_k^B\rangle\}_k$. What is the probability that we obtain outcome "$|b_j^A\rangle$" for A, and outcome "$|b_k^B\rangle$" for B? To find out, we first write down a basis for the joint state space of qudits A and B: $\{\{|b_j^A\rangle|b_k^B\rangle\}_j\}_k$. We then apply the usual measurement rule to compute the probability

$$p_{jk} = |\langle b_j^A|\langle b_k^B||\psi\rangle_{AB}|^2 . \tag{1.55}$$

Example 1.5.7 Consider two qubits in an EPR pair:

$$|\text{EPR}\rangle = \frac{1}{\sqrt{2}}(|00\rangle + |11\rangle) . \tag{1.56}$$

Suppose each qubit is measured in the standard basis. Then the probabilities of obtaining outcomes 00, 01, 10, and 11 are given by

$$p_{00} = p_{11} = \frac{1}{2} , \quad p_{01} = p_{10} = 0 . \tag{1.57}$$

■

1.5.5   Post-Measurement States

In general, when a state $|\psi\rangle \in \mathbb{C}^d$ is measured in a basis $\{|b_i\rangle\}_i$ of $\mathbb{C}^d$, once the measurement outcome "b_i" is obtained the state $|\psi\rangle$ automatically "collapses" to the basis state that is consistent with the outcome: it becomes the state $|b_i\rangle$. We will discuss the formalism associated with post-measurement states in more detail in the next chapter, when we consider generalized measurements.

1.6   Unitary Transformations and Gates

 Just like it is possible to manipulate classical bits, such as by flipping a bit or adding two bits, it is possible to perform operations on qubits. However, the laws of quantum mechanics do not allow every possible operation: some operations are physically impossible.

1.6.1   Unitary Transformations

First, consider operations that transform the state of some qubits to a different state of the same qubits. Mathematically, we are interested in operations that transform normalized states in $\mathbb{C}^d$ to normalized states in the same space. According to the laws of quantum mechanics, a necessary condition on any such transformation for it to be a valid quantum operation is that it should be *linear*: any quantum map U that performs a transformation

$$U : |\psi_{\text{in}}\rangle \mapsto |\psi_{\text{out}}\rangle = U(|\psi_{\text{in}}\rangle) \tag{1.58}$$

must satisfy

$$U(\alpha|\psi_1\rangle + \beta|\psi_2\rangle) = \alpha U(|\psi_1\rangle) + \beta U(|\psi_2\rangle) \,.$$

This allows us to deduce that any quantum operation that acts on d-dimensional qudits can be represented by some $d \times d$ matrix U with complex coefficients. This is because any linear map on $\mathbb{C}^d$ has a matrix representation. Furthermore, since we want the operation to map quantum states to quantum states, it should preserve lengths: for all possible states $|\psi_{\text{in}}\rangle$,

$$\langle\psi_{\text{out}}|\psi_{\text{out}}\rangle = \langle\psi_{\text{in}}|U^\dagger U|\psi_{\text{in}}\rangle = 1 \,, \tag{1.59}$$

where recall that for matrices the "dagger" notation $U^\dagger$ designates the conjugate-transpose: $U^\dagger = (U^*)^T$. Observe that $(U|\psi\rangle)^\dagger = \langle\psi|U^\dagger$. Similarly, the same should be true for the operation $U^\dagger$,

$$\langle\psi_{\text{out}}|\psi_{\text{out}}\rangle = \langle\psi_{\text{in}}|UU^\dagger|\psi_{\text{in}}\rangle = 1 \,. \tag{1.60}$$

This shows that the condition that the operation U preserves the length of any vector is equivalent to the condition that $U^\dagger U = UU^\dagger = \mathbb{I}$, where $\mathbb{I}$ is the identity matrix.

Definition 1.6.1 (Identity). *The identity matrix is a diagonal, square matrix with all diagonal entries equal to 1:*

$$\mathbb{I} = \begin{pmatrix} 1 & 0 & \cdots & \cdots & 0 \\ 0 & 1 & \cdots & \cdots & 0 \\ \vdots & \vdots & \ddots & \ddots & \vdots \\ 0 & 0 & \cdots & 0 & 1 \end{pmatrix}. \tag{1.61}$$

For any dimension d, we denote the $d \times d$ identity matrix as $\mathbb{I}_d$. We sometimes leave the dimension implicit and simply write $\mathbb{I}$.

Remark 1.6.1 The identity matrix leaves all quantum states invariant, i.e. for any quantum state $|\psi\rangle$, $\mathbb{I}|\psi\rangle = |\psi\rangle$.

Definition 1.6.2 (Unitary operation). *An operation U is unitary if and only if $U^\dagger U = UU^\dagger = \mathbb{I}$.*

The allowed operations on quantum states $|\psi\rangle$ are precisely the unitary operations. Note that $\mathbb{I}$ is itself a unitary operation, called the *identity operation*. This just means that the state is not transformed at all. Note that since $U^\dagger U = \mathbb{I}$, any operation U is reversible: if $|\psi\rangle$ has been transformed to $U|\psi\rangle$ we can undo U by applying $U^\dagger$, which is also unitary, to obtain $U^\dagger U|\psi\rangle = \mathbb{I}|\psi\rangle = |\psi\rangle$. To gain some intuition on this, let us have a look at some examples.

Example 1.6.1 Consider the matrix

$$H = \frac{1}{\sqrt{2}} \begin{pmatrix} 1 & 1 \\ 1 & -1 \end{pmatrix}. \tag{1.62}$$

You can verify that $H^\dagger = H$ and thus

$$H^\dagger H = HH = \begin{pmatrix} 1 & 0 \\ 0 & 1 \end{pmatrix} = \mathbb{I}. \tag{1.63}$$

That is, H is unitary. We have that

$$H|0\rangle = \frac{1}{\sqrt{2}} \begin{pmatrix} 1 & 1 \\ 1 & -1 \end{pmatrix} \begin{pmatrix} 1 \\ 0 \end{pmatrix} = \frac{1}{\sqrt{2}} \begin{pmatrix} 1 \\ 1 \end{pmatrix} = |+\rangle. \tag{1.64}$$

Similarly, you can verify that $H|1\rangle = |-\rangle$. We thus see that H transforms the computational basis $\{|0\rangle, |1\rangle\}$ into the Hadamard basis $\{|+\rangle, |-\rangle\}$. The transformation H is called the *Hadamard transform*, or *Hadamard gate*. ∎

Let us now consider a somewhat more complicated operation.

Example 1.6.2 For any $\theta \in \mathbb{R}$, consider the matrix

$$R(\theta) = \begin{pmatrix} \cos\frac{\theta}{2} & -\sin\frac{\theta}{2} \\ \sin\frac{\theta}{2} & \cos\frac{\theta}{2} \end{pmatrix}. \tag{1.65}$$

The conjugate-transpose of this matrix is given by

$$R^{\dagger}(\theta) = \begin{pmatrix} \cos\frac{\theta}{2} & \sin\frac{\theta}{2} \\ -\sin\frac{\theta}{2} & \cos\frac{\theta}{2} \end{pmatrix}, \tag{1.66}$$

and therefore

$$R(\theta)R^{\dagger}(\theta) = \begin{pmatrix} \cos\frac{\theta}{2} & -\sin\frac{\theta}{2} \\ \sin\frac{\theta}{2} & \cos\frac{\theta}{2} \end{pmatrix} \cdot \begin{pmatrix} \cos\frac{\theta}{2} & \sin\frac{\theta}{2} \\ -\sin\frac{\theta}{2} & \cos\frac{\theta}{2} \end{pmatrix}$$

$$= \begin{pmatrix} \cos^2\frac{\theta}{2} + \sin^2\frac{\theta}{2} & 0 \\ 0 & \sin^2\frac{\theta}{2} + \cos^2\frac{\theta}{2} \end{pmatrix} = \begin{pmatrix} 1 & 0 \\ 0 & 1 \end{pmatrix}.$$

You can check that $R^{\dagger}(\theta)R(\theta) = \mathbb{I}$ as well, therefore $R(\theta)$ is unitary. Moreover,

$$R(\theta)\,|0\rangle = \begin{pmatrix} \cos\frac{\theta}{2} & -\sin\frac{\theta}{2} \\ \sin\frac{\theta}{2} & \cos\frac{\theta}{2} \end{pmatrix} \cdot \begin{pmatrix} 1 \\ 0 \end{pmatrix} = \begin{pmatrix} \cos\frac{\theta}{2} \\ \sin\frac{\theta}{2} \end{pmatrix},$$

$$R(\theta)\,|1\rangle = \begin{pmatrix} \cos\frac{\theta}{2} & -\sin\frac{\theta}{2} \\ \sin\frac{\theta}{2} & \cos\frac{\theta}{2} \end{pmatrix} \cdot \begin{pmatrix} 0 \\ 1 \end{pmatrix} = \begin{pmatrix} -\sin\frac{\theta}{2} \\ \cos\frac{\theta}{2} \end{pmatrix}.$$

If we take $\theta = \frac{\pi}{2}$, then $\cos\frac{\theta}{2} = \sin\frac{\theta}{2} = \cos\frac{\pi}{4} = \frac{1}{\sqrt{2}}$ and therefore

$$R\left(\frac{\pi}{2}\right)|0\rangle = |+\rangle \qquad \text{and} \qquad R\left(\frac{\pi}{2}\right)|1\rangle = -|-\rangle. \tag{1.67}$$

∎

QUIZ 1.6.1 *What is the action of the Hadamard transformation on $|+\rangle$?*

(a) $H|+\rangle = |0\rangle$
(b) $H|+\rangle = |1\rangle$
(c) $H|+\rangle = |-\rangle$

Since we will be working with unitaries a lot, it is useful to have multiple ways of recognizing them. Definition 1.6.2 provides one such way. Here is another.

Lemma 1.6.2 *Let U be a linear map on $\mathbb{C}^d$ represented by a $d \times d$ matrix. Then U is unitary if and only if the columns of U form an orthonormal basis of $\mathbb{C}^d$. Equivalently, U is unitary if and only if it sends the canonical basis $\{|e_0\rangle, \ldots, |e_{d-1}\rangle\}$ to $\{|u_0\rangle = U|e_0\rangle, \ldots, |u_{d-1}\rangle = U|e_{d-1}\rangle\}$ such that $\{|u_0\rangle, \ldots, |u_{d-1}\rangle\}$ is also an orthonormal basis of $\mathbb{C}^d$.*

More generally, U is unitary if and only if it transforms any orthonormal basis of $\mathbb{C}^d$ into an orthonormal basis.

Proof The condition that the columns $|u_0\rangle,\ldots,|u_{d-1}\rangle$ of U are orthonormal is equivalent to the condition $U^\dagger U = \mathbb{I}$. The latter condition is equivalent to

$$\||U|v\rangle\|^2 = \langle v| U^\dagger U |v\rangle = \langle v| v\rangle = \||v\rangle\|^2$$

for any vector $|v\rangle$. By taking the conjugate, this is equivalent to $\|U^\dagger |v\rangle\| = \||v\rangle\|$ for any vector $|v\rangle$, hence $UU^\dagger = \mathbb{I}$ as well.

For the "more generally" part, note that if $U^\dagger U = \mathbb{I}$ then U transforms any orthonormal basis into an orthonormal basis. Conversely, if U transforms any orthonormal basis into an orthonormal basis then it transforms the standard basis into an orthonormal basis, so using the first part U is unitary. ∎

Remark 1.6.3 A useful consequence is that if one fixes k orthonormal vectors $|v_0\rangle,\ldots,|v_k\rangle$ in $\mathbb{C}^d$, for $0 \le k \le d-1$, then there always exists a unitary U that sends $|e_i\rangle$ to $|v_i\rangle = U|e_i\rangle$ for all $i \in \{0,\ldots,k\}$. (In fact, as soon as $k < d-1$ there are many such operations!) To see this, simply complete $|v_0\rangle,\ldots,|v_k\rangle$ to an orthonormal basis $\{|v_0\rangle,\ldots,|v_k\rangle,\ldots,|v_{d-1}\rangle\}$ of $\mathbb{C}^d$ and define U to be the matrix whose columns are given by $|v_0\rangle,\ldots,|v_{d-1}\rangle$. By Lemma 1.6.2, U is a unitary map, and it sends $|e_i\rangle$ to $|v_i\rangle = U|e_i\rangle$, as desired.

QUIZ 1.6.2 *Is $U = \frac{1}{\sqrt{2}}\begin{pmatrix} 1 & 1 \\ 1 & 1 \end{pmatrix}$ a valid unitary transformation?*

(a) *Yes*

(b) *No*

QUIZ 1.6.3 *Consider a unitary transformation $U = \begin{pmatrix} 1 & 0 \\ 0 & i \end{pmatrix}$. Which operation corresponds to $U^\dagger$?*

(a) $U^\dagger = \begin{pmatrix} -1 & 0 \\ 0 & -i \end{pmatrix}$

(b) $U^\dagger = \begin{pmatrix} -i & 0 \\ 0 & 1 \end{pmatrix}$

(c) $U^\dagger = \begin{pmatrix} i & 0 \\ 0 & -1 \end{pmatrix}$

(d) $U^\dagger = \begin{pmatrix} 1 & 0 \\ 0 & -i \end{pmatrix}$

QUIZ 1.6.4 *Consider a unitary operation U that has the following action: $U|0\rangle|0\rangle = |0\rangle|0\rangle$ and $U|1\rangle|0\rangle = |1\rangle|1\rangle$. What is the action of U on $|-\rangle|0\rangle$?*

(a) $U\,|-\rangle\,|0\rangle = \frac{1}{\sqrt{2}}\,(|0\rangle\,|0\rangle - |1\rangle\,|1\rangle)$

(b) $U\,|-\rangle\,|0\rangle = |-\rangle\,|-\rangle$

1.6.2 Pauli Matrices as Unitary Operations

The Pauli matrices are unitary 2×2 matrices defined as

$$X = \begin{pmatrix} 0 & 1 \\ 1 & 0 \end{pmatrix},$$

$$Z = \begin{pmatrix} 1 & 0 \\ 0 & -1 \end{pmatrix},$$

$$Y = iXZ.$$

(The i is there to make Y Hermitian, that is, $Y^{\dagger} = Y$.) These matrices play a prominent role in physics, because they model simple observables (see Box 1.1) that can be performed on a single qubit. As we will see below, they also have an interesting interpretation as operations or *gates* in quantum computing.

Exercise 1.6.1   Verify that the Pauli matrices X, Z, and Y are unitary.

The Pauli X matrix acts on the standard basis vectors by interchanging them:

$$X\,|0\rangle = |1\rangle\,,$$
$$X\,|1\rangle = |0\rangle\,.$$

In analogy to classical computation X is often referred to as the NOT gate, since it changes 0 to 1 and vice versa. This is also known as a *bit-flip* operation. On the other hand, the Pauli Z matrix acts on the standard basis by introducing a *phase flip*:

$$Z\,|0\rangle = |0\rangle\,,$$
$$Z\,|1\rangle = -\,|1\rangle\,.$$

The Pauli Z matrix has the effect of interchanging the vectors $|+\rangle$ and $|-\rangle$. To be precise, we have

$$Z\,|+\rangle = Z\left(\frac{1}{\sqrt{2}}(|0\rangle + |1\rangle)\right) = \frac{1}{\sqrt{2}}(Z\,|0\rangle + Z\,|1\rangle) = \frac{1}{\sqrt{2}}(|0\rangle - |1\rangle) = |-\rangle. \quad (1.68)$$

Similarly, $Z\,|-\rangle = |+\rangle$. We thus see that Z acts like a bit flip on the Hadamard basis, while it acts like a phase flip on the standard basis. Applying both a bit and a phase flip gives $Y = iXZ$. This matrix, when acting on the standard basis vectors, introduces a bit flip and a phase flip:

$$Y\,|0\rangle = iXZ\,|0\rangle = iX\,|0\rangle = i\,|1\rangle, \quad (1.69)$$
$$Y\,|1\rangle = -iXZ\,|0\rangle = -iX\,|1\rangle = -i\,|0\rangle. \quad (1.70)$$

1.6.3 No Cloning!

We now use our understanding of unitaries U to show that arbitrary qubits, unlike classical bits, cannot be copied! We will see throughout the book that this fundamental limitation of quantum mechanics plays an essential role in quantum cryptography. To see why we cannot copy arbitrary qubits $|\psi\rangle$, suppose that there existed a copying unitary C. Such a unitary would have the property that

$$C(|\psi\rangle \otimes |0\rangle) = |\psi\rangle \otimes |\psi\rangle , \tag{1.71}$$

for *any* input qubit $|\psi\rangle$. That is, it would produce a copy of $|\psi\rangle$. Then for any $|\psi_1\rangle$ and $|\psi_2\rangle$,

$$C(|\psi_1\rangle \otimes |0\rangle) = |\psi_1\rangle \otimes |\psi_1\rangle ,$$
$$C(|\psi_2\rangle \otimes |0\rangle) = |\psi_2\rangle \otimes |\psi_2\rangle .$$

Since C is a unitary, we have $C^\dagger C = \mathbb{I}$ and hence

$$\begin{aligned}
\langle \psi_1 | \psi_2 \rangle &= \langle \psi_1 | C^\dagger C | \psi_2 \rangle \\
&= \langle \psi_1 | \psi_2 \rangle \langle 0 | 0 \rangle \\
&= ((\langle \psi_1 | \otimes \langle 0 |)(|\psi_2\rangle \otimes |0\rangle)) \\
&= ((\langle \psi_1 | \otimes \langle 0 |) C^\dagger C (|\psi_2\rangle \otimes |0\rangle)) \\
&= ((\langle \psi_1 | \otimes \langle \psi_1 |)(|\psi_2\rangle \otimes |\psi_2\rangle)) = (\langle \psi_1 | \psi_2 \rangle)^2 .
\end{aligned}$$

Clearly, whenever $0 < |\langle \psi_1 | \psi_2 \rangle| < 1$ the above cannot hold and hence such a copying unitary C does not exist. Note that $|\psi_1\rangle = |0\rangle$ and $|\psi_2\rangle = |+\rangle$, for example, have precisely this property. Hence there does not even exist a unitary that satisfies (1.71) just for these two states. Note that if we have only classical bits $|0\rangle$ and $|1\rangle$, then these can be copied. Indeed, $0^2 = 0$ and so for this restricted case there is no contradiction in (1.71).

An interesting consequence of the no-cloning principle, which distinguishes quantum information from classical information, is that in general given an unknown qubit $|\psi\rangle = \alpha |0\rangle + \beta |1\rangle$ it is not possible to determine the amplitudes α and β exactly. Indeed, if it were possible to measure α and β, then one would be able to clone $|\psi\rangle$ by first determining the amplitudes α and β and then building a machine that repeatedly prepares a qubit in the state $\alpha |0\rangle + \beta |1\rangle$. As a result, qubits are very precious. For example, when trying to send a qubit $|\psi\rangle$ through a communication channel it is generally not possible to simply "try again" in case the communication fails.

1.7   The Bloch Sphere

Single-qubit states can be represented in a very convenient visual way in terms of the so-called *Bloch sphere*. To see how this works, write an arbitrary qubit state as

$$|\psi\rangle = e^{i\gamma} \left(\cos \frac{\theta}{2} |0\rangle + e^{i\phi} \sin \frac{\theta}{2} |1\rangle \right), \tag{1.72}$$

where γ, θ, and ϕ are real numbers that can always be taken in $[0, 2\pi)$. As a first step we observe that the global phase $e^{i\gamma}$ can be neglected, as it has no effect at all on the outcome distribution of any measurement that could be performed on the state. To see this, consider the states

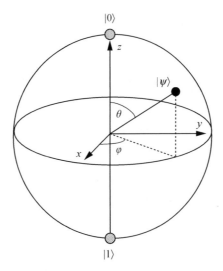

Fig. 1.3   The Bloch sphere. The qubit $|\psi\rangle$ is represented by its Bloch vector
$\vec{r} = (\cos\phi\sin\theta, \sin\phi\sin\theta, \cos\theta)$.

$$|\psi_1\rangle = e^{i\gamma_1}\left(\cos\frac{\theta}{2}|0\rangle + e^{i\phi}\sin\frac{\theta}{2}|1\rangle\right), \tag{1.73}$$

$$|\psi_2\rangle = e^{i\gamma_2}\left(\cos\frac{\theta}{2}|0\rangle + e^{i\phi}\sin\frac{\theta}{2}|1\rangle\right), \tag{1.74}$$

for some real numbers γ_1, γ_2. Note that $|\psi_1\rangle = e^{i(\gamma_1 - \gamma_2)}|\psi_2\rangle$. Then for any measurement with respect to a basis $\{|b\rangle\}_b$, the probability of obtaining an outcome b is equal for both states, since

$$|\langle\psi_1|b\rangle|^2 = \langle b|\psi_1\rangle\langle\psi_1|b\rangle = e^{i(\gamma_1 - \gamma_2)}e^{-i(\gamma_1 - \gamma_2)}\langle b|\psi_2\rangle\langle\psi_2|b\rangle = |\langle\psi_2|b\rangle|^2. \tag{1.75}$$

Also, note that this parametrization preserves the normalization condition since $|\alpha|^2 + |\beta|^2 = \cos^2(\theta/2) + \sin^2(\theta/2) = 1$. Thus the state can be characterized using the real numbers (θ, ϕ) only. This allows us to think of the qubit as a point on a 3-dimensional sphere, as in Figure 1.3. It should be emphasized that this sphere does not follow the same coordinates as we have used for the vectors $|v\rangle \in \mathbb{C}^2$; we need to translate to the new coordinate system.

Definition 1.7.1. *The parametrization* (θ, ϕ) *of*

$$|\psi\rangle = e^{i\gamma}\left(\cos\frac{\theta}{2}|0\rangle + e^{i\phi}\sin\frac{\theta}{2}|1\rangle\right) \tag{1.76}$$

is called the Bloch sphere *representation (Figure 1.3). Any single-qubit state* (1.76) *can be represented by a* Bloch vector $\vec{r} = (\cos\phi\sin\theta, \sin\phi\sin\theta, \cos\theta)$.

Consider a qubit in the representation of Eq. (1.72) where $\gamma = \phi = 0$. Then the Bloch sphere representation of such a qubit lies on the xz-plane. The usefulness of this representation becomes immediately apparent when we consider the effects of the Hadamard gate on a qubit. Note that $(|0\rangle + |1\rangle)/\sqrt{2}$ can be found in Figure 1.3 at the intersection of the positive x-axis and the sphere. It is then easy to see that we can describe the effect of H on $(|0\rangle + |1\rangle)/\sqrt{2}$ as a rotation around the y-axis towards $|1\rangle$, followed by a reflection in the xy-plane. In fact, the Bloch sphere representation allows one to view all single-qubit operations as rotations on this sphere. For the sake of building intuition about quantum operations, it is useful to see how this can be done. A rotation matrix $R_s(\theta)$ is a unitary operation that rotates a qubit Bloch vector around the axes $s \in \{x, y, z\}$ by an angle θ. Such matrices have the following form:

$$R_x(\theta) = e^{-i\theta X/2}, \ R_y(\theta) = e^{-i\theta Y/2}, \ \text{and} \ R_z(\theta) = e^{-i\theta Z/2}, \quad (1.77)$$

where X, Y, Z are the Pauli matrices introduced in the previous section. Especially important is the rotation around the z-axis. We can express it in more detail as

$$R_z(\theta) = e^{-i\theta Z/2} = \begin{pmatrix} e^{-i\theta/2} & 0 \\ 0 & e^{i\theta/2} \end{pmatrix} = e^{-i\theta/2} \begin{pmatrix} 1 & 0 \\ 0 & e^{i\theta} \end{pmatrix}.$$

It can be shown that any arbitrary single-qubit operation U can be expressed in terms of these rotations as

$$U = e^{i\alpha} R_z(\beta) R_y(\gamma) R_z(\delta)$$

for some real numbers α, β, γ, and δ.

Remark 1.7.1 It would be natural to think that more generally for n-qubit states $|\psi\rangle = \sum_x \alpha_x |x\rangle$ the coefficients α_x can be reparametrized using $2^{n+1} - 1$ real parameters and plotted on some form of higher-dimensional analogue of the Bloch sphere. Unfortunately this is not the case, and the Bloch sphere representation is only used for a single qubit, where it forms a useful visualization tool.

QUIZ 1.7.1 *Which of the following states lies on the x-axis of the Bloch sphere?*

(a) $|\psi_1\rangle = |0\rangle$
(b) $|\psi_2\rangle = \frac{1}{\sqrt{2}}(|0\rangle + |1\rangle)$
(c) $|\psi_3\rangle = \frac{1}{\sqrt{2}}(|0\rangle + i|1\rangle)$

QUIZ 1.7.2 *Which of the following states lies on the equator, i.e. the xy-plane, of the Bloch sphere for all values of θ in the indicated range?*

(a) $|\psi_1\rangle = \cos\left(\frac{\theta}{2}\right)|0\rangle + e^{i\frac{\pi}{2}}\sin\left(\frac{\theta}{2}\right)|1\rangle, \ \theta \in [0, \pi]$
(b) $|\psi_2\rangle = \frac{1}{\sqrt{2}}\left(|0\rangle + e^{i\theta}|1\rangle\right), \ \theta \in [0, 2\pi]$
(c) $|\psi_3\rangle = e^{i\theta}|1\rangle, \ \theta \in [0, 2\pi]$

1.8 Implementing Quantum Cryptography

The goal of this book is to teach you the theory needed to become a quantum cryptographer who is capable of analyzing – and maybe even designing! – quantum cryptographic protocols. As such, we will adopt a mathematical approach throughout the remainder of this book, and not consider how quantum cryptographic protocols can be implemented in the real world. That is, we will simply assume that qubits can be manipulated, exchanged, and measured by the protocol participants. If you are nevertheless interested in understanding physical implementations, we provide a very brief introduction to such implementations in this section. Indeed, you may wish to know how far quantum cryptography actually is from a practical reality. Or, if you want to embark on designing your own protocols, which protocols may be easier to realize in practice.

1.8.1 Ingredients

As you might imagine, there is no easy answer to such questions: it depends! To get ourselves closer to answering them, it is useful to examine what is actually needed to implement a quantum cryptographic protocol in the real world. First, we need a device that a user can use to manipulate quantum information locally in order to play their part in a quantum cryptographic protocol. That is, we need a device that can perform quantum measurements, or even more general quantum operations. In full analogy to the classical world, you can think of this device as the quantum laptop that a user might use in order to execute their part of a protocol. In the context of quantum communication, such a device is generally called an end node. In the quantum domain, very simple end nodes capable of preparing and measuring one quantum bit at a time can already be used to realize quantum cryptographic functionality that is impossible to replicate classically. Indeed, we will see one example in Chapter 8! Such end nodes may be realized using relatively simple photonic quantum devices that do not require a quantum memory.

At first glance, it may be surprising that one can do things that are impossible classically with an end node that can only deal with a single qubit at a time. After all, in quantum computing one needs a quantum computer capable of manipulating more quantum bits than can be simulated on a classical supercomputer in order to gain a quantum advantage. Intuitively, the reason why such simple end nodes suffice to gain a quantum advantage in quantum cryptography is the fact that already one qubit suffices to observe some of the properties of quantum mechanics that are essential for cryptography, such as the non-cloning principle and uncertainty relations. What's more, two quantum bits – one for each end node – can share a property called quantum entanglement, which we will learn about in Chapter 4. Since it is impossible to simulate all the properties of quantum entanglement using any amount of classical communication, we can unlock many of the benefits of quantum cryptography using two simple end nodes that share an entangled pair of qubits. Of course, using more sophisticated end nodes one may hope to realize more complex quantum cryptographic functionality. We will see some examples of this later in Chapters 10 and 13.

End nodes themselves are of course not enough: they need a way to talk to each other! The second ingredient that we need is a means to transmit quantum states from one end node to the other, or to create quantum entanglement between quantum devices. That is, we need a way to communicate quantum information between end nodes. This is analogous to the classical communication channel between laptops or phones that is needed in order to execute classical cryptographic protocols. Quantum communication can be performed over physical media that are able to carry light, such as commercial telecom fibers, or through the air, such as freespace connections from a quantum satellite to a ground station.

The third ingredient needed to make quantum cryptographic technologies broadly usable is a bit more subtle to express and relates to the cost and reliability of quantum communication technologies. Such cost could be lowered, for example, by an ability to share parts of such technologies between many users. Classically, it is often cost-effective for many users to share a network such as the internet, which introduces an extra layer of complexity. This motivates the creation of quantum communication networks that can be shared by many users.

When assessing how difficult it is to realize a quantum cryptographic protocol in practice, it is important to examine all ingredients. First, we thus want to examine the requirements of the end nodes that are needed in order to realize a protocol. That is, we want to answer questions such as how many qubits do the protocol participants need to manipulate at once? Does a protocol only require quantum measurements? Or, do we need to store qubits for some time in a quantum memory? Second, we need to be able to allow for quantum communication between the end nodes. This leads to a number of questions, including how far the protocol participants should be from each other. Whether a quantum cryptographic protocol is of interest in practice often depends on the allowed distance between the protocol participants. For example, secret communication using quantum cryptography is generally a lot more interesting if our protagonists Alice and Bob are at least in two separate buildings!

Due to the various different ingredients necessary to realize quantum cryptography in practice, progress in quantum communication technologies is often thought about along three axes (Figure 1.4). Each of these axes corresponds to one of the three ingredients above. Progress may be made independently along one of the axes: (1) accessibility measures how available the technology is to users in terms of practical measures such as cost, reliability, and ease of use. When designing your own quantum cryptographic protocols to be put into industrial use, this criterion is evidently important to answer the question of whether the benefits of using the protocol (presently) justify the costs for the users. The second criterion, (2) distance, measures the distance over which end-to-end quantum communication or entanglement generation can be performed. This is important for understanding how far protocol participants can be away from each other, and whether a quantum cryptographic protocol leads to an interesting use case for achievable distances. Finally, (3) functionality measures the capabilities of the end nodes, matched by the quantum communication channel connecting them. We already mentioned some such capabilities: for example, our protocol might ask to perform quantum measurements, or require storing information in a quantum memory, or maybe even ask for quantum operations on many qubits at once.

Trying to understand progress along such axes is evidently in general quite a complicated endeavor, requiring us to consider many aspects, and indeed many parameters describing the properties of quantum hardware. Delving into the details

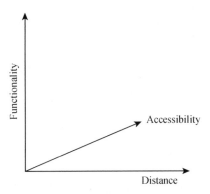

Fig. 1.4   Three axes along which one may measure progress in quantum communication technologies.

of quantum hardware implementations certainly deserves a book on its own, and we thus focus here on one simplifying classification. This classification does, however, allow us to get some initial insights into which types of protocols may be more easy to realize in practice. Specifically, functionality has been characterized by stages of development of a quantum network (see S. Wehner, D. Elkouss, and R. Hanson. Quantum internet: A vision for the road ahead. *Science*, **362**(6412):eaam9288, 2018). Each subsequent stage is more difficult to build from the perspective of quantum hardware, but offers a higher level of functionality to the end user. Examples of existing quantum protocols may be classified into such stages of functionality (Figure 1.5), which we briefly sketch here for completeness. To create a baseline, the stages of functionality of a quantum network include a trusted repeater stage. At this stage, no end-to-end quantum communication or cryptographic security is possible. Instead, the communication channel is chopped up into segments connected via a trusted repeater, often also called trusted node. Quantum cryptography may be used to secure communication on each segment, but interception at each trusted node on the segment is possible (in Chapter 8 we will figure out how to do this as an exercise!). The first true quantum network stage is the prepare-and-measure stage, where each end node may send a one-qubit state to any other end node in such a way that either the state is measured, or the transmission is declared lost. In the entanglement network stage, end nodes are able to produce quantum entanglement between them, while the end nodes themselves remain simple devices capable of measuring only one qubit at a time. In the quantum network stage, end nodes are for the first time able to store and manipulate a small number of quantum bits, unlocking more advanced protocols. Higher stages demand significant advances in end nodes, achieving essentially noise-free computation on at least a few quantum bits (Few qubit fault tolerant stage), and finally connecting end nodes that are large-scale quantum computers (Quantum Computing stage). You can find examples of some protocols we will discuss later in this book in Figure 1.5. We would like to emphasize that it is unknown whether there is a better analysis, or an altogether different protocol in order to solve the same tasks in a lower stage of functionality. We hope that by the end of this book, you are well on your way to both designing and analyzing your own quantum protocols to tackle this question!

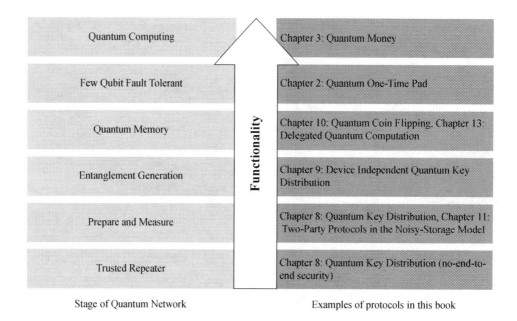

Stage of Quantum Network	Examples of protocols in this book
Quantum Computing	Chapter 3: Quantum Money
Few Qubit Fault Tolerant	Chapter 2: Quantum One-Time Pad
Quantum Memory	Chapter 10: Quantum Coin Flipping, Chapter 13: Delegated Quantum Computation
Entanglement Generation	Chapter 9: Device Independent Quantum Key Distribution
Prepare and Measure	Chapter 8: Quantum Key Distribution, Chapter 11: Two-Party Protocols in the Noisy-Storage Model
Trusted Repeater	Chapter 8: Quantum Key Distribution (no-end-to-end security)

Fig. 1.5   Stages of Quantum Network Development. Each stage is more difficult to build, but allows access to a larger set of possible application protocols. Examples from this book can be found for all stages. Technological advances or an improved analysis can lead to a specific protocol being realized at a lower stage than the one that is obvious from the protocol description.

1.8.2   State of the Art

So now that we have gathered some background knowledge on quantum communication technologies, let us return to the pressing question: Where are we at in putting quantum cryptography into practice? Can we already realize all the protocols presented in this book?

Let us examine such questions along the three axes above. At short distances, and limited functionality, quantum communication is already a commercial reality using relatively easy to use and reliable devices. Devices that realize quantum key distribution (QKD) for secure communication (see Chapter 8) over short distances are commercially available from many vendors around the globe (see e.g. Lux Quanta, http://luxquanta.com; ID Quantique, www.idquantique.com; QuantumCTek, www.quantum-info.com/English/; QBird BV, www.q-bird.nl). Short presently refers to around 100 km in deployed telecom fiber, whereas longer distances of several hundred kilometers have been realized in research labs (see e.g. Toshiba Labs, www.toshiba.eu/pages/eu/Cambridge-Research-Laboratory/quantum-information). These connections are generally point-to-point, and require a dedicated fiber connection between users. At present, many systems require a dedicated dark fiber, that is, a fiber that is not used for any other communication at the same time. However, some systems can already share a fiber with conventional classical fiber communication. Recent years have seen implementations of a variant of QKD called MDI-QKD (see e.g. Figure 1.6), which we will explore in the Julia sheet accompanying Chapter 8. Here many users are connected to a central hub that can

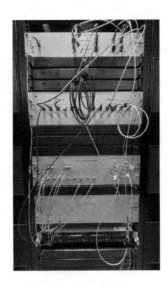

Fig. 1.6   Commercial MDI-QKD system by QBird. Several users are connected via one central hub. [Photo QBird B.V.]

be used to perform QKD between any two users in a metropolitan area connected to it. In principle, one might imagine later connecting such hubs via long-distance backbones to realize a large quantum network connecting together many metropolitan networks.

How about longer distances? Long-distance quantum communication is a highly active area of research. You may be wondering why long-distance quantum communication is actually difficult. After all, we have become quite advanced in terms of classical communication over worldwide distances. Qubits are generally transmitted using light, for example as photons over an optical fiber. It turns out, however, that the transmission of photons over fiber is highly lossy (in fact, exponential in the length of the fiber!). Fiber connections alone thus cannot help us transmit qubits over large distances. In the classical domain, one uses signal amplifiers along fibers in order to mitigate the loss of light in the fiber. Unfortunately – or maybe fortunately for cryptography! – we cannot use such amplifiers in the quantum domain, since they effectively create a copy of some of the quantum information we are trying to send. Moreover, as we saw above, it is impossible to make a copy of an arbitrary quantum bit. The fact that we cannot copy arbitrary qubits thus makes it quite challenging to send quantum information over long distances. However, it is also the very same feature of quantum communication that makes it so suitable for solving cryptographic tasks.

Broadly speaking, two complementary approaches exist in developing long-distance quantum communication in the future. While outside the scope of this book, we provide a number of references that allow you to start reading about such developments. One approach is to use (telecom) fibers in the ground. This necessitates the development of a quantum repeater that allows us to overcome the exponential loss in fiber. Such a quantum repeater would be inserted at specific intervals into the ground. Such a fiber-based approach would allow the connection of potentially many users via the existing telecom grid. Several possible repeater designs are envisioned and we refer to the work of S. Muralidharan, et al. (Optimal architectures for long distance quantum communication. *Scientific Reports*, **6**(20463), 2016) and N. Sangouard, et al. (Quantum repeaters based on atomic ensembles and linear optics. *Reviews of Modern Physics*, **83**:33–80, 2011) for some surveys introducing these fascinating ideas. At present, no quantum repeaters exist that can bridge significant distances, and only proof of principle experiments have been performed (see e.g. M. K. Bhaskar, et al. Experimental demonstration of memory-enhanced quantum communication. *Nature*, **580**:60–64, 2020; D. Lago-Rivera, et al. Telecom-heralded entanglement between multimode solid-state quantum memories. *Nature*, **594**:37–40, 2021; S. Langenfeld, et al. Quantum repeater node demonstrating unconditionally secure key distribution. *Physical Review Letters*, **126**(230506), 2021). Important for us here in this book is the fact that a quantum repeater would enable end-to-end quantum communication, and hence all the protocols in this book can in principle be realized with end-to-end security once such a device has been built. Right now, several networks exist that chain together short QKD links using a trusted node (see above). A trusted node, however, does not enable end-to-end quantum communication, and also no end-to-end security.

Another approach to bridging long distances is to use quantum satellites. Proof of principle demonstrations have been performed, including generating quantum entanglement over more than 1200 km (post-selected on successful detection events; see Juan Yin, et al. Satellite-based entanglement distribution over 1200 kilometers. *Science*, **356**(6343):1140–1144, 2017). Quantum satellites thus promise to bridge very long distances. Yet they typically require large-scale telescopes on the ground, which may make them less suitable for connecting very many users on the ground. Depending on the orbit of the satellite, quantum communication may also only be possible for a small part of the day. Quite conceivably, the two approaches may go hand-in-hand in the future: quantum satellites might be used to create very long-distance backbones for quantum communication, while fiber-based communication may be used for medium-scale distances to achieve high connectivity on the ground.

How about achieving higher stages of functionality? When considering protocols that ask for more than preparing and measuring single qubits, or producing entanglement between end nodes, we need to move to a higher stage of functionality in order to put them into practice. We again provide a very brief overview including some references to help you get started. Achieving higher stages requires an advancement not only of the quantum communication network connecting users, but crucially also of the end nodes that the users use to run applications. Starting with the quantum memory stage, the end nodes are expected to have a quantum memory, and the ability to execute general quantum operations on the qubits. This enables them to execute protocols that require the protocol participants to store qubits for some period of time. From this stage onwards, end nodes are thus no longer simple photonic devices as presently used in QKD systems, but processing nodes, i.e. small quantum computers capable of manipulating qubits, not necessarily in a fault-tolerant manner as desired for general quantum computation. Small means that the processing nodes have only a small number of qubits, possibly not more than one or two. Crucial for

Fig. 1.7   Alice, one of the three nodes of the Delft processing node quantum network. Inside the black aluminum cylinder, the diamond sample is cooled to $-270\,^{\circ}$C, to reduce the noise from the environment and enable the quantum control. [Hanson Lab, Photo Marieke de Lorijn for QuTech]

the use of such processing nodes as end nodes is that they must possess an optical interface capable of connecting to a quantum network, and storage times that are long enough to allow for (classical) communication to be exchanged between the users while still retaining sufficient information inside the quantum memory. As with quantum repeaters, the development of such processing nodes is an active area of research and we refer to S. Wehner, D. Elkouss, and R. Hanson (Quantum internet: A vision for the road ahead. *Science*, **362**(6412):eaam9288, 2018) for an overview. As of now, the ability to link multiple processing nodes has been demonstrated by M. Pompili, et al. (Realization of a multinode quantum network of remote solid-state qubits. *Science*, **372**(6539):259–264, 2021) by creating a three-node quantum network depicted in Figure 1.7.

As you can see, quantum communication is, on the one hand, already a commercial reality. On the other hand, much of it is still at the forefront of cutting-edge quantum research. There is much to do, not only in understanding existing quantum protocols, but also in exploring completely new quantum application protocols. We hope that this book will prepare you for contributing to this existing field of research.

CHAPTER NOTES

For more extensive background than we provide here, we recommend the standard textbook on quantum information by A. Nielsen and I. L. Chuang, *Quantum Computation and Quantum Information* (Cambridge University Press, 2000). Another classic, which also makes the connection with physical implementations, is by B. Schumacher and M. Westmoreland, *Quantum Processes Systems and Information* (Cambridge University Press, 2010). For a far more extensive introduction to linear algebra, without reference to quantum information, we can recommend the book by G. Strang, *Introduction to Linear Algebra*, 3rd edition (Wellesley-Cambridge Press, 1993). If you want to learn much more about quantum mechanics (far more than needed for this book), a standard textbook is by D. J. Griffiths and D. F. Schroeter, *Introduction to Quantum Mechanics* (Cambridge University Press, 2018). For a more light-hearted introduction, focusing on the intuition, we recommend the small book by L. Susskind and A. Friedman, *Quantum Mechanics: The Theoretical Minimum* (Basic Books, 2014).

PROBLEMS

1.1 The simplest quantum communication task

In this problem we investigate the simplest of quantum communication tasks: sending a classical bit using a qubit. Let's recruit two protagonists: Alice and Bob. Alice, a PhD student at Caltech, wants to send some information to Bob, a postdoc at TU Delft. Bob only accepts messages coming through their shared quantum communication device, which can prepare, send, receive, and measure qubits. Imagine Alice wants to send a very simple message that consists of a single bit $a \in \{0,1\}$. In order to do this she encodes her bit by preparing a qubit in the standard basis according to the encoding scheme

$$0 \longrightarrow |0\rangle,$$
$$1 \longrightarrow |1\rangle.$$

Let's further suppose, for now, that Bob knows that Alice sent a qubit encoded according to this scheme. Upon reception of Alice's qubit Bob measures it in the standard basis. Let $b \in \{0,1\}$ denote Bob's outcome. Let p_0 be the probability that $b = 0$ and p_1 the probability that $b = 1$.

1. Compute p_0 and p_1, first in the case that Alice's bit is $a = 0$ and then in the case that it is $a = 1$.

Suppose now that instead of encoding her bit a in the computational basis Alice chooses to encode it in the Hadamard basis:

$$0 \longrightarrow |+\rangle = \frac{1}{\sqrt{2}}(|0\rangle + |1\rangle),$$
$$1 \longrightarrow |-\rangle = \frac{1}{\sqrt{2}}(|0\rangle - |1\rangle).$$

This means that if Alice wants to send the bit 0 she sends the state $|+\rangle$ to Bob.

2. Assume that Bob is unaware of Alice's change of encoding scheme, so that he still measures in the standard basis. Compute p_0 and p_1 in both cases, $a = 0$ and $a = 1$.

3. In both scenarios we considered Alice attempts to send a classical bit to Bob by encoding it in a quantum state. However, in only one of the scenarios could Bob reliably retrieve Alice's bit from the measurement he makes. Which of the two scenarios is this?

4. Suppose Bob knows that Alice is encoding her bit in the Hadamard basis. Describe a unitary operation U such that if Bob applies U to the qubit he receives from Alice, and then measures it in the computational basis, he always obtains the correct outcome (i.e. the outcome 0 when Alice sends a $|+\rangle$ state, and the outcome 1 when Alice sends a $|-\rangle$ state).

Lastly, imagine that Alice's qubit preparation machine is somewhat broken and, when she asks it to prepare her qubit in the state $|0\rangle$, it actually prepares the state

$$|\phi\rangle = \frac{\sqrt{2}}{\sqrt{3}} |0\rangle + \frac{1}{\sqrt{3}} |1\rangle \ .$$

Now imagine that Bob knows this, but the only thing he can do is decide to measure in either the standard basis or the Hadamard basis.

5. Which of Bob's two possible basis choices gives him the highest probability of obtaining the outcome 0, and what is the associated probability?

1.2 The EPR pair

In this problem we show that the EPR state cannot be written as the tensor product of two single-qubit states. There are many possible proofs of this, and the problem indicates one of them. We encourage you to find others! Let

$$|\psi\rangle_{AB} = \alpha_{00} |00\rangle_{AB} + \alpha_{01} |01\rangle_{AB} + \alpha_{10} |10\rangle_{AB} + \alpha_{00} |11\rangle_{AB}$$

be an arbitrary two-qubit state expressed as a linear combination of basis states. Suppose that there exist two single-qubit states

$$|\phi\rangle_A = \beta_{0.A} |0\rangle_A + \beta_{1.A} |1\rangle_A \quad \text{and} \quad |\psi\rangle_B = \beta_{0.B} |0\rangle_B + \beta_{1.B} |1\rangle_B$$

such that $|\psi\rangle_{AB} = |\phi\rangle_A \otimes |\psi\rangle_B$.

1. Show that if $\alpha_{xy} = 0$ for some $x, y \in \{0,1\}$ then necessarily $\beta_{x.A} = 0$ or $\beta_{y.B} = 0$. Remember the definition of the EPR state

$$|\text{EPR}\rangle_{AB} = \frac{1}{\sqrt{2}} |0\rangle_A |0\rangle_B + \frac{1}{\sqrt{2}} |1\rangle_A |1\rangle_B.$$

2. Show that there do not exist any two single-qubit states $|\phi\rangle_A$ and $|\psi\rangle_B$ such that $|\text{EPR}\rangle_{AB} = |\phi\rangle_A \otimes |\psi\rangle_B$.

QUIZ SOLUTIONS

Quiz 1.2.1 (b); Quiz 1.2.2 (a); Quiz 1.3.1 (b); Quiz 1.4.1 False; Quiz 1.4.2 (c); Quiz 1.4.3 False; Quiz 1.5.1 (a); Quiz 1.5.2 (c); Quiz 1.6.1 (a); Quiz 1.6.2 (b); Quiz 1.6.3 (d); Quiz 1.6.4 (a); Quiz 1.7.1 (b); Quiz 1.7.2 (b)

CHEAT SHEET

Given two vectors $|v_1\rangle = \begin{pmatrix} a_1 & \cdots & a_d \end{pmatrix}^T$ and $|v_2\rangle = \begin{pmatrix} b_1 & \cdots & b_d \end{pmatrix}^T$,

1. (Inner product)

$$\langle v_1 | v_2 \rangle := \langle v_1 | \, | v_2 \rangle = \sum_{i=1}^{d} a_i^* b_i.$$

2. (Tensor product)

$$|v_1\rangle \otimes |v_2\rangle := \begin{pmatrix} a_1 b_1 & a_1 b_2 & \cdots & a_1 b_d & a_2 b_1 & \cdots & a_2 b_d & \cdots & a_d b_d \end{pmatrix}^T.$$

Commonly used orthonormal bases for qubits

Standard basis for one qubit: $\mathcal{S} = \{|0\rangle, |1\rangle\}$ where $|0\rangle = \begin{pmatrix} 1 \\ 0 \end{pmatrix}$ and $|1\rangle = \begin{pmatrix} 0 \\ 1 \end{pmatrix}$.

Standard basis for n qubits: $\mathcal{S}_n = \{|x\rangle\}_{x \in \{0,1\}^n}$ where for any string $x = x_1 x_2 \cdots x_n$, $|x\rangle = |x_1\rangle \otimes |x_2\rangle \otimes \cdots \otimes |x_n\rangle$.

Hadamard basis for one qubit: $\mathcal{H} = \{|+\rangle, |-\rangle\}$ where $|\pm\rangle = \frac{1}{\sqrt{2}}(|0\rangle \pm |1\rangle)$. Since these are orthonormal bases, the following hold:

$$\langle 0|1 \rangle = \langle 1|0 \rangle = 0, \qquad \langle 0|0 \rangle = \langle 1|1 \rangle = 1,$$

$$\langle +|- \rangle = \langle -|+ \rangle = 0, \qquad \langle +|+ \rangle = \langle -|- \rangle = 1,$$

$$\langle x|x' \rangle = \delta_{xx'}, \text{ where } x, x' \in \{0,1\}^n \text{ and } \delta_{xx'} \text{ is the Kronecker delta function.}$$

Common representations of a qubit

Standard representation: $|\psi\rangle = \alpha |0\rangle + \beta |1\rangle$, where $\alpha, \beta \in \mathbb{C}$.

Bloch sphere representation: $|\psi\rangle = e^{i\gamma}\left(\cos\frac{\theta}{2}|0\rangle + e^{i\phi}\sin\frac{\theta}{2}|1\rangle\right)$, where $\gamma, \theta, \phi \in \mathbb{R}$.

Properties of the tensor product

For any $|v_1\rangle, |v_2\rangle$, and $|v_3\rangle$,

1. Distributive: $|v_1\rangle \otimes (|v_2\rangle + |v_3\rangle) = |v_1\rangle \otimes |v_2\rangle + |v_1\rangle \otimes |v_3\rangle$
   Also, $|v_1\rangle \otimes (|v_2\rangle + |v_3\rangle) = |v_2\rangle \otimes |v_1\rangle + |v_3\rangle \otimes |v_1\rangle$.
2. Associative: $|v_1\rangle \otimes (|v_2\rangle \otimes |v_3\rangle) = (|v_1\rangle \otimes |v_2\rangle) \otimes |v_3\rangle$.

Similarly, these relations hold for any $\langle v_1|, \langle v_2|$, and $\langle v_3|$.

Probability of measurement outcomes

Consider measuring a quantum state $|\psi\rangle$ in an orthonormal basis $\mathcal{B} = \{|b_i\rangle\}_{i=1}^{d}$. The probability of measuring a particular outcome "b_i" is $p_i = |\langle b_i|\psi\rangle|^2$. After the measurement, if a certain outcome "b_i" is observed, then the state $|\psi\rangle$ collapses to $|b_i\rangle$.

Pauli matrices

The Pauli matrices are 2×2 matrices,

$$X = \begin{pmatrix} 0 & 1 \\ 1 & 0 \end{pmatrix}, \quad Z = \begin{pmatrix} 1 & 0 \\ 0 & -1 \end{pmatrix}, \quad Y = iXZ,$$

and the following relations hold:

$$X\left|0\right\rangle = \left|1\right\rangle, \; X\left|1\right\rangle = \left|0\right\rangle \qquad X\left|+\right\rangle = \left|+\right\rangle, \; X\left|-\right\rangle = -\left|-\right\rangle$$

$$Z\left|0\right\rangle = \left|0\right\rangle, \; Z\left|1\right\rangle = -\left|1\right\rangle \qquad Z\left|+\right\rangle = \left|-\right\rangle, \; Z\left|-\right\rangle = \left|+\right\rangle$$

$$Y\left|0\right\rangle = i\left|1\right\rangle, \; Y\left|1\right\rangle = -i\left|0\right\rangle \qquad Y\left|+\right\rangle = -i\left|-\right\rangle, \; Y\left|-\right\rangle = i\left|+\right\rangle$$

2

Quantum Tools and a First Protocol

This chapter covers our first cryptographic task: we will learn how to encrypt quantum states! To prepare our entry into quantum communication and cryptography, we first need to learn a little more about quantum information. Before proceeding, make sure you are comfortable with the notions introduced in Chapter 1. In this chapter we extend these notions in several ways that will be essential to model interesting cryptographic scenarios.

2.1 Probability Notation

We start by recalling standard notions of probability theory, and defining associated notation which we use frequently. Consider a discrete random variable X taking values in a finite set $\mathcal{X}$. We often write $|X|$ for the size of the set $\mathcal{X}$ over which X ranges. The probability distribution of X is specified by a function $P_X(\cdot) : \mathcal{X} \to [0,1]$ such that for any $x \in \mathcal{X}$, $P_X(x)$ denotes the probability that X takes on a specific value $x \in \mathcal{X}$. Recall that for a probability distribution, the normalization condition $\sum_{x \in \mathcal{X}} P_X(x) = 1$ always holds. When the distribution is clear from context we use the shorthand

$$p_x = \Pr(X = x) = P_X(x)$$

for the probability that x occurs. If P is a distribution and X a random variable, we will write $X \sim P$ to indicate that the distribution of X is P. We sometimes extend this notation and write $X \sim Y$, where X, Y are random variables, to indicate that they have the same distribution.

Example 2.1.1 Let $\mathcal{X} = \{1,2,3,4,5,6\}$ correspond to the faces of a six-sided die. If the die is fair, i.e. all sides have equal probability of occurring, then $P_X(x) = 1/6$ for all $x \in \mathcal{X}$. Using our shorthand notation this can also be written as $p_x = 1/6$. The size of the range of X is $|X| = 6$. ∎

A random variable X ranging over a set $\mathcal{X}$ can be correlated with another random variable Y ranging over $\mathcal{Y}$. This means that they have a joint distribution $P_{XY}(\cdot, \cdot) : \mathcal{X} \times \mathcal{Y} \to [0,1]$ which is not necessarily a product. That is, $P_{XY}(x,y) \neq P_X(x)P_Y(y)$ in general, where P_X (resp. P_Y) is the marginal distribution of X (resp. Y), defined by $P_X(x) = \sum_{y \in \mathcal{Y}} P_{XY}(x,y)$ (and similarly for Y). This leads to the notion of *conditional probabilities* $P_{X|Y}(x|y)$, where $P_{X|Y}(x|y)$ is the probability that X takes on the value x, conditioned on the event that Y takes on the value y. Bayes' rule relates this conditional probability to the joint probabilities:

$$P_{X|Y}(x|y) = \frac{P_{XY}(x,y)}{P_Y(y)} \ ,$$

whenever $P_Y(y) > 0$.[1] We use the following shorthand when it is clear which random variable we refer to:

$$p_{x|y} = \Pr(X = x|Y = y) = P_{X|Y}(x|y) \ .$$

Example 2.1.2 Let $Y \in \mathcal{Y} = \{\text{"fair"}, \text{"unfair"}\}$ refer to the choice of either a fair or an unfair die, each chosen with equal probability: $P_Y(\text{fair}) = 1/2$ and $P_Y(\text{unfair}) = 1/2$. If X denotes the fair or unfair die, where the unfair die always rolls a "6" (that is, $\mathcal{X} = \{1,2,3,4,5,6\}$, with $P_X(6) = 1$ and $P_X(x) = 0$ for $x \neq 6$), then $P_{X|Y}(x|\text{fair}) = 1/6$ for all x, but $P_{X|Y}(6|\text{unfair}) = 1$ and $P_{X|Y}(x|\text{unfair}) = 0$ for $x \neq 6$. ∎

Exercise 2.1.1 Compute explicitly the joint probability $P_{XY}(x,y)$ for the random variables in Example 2.1.2.

Exercise 2.1.2 Suppose that Alice chooses between the fair or unfair die from Example 2.1.2 with probability $P_Y(\text{fair}) = P_Y(\text{unfair}) = 1/2$, but does not reveal to us which choice was made. Imagine that we roll the (fair or unfair) die and obtain the outcome X. Suppose that we see $X = 3$. Can we guess what die Alice used? That is, what is the most likely value of Y, "fair" or "unfair"? Answer the same question in the case when we observe that $X = 6$.

2.2 Density Matrices

The quantum generalizations of probability distributions, i.e. probability distributions over quantum states, are called *density matrices*. There are two main motivations for working with density matrices. The first motivation is to model the kind of scenario described above. Suppose, for example, that we build a device that prepares either a state $|\psi_1\rangle$, with some probability p_1, or a state $|\psi_2\rangle$, with probability p_2. Wouldn't it be nice to have a concise mathematical way to describe the "average" quantum state returned by this device, without having to resort to words as in the previous sentence? We will call such a state a *mixed state*, in contrast to the pure states we have studied so far.

There is a second motivation for introducing density matrices, which is that they are necessary to describe the quantum state of a subsystem of a general system. To understand why this is the case, imagine having two quantum systems A and B. For example, A and B are two qubits such that the state of A and B is a normalized vector $|\psi\rangle_{AB} \in \mathbb{C}^4$. Given this situation, how can we mathematically describe the state of qubit A? Note that physically speaking, if we imagine qubits A and B as being in

1 The marginal distribution of X given $Y = y$ is undefined if y cannot occur, i.e. whenever $P_Y(y) = 0$.

very far-away locations, then intuitively there must be a way to describe the state of A without referring to B at all. So how do we do it?

Consider first an easy case. Suppose that the joint state of A and B takes the form

$$|\psi\rangle_{AB} = |\psi_1\rangle_A \otimes |\psi_2\rangle_B .$$

Then the answer is clear: the state of A is the normalized vector $|\psi_1\rangle_A$. However, remember from Chapter 1 that there exist quantum states $|\psi\rangle_{AB}$ that cannot be written as a simple tensor product like this! A good example of such a state is the EPR pair

$$|\text{EPR}\rangle_{AB} = \frac{1}{\sqrt{2}} |0\rangle_A |0\rangle_B + \frac{1}{\sqrt{2}} |1\rangle_A |1\rangle_B .$$

As shown in Problem 1.2, it is impossible to express $|\text{EPR}\rangle_{AB} = |\psi_1\rangle_A \otimes |\psi_2\rangle_B$ for some states $|\psi_1\rangle_A$ and $|\psi_2\rangle_B$. In this case, how can we describe the state of A? It seems like we dug ourselves into a mathematical rabbit-hole. Either we find a way to describe the state of A, or there is a problem with our formalism. As we will see, the answer to this question is the same as the previous one: the notion of a *density matrix* will help us save the day.

2.2.1 Introduction to the Formalism

We start by giving a different way to represent pure quantum states: as matrices. Recall that a ket $|\psi\rangle$ is a column vector, while the bra $\langle\psi|$ is a row vector. Therefore, $\rho = |\psi\rangle\langle\psi|$ is a rank-1 matrix: it has precisely 1 nonzero eigenvalue (equal to 1), with associated eigenstate $|\psi\rangle$. The matrix ρ is called the *density matrix* of the quantum state.

Example 2.2.1 For the states $|0\rangle$ and $|+\rangle = (|0\rangle + |1\rangle)/\sqrt{2}$ we obtain the density matrices

$$|0\rangle\langle 0| = \begin{pmatrix} 1 \\ 0 \end{pmatrix} (1\ 0) = \begin{pmatrix} 1 & 0 \\ 0 & 0 \end{pmatrix},$$

$$|+\rangle\langle +| = \frac{1}{2} \begin{pmatrix} 1 \\ 1 \end{pmatrix} (1\ 1) = \frac{1}{2} \begin{pmatrix} 1 & 1 \\ 1 & 1 \end{pmatrix}. \qquad\blacksquare$$

How does writing down states as matrices help us resolve the questions above? To see how, let us first consider the first motivation that we gave: the need for a formalism that can represent probabilistic combinations of pure quantum states. But before that, let us remember that physically the only information that we can obtain about a quantum state is obtained by performing a measurement. Moreover, if a state $|\psi\rangle$ is measured in a basis that contains the vector $|b\rangle$, then the probability of obtaining the outcome "$|b\rangle$" is given by

$$|\langle b| \psi\rangle|^2 = \langle b| \psi\rangle\langle\psi |b\rangle = \langle b| \rho |b\rangle , \qquad (2.1)$$

where as earlier $\rho = |\psi\rangle\langle\psi|$ is the density matrix representation of the pure state $|\psi\rangle$. In words: the probability of obtaining the outcome "$|b\rangle$" is obtained by computing the *overlap* of $|b\rangle$ with ρ, which is defined as the quantity $\langle b| \rho |b\rangle$.

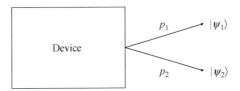

Fig. 2.1 Device that prepares two possible states with equal probability.

 Remark 2.2.1 While $|\psi\rangle \neq -|\psi\rangle$ as vectors, as matrices we have $|\psi\rangle\langle\psi| = (-|\psi\rangle)$ $(-\langle\psi|)$. Thanks to the modulus squared in the computation of probabilities we see that the (-1) phase has no observable consequence, and so representing the state vector as a matrix does not lose information.

Moving on, let us consider the case where our preparation device prepares one of two possible states, $|\psi_1\rangle$ or $|\psi_2\rangle$, with equal probability $p_1 = p_2 = 1/2$ as in Figure 2.1. We claim that an accurate matrix representation of the state produced by the device can be obtained by taking the linear combination

$$\rho = \frac{1}{2}|\psi_1\rangle\langle\psi_1| + \frac{1}{2}|\psi_2\rangle\langle\psi_2| .$$

More generally, if the device prepares $|\psi_x\rangle$ with probability p_x, the density matrix representation of the resulting state is

$$\rho = \sum_x p_x |\psi_x\rangle\langle\psi_x| . \tag{2.2}$$

To verify that this choice of representation is accurate, consider what happens if we measure the state output by the device in a basis that contains the vector $|b\rangle$. If the state is $|\psi_x\rangle$ for some x, then the probability of obtaining the outcome "$|b\rangle$" is

$$q_{b|x} = |\langle b | \psi_x\rangle|^2 = \langle b|\psi_x\rangle\langle\psi_x|b\rangle .$$

Since the state $|\psi_x\rangle$ is prepared with probability p_x we expect the overall probability of obtaining the outcome "$|b\rangle$" to be

$$q_b = \sum_x p_x q_{b|x} .$$

Observe that

$$q_b = \sum_x p_x q_{b|x} = \sum_x p_x \langle b|\psi_x\rangle\langle\psi_x|b\rangle = \langle b| \left(\sum_x p_x |\psi_x\rangle\langle\psi_x| \right) |b\rangle = \langle b|\rho|b\rangle ,$$

which is precisely the same rule as (2.1). This means that the density matrix representation (2.2) captures the right amount of information about the state of the system so that the distribution of outcomes of any measurement on the state can be recovered using the generalized measurement rule (2.1).

Example 2.2.2 Suppose more generally that a device prepares a state with density matrix ρ_x with probability p_x. Then the density matrix that describes the overall state prepared by the device is given by

$$\rho = \sum_x p_x \rho_x \, .$$

The set of probabilities and density matrices $\mathcal{E} = \{(p_x, \rho_x)\}_x$ is called an *ensemble* of states. ∎

Example 2.2.3 Suppose that a device prepares $|0\rangle\langle 0|$ with probability $1/2$, and $|+\rangle\langle +|$ with probability $1/2$ (Figure 2.1). Then the resulting density matrix is given by

$$\rho = \frac{1}{2}|0\rangle\langle 0| + \frac{1}{2}|+\rangle\langle +| = \frac{1}{2}\begin{pmatrix} 1 & 0 \\ 0 & 0 \end{pmatrix} + \frac{1}{4}\begin{pmatrix} 1 & 1 \\ 1 & 1 \end{pmatrix} = \frac{1}{4}\begin{pmatrix} 3 & 1 \\ 1 & 1 \end{pmatrix} .$$

∎

Be careful that *superposition* is not the same as a *probabilistic combination*! Intuitively, the difference is that a probabilistic combination, often called a *mixture*, is an inherently classical object: there is a process that prepares one *or* the other state with some probability. In contrast, a state in a superposition is in some sense one *and* the other; it is a truly quantum phenomenon. The following example illustrates the difference between the two.

Example 2.2.4 Consider the difference between preparing a *mixture* of $|0\rangle\langle 0|$ and $|1\rangle\langle 1|$, and creating a *superposition* over $|0\rangle$ and $|1\rangle$. First, consider a source that prepares the states $|0\rangle\langle 0|$ and $|1\rangle\langle 1|$ with probabilities $p_0 = p_1 = 1/2$. Suppose we measure the resulting density matrix

$$\rho = \frac{1}{2}|0\rangle\langle 0| + \frac{1}{2}|1\rangle\langle 1| = \frac{1}{2}\mathbb{I}$$

in the Hadamard basis $\{|+\rangle, |-\rangle\}$. Then the probability of each possible outcome is given by

$$q_+ = \langle +|\rho|+\rangle = \frac{1}{2} \, ,$$
$$q_- = \langle -|\rho|-\rangle = \frac{1}{2} \, .$$

In contrast, consider now a state that is an equal superposition of $|0\rangle$ and $|1\rangle$, i.e. the state $|+\rangle = \frac{1}{\sqrt{2}}(|0\rangle + |1\rangle)$. Measuring $|+\rangle$ in the Hadamard basis results in $q_+ = 1$ and $q_- = 0$. The probabilities are different, so the two states are different! Indeed,

$$|+\rangle\langle +| = \frac{1}{2}\begin{pmatrix} 1 & 1 \\ 1 & 1 \end{pmatrix} \neq \frac{1}{2}\begin{pmatrix} 1 & 0 \\ 0 & 1 \end{pmatrix} = \frac{1}{2}\mathbb{I} \, .$$

∎

Remark 2.2.2 Note that the same density matrix ρ can be obtained from different ensembles $\{(p_x, \rho_x)\}_x$. A simple example is provided by the density matrix

$$\rho = \frac{\mathbb{I}}{2},$$

which is also called the *maximally mixed* state. You can verify that

$$\frac{\mathbb{I}}{2} = \frac{1}{2}(|0\rangle\langle 0| + |1\rangle\langle 1|) = \frac{1}{2}(|+\rangle\langle +| + |-\rangle\langle -|),$$

and many other equivalent decompositions are possible. (The maximally mixed state arises very frequently in cryptography, because it represents a state of *complete uncertainty*.) What this means is that the two processes, generating the states $|0\rangle$ or $|1\rangle$ with probability 1/2 each, or generating the states $|+\rangle$ or $|-\rangle$ with probability 1/2 each, return quantum states that are physically indistinguishable: they have the same density matrix representation $\rho = (1/2)\mathbb{I}$.

> **QUIZ 2.2.1** *Suppose a system is produced in state $|0\rangle$ with probability $p_0 = 1/2$ and in state $|-\rangle$ with probability $p_1 = 1/2$. What is the resulting density matrix?*
>
> **(a)** $\rho = \frac{1}{2}\begin{pmatrix} 1 & 0 \\ 0 & 1 \end{pmatrix}$
>
> **(b)** $\rho = \frac{1}{4}\begin{pmatrix} 3 & 1 \\ 1 & 1 \end{pmatrix}$
>
> **(c)** $\rho = \frac{1}{4}\begin{pmatrix} 3 & -1 \\ -1 & 1 \end{pmatrix}$
>
> **(d)** $\rho = \frac{1}{2}\begin{pmatrix} 1 & 1 \\ 1 & 1 \end{pmatrix}$

2.2.2 A Little Bit of Math

To formally define density matrices and their properties we recall some important notions from linear algebra. We start with Hermitian, and then positive semidefinite, matrices.

> **Definition 2.2.1 (Hermitian matrix M).** *A $d \times d$ complex matrix M is Hermitian if it satisfies $M^\dagger = M$, where recall from Definition 1.1.2 that $M^\dagger$ denotes the conjugate-transpose.*

To define density matrices formally, we need a few more mathematical concepts. The spectral theorem states that any Hermitian matrix M can be diagonalized with real eigenvalues. This means that there exists an orthonormal basis $\{|v_j\rangle\}_j$ of $\mathbb{C}^d$ (the *eigenvectors*) and real numbers λ_j (the *eigenvalues*) such that $M = \sum_j \lambda_j |v_j\rangle\langle v_j|$.

> **Definition 2.2.2 (Positive semidefinite matrix).** *A Hermitian matrix M is positive semidefinite if all its eigenvalues $\{\lambda_i\}_i$ are non-negative. This condition is denoted $M \geq 0$.*

Exercise 2.2.1 Show that a matrix M is positive semidefinite if and only if $\langle v| M |v\rangle \geq 0$ for all unit vectors $|v\rangle$. In particular, the diagonal coefficients $\langle i| M |i\rangle$ of M in any basis are non-negative.

Exercise 2.2.2 Show that the diagonal coefficients being positive is not a sufficient condition for M to be positive semidefinite: find an M such that the diagonal coefficients of M are all positive, but M itself is not positive semidefinite.

Exercise 2.2.3 Even worse: find an M such that *all* coefficients (i.e. entries) of M are non-negative, but M is not positive semidefinite.

An important operation on matrices is the *trace*, which is simply the sum of the diagonal elements. It is convenient to note that the trace can also be expressed as follows.

Definition 2.2.3 (Trace of a matrix). *The trace of a $d \times d$ matrix M is*

$$\mathrm{tr}(M) = \sum_{i=0}^{d-1} \langle i| M |i\rangle \,,$$

where $\{|i\rangle\}_i$ is any *orthonormal basis of $\mathbb{C}^d$.*

The definition implicitly assumes that the definition of the trace does not depend on the choice of orthonormal basis. Let's verify that this is indeed the case. First, in the following exercise we verify an important property of the trace, which is that it is *cyclic*. We will frequently make use of this property in our calculations.

Exercise 2.2.4 Show that for any matrices M, N (such that both products MN and NM are well-defined) we have $\mathrm{tr}(MN) = \mathrm{tr}(NM)$. We will often use this property to perform manipulations such as

$$\langle i|A|i\rangle = \mathrm{tr}(\langle i|A|i\rangle) = \mathrm{tr}(A|i\rangle\langle i|) \,,$$

where we made use of the fact that the trace is cyclic with $M = \langle i|$ and $N = A|i\rangle$. (Make sure you can follow all the kets and bras!) It is worth noting that in general a noncyclic permutation of the matrices does not preserve the trace. More precisely, for matrices M, N, P, in general

$$\mathrm{tr}(MNP) \neq \mathrm{tr}(NMP) \,.$$

Now, if $\{|u_i\rangle\}_i$ is any orthonormal basis of $\mathbb{C}^d$, we know that there exists a unitary transformation U such that $U|i\rangle = |u_i\rangle$ for all $i = 0, \ldots, d-1$. So given a $d \times d$ matrix M,

$$\sum_i \langle u_i| M |u_i\rangle = \sum_i \langle i| U^* M U |i\rangle$$

$$= \mathrm{tr}(U^\dagger M U)$$

$$= \mathrm{tr}(M U U^\dagger)$$

$$= \mathrm{tr}(M) \,,$$

where for the second line we used the cyclicity property and for the last line we used $UU^\dagger = \mathbb{I}$. This shows that our definition of the trace is indeed independent of the choice of orthonormal basis! In particular, by choosing the basis of eigenvectors of M, you can verify that for any Hermitian matrix M, $\mathrm{tr}(M)$ is the sum of its eigenvalues (counted with multiplicity).

2.2.3 Density Matrices and Their Properties

Before we take the density matrix ρ as our new definition for a general quantum state, let us investigate when an arbitrary matrix ρ can be considered a valid density matrix, that is, a valid representation of a quantum state. It turns out that two properties are necessary and sufficient: the matrix ρ should be *positive semidefinite* and have *trace equal to 1*.

To see why this is true, consider the diagonalized representation of a density matrix ρ as a function of its eigenvalues $\{\lambda_j\}_j$ and corresponding eigenvectors $\{|v_j\rangle\}_j$:

$$\rho = \sum_j \lambda_j |v_j\rangle\langle v_j| .$$

Imagine that we measure ρ in an orthonormal basis $\{|w_k\rangle\}_k$. Based on (2.1) we know that the probability of obtaining the measurement outcome k is given by

$$q_k = \langle w_k| \rho |w_k\rangle . \tag{2.3}$$

For this to specify a proper distribution, it must be that $q_k \geq 0$ and $\sum_k q_k = 1$. By performing the measurement in the eigenbasis of ρ, $|w_j\rangle = |v_j\rangle$, we obtain the necessary conditions $\lambda_j \geq 0$, that is, ρ is a *positive semidefinite* matrix, and $\mathrm{tr}(\rho) = 1$, since

$$1 = \sum_j q_j = \sum_j \lambda_j \, \mathrm{tr}(|v_j\rangle\langle v_j|) = \mathrm{tr}(\rho) .$$

This shows that the two conditions are necessary for ρ to lead to well-defined distributions on measurement outcomes when using the rule (2.1). The following exercise asks you to show that the conditions are also sufficient.

Exercise 2.2.5 Show that for any positive semidefinite matrix ρ with trace 1, and any orthonormal basis $\{|w_k\rangle\}_k$, the numbers $q_k = \langle w_k| \rho |w_k\rangle$ are real, non-negative, and sum to 1.

We give a formal definition of a density matrix, which is the most general way of representing a quantum state.

Definition 2.2.4 (Density matrix). *A* density matrix *on* $\mathbb{C}^d$ *is a* $d \times d$ *matrix* ρ *such that* $\rho \geq 0$ *and* $\mathrm{tr}(\rho) = 1$. *If furthermore* ρ *is of rank 1, then* ρ *is called a* pure density matrix. *Otherwise it is called a* mixed *density matrix*.

Note that by definition a pure density matrix is of the form $\rho = \lambda_1 |u_1\rangle\langle u_1|$, where the trace condition implies that necessarily $\lambda_1 = 1$. Thus, for the case of pure states, density matrices and the vector representation we got used to before are in one-to-one correspondence. (Except for the phase, which as we pointed out is not relevant since there is no observation on the state that can determine it.)

We also summarize the rule for computing outcome probabilities when measuring a quantum system described by the density matrix ρ.

Definition 2.2.5 (Measuring a density matrix in a basis). *Consider a density matrix ρ. Measuring ρ in the orthonormal basis $\{|b_j\rangle\}_j$ results in outcome j with probability*

$$q_j = \langle b_j|\rho|b_j\rangle .$$

QUIZ 2.2.2 *Is $\rho = \begin{pmatrix} 1 & 0 \\ 0 & 1 \end{pmatrix}$ a valid density matrix?*

(a) *Yes*

(b) *No*

QUIZ 2.2.3 *Is there always a unique way of preparing the state described by a given density matrix?*

(a) *Yes*

(b) *No*

2.2.4 Bloch Representation for One-Qubit Mixed States

In Chapter 1 we saw that single-qubit states have a convenient graphical representation in terms of a vector on the Bloch sphere. In particular, any pure quantum state can be described by a *Bloch vector* $\vec{r} = (\cos\phi\sin\theta, \sin\phi\sin\theta, \cos\theta)$. Rather conveniently, the representation extends to mixed states. Concretely, it is always possible to write a single-qubit density matrix as

$$\rho = \frac{1}{2}\left(\mathbb{I} + v_x X + v_z Z + v_y Y\right) , \tag{2.4}$$

where X, Y, Z are the Pauli matrices defined in Chapter 1 and v_x, v_y, v_z are real coefficients. The fact that such an expansion always exists follows from the fact that the matrices $\mathcal{P} = \{\mathbb{I}, X, Y, Z\}$ form a basis for the space of 2×2 density matrices that correspond to a qubit.

Exercise 2.2.6 Use the fact that all matrices $M, N \in \mathcal{P}$ with $M \neq N$ anti-commute, i.e. $\{M,N\} = MN + NM = 0$, to show that $\text{tr}(MN) = 0$ whenever $M \neq N \in \mathcal{P}$.

Exercise 2.2.7 Using the orthogonality condition (2.5), show that

$$|0\rangle\langle 0| = \frac{1}{2}\left(\mathbb{I} + Z\right) ,$$

$$|1\rangle\langle 1| = \frac{1}{2}\left(\mathbb{I} - Z\right) .$$

The exercise shows that the matrices $\mathbb{I}, X, Y, Z$ are orthogonal under the Hilbert–Schmidt inner product $\langle A, B \rangle = \text{tr}(A^\dagger B)$. That is,

$$\text{tr}(X^\dagger Y) = \text{tr}(X^\dagger Z) = \text{tr}(X^\dagger \mathbb{I}) = 0 \, , \tag{2.5}$$

and similarly for all other pairs of matrices. This is why we can refer to them as an orthonormal basis.

If ρ is pure you can verify that the vector $\vec{v} = (v_x, v_y, v_z)$ is precisely the Bloch vector $\vec{r}$ defined in Chapter 1. For pure states $\|\vec{v}\|_2^2 = v_x^2 + v_y^2 + v_z^2 = 1$. In other words, pure states live on the surface of the Bloch sphere. For mixed states, however, we can have $\|\vec{v}\|_2^2 \leq 1$. Mixed states thus lie in the interior of the Bloch sphere. For the case of 2×2 matrices, the vector $\vec{v}$ tells us immediately whether the matrix ρ is a valid one-qubit quantum state: this is the case if and only if $\|\vec{v}\|_2^2 \leq 1$.

QUIZ 2.2.4 *A qubit density matrix with Bloch vector $v = (0.8, 0, 0.8)$ is*

(a) *A pure state*

(b) *A mixed state*

(c) *Not a valid quantum state*

QUIZ 2.2.5 *The matrix $\rho = \frac{1}{2}\mathbb{I}$ is*

(a) *A pure state*

(b) *A mixed state*

(c) *Not a valid quantum state*

2.2.5 Combining Density Matrices

Suppose we are given two quantum systems A and B, described by density matrices ρ_A and ρ_B respectively. How should their joint state ρ_{AB} be defined? In the previous chapter we saw that two pure quantum states $|v_1\rangle \in \mathbb{C}^{d_1}$ and $|v_2\rangle \in \mathbb{C}^{d_2}$ can be combined by taking their tensor product $|v_1\rangle \otimes |v_2\rangle \in \mathbb{C}^{d_1} \otimes \mathbb{C}^{d_2}$. It turns out that the rule for mixed states is very similar.

Let us start with the simple case where ρ_A, ρ_B are 2×2-dimensional matrices. Then

$$\rho_A \otimes \rho_B = \begin{pmatrix} m_{11} & m_{12} \\ m_{21} & m_{22} \end{pmatrix} \otimes \begin{pmatrix} n_{11} & n_{12} \\ n_{21} & n_{22} \end{pmatrix} = \begin{pmatrix} m_{11}\begin{pmatrix} n_{11} & n_{12} \\ n_{21} & n_{22} \end{pmatrix} & m_{12}\begin{pmatrix} n_{11} & n_{12} \\ n_{21} & n_{22} \end{pmatrix} \\ m_{21}\begin{pmatrix} n_{11} & n_{12} \\ n_{21} & n_{22} \end{pmatrix} & m_{22}\begin{pmatrix} n_{11} & n_{12} \\ n_{21} & n_{22} \end{pmatrix} \end{pmatrix}$$

$$= \begin{pmatrix} m_{11}n_{11} & m_{11}n_{12} & m_{12}n_{11} & m_{12}n_{12} \\ m_{11}n_{21} & m_{11}n_{22} & m_{12}n_{21} & m_{12}n_{22} \\ m_{21}n_{11} & m_{21}n_{12} & m_{22}n_{11} & m_{22}n_{12} \\ m_{21}n_{21} & m_{21}n_{22} & m_{22}n_{21} & m_{22}n_{22} \end{pmatrix} .$$

For example, if we have two density matrices $\rho_A = \begin{pmatrix} 1 & 0 \\ 0 & 0 \end{pmatrix}$ and $\rho_B = \begin{pmatrix} 0 & 0 \\ 0 & 1 \end{pmatrix}$, then

$$\rho_{AB} = \rho_A \otimes \rho_B = \begin{pmatrix} 1 \cdot \rho_B & 0 \cdot \rho_B \\ 0 \cdot \rho_B & 0 \cdot \rho_B \end{pmatrix} = \begin{pmatrix} 0 & 0 & 0 & 0 \\ 0 & 1 & 0 & 0 \\ 0 & 0 & 0 & 0 \\ 0 & 0 & 0 & 0 \end{pmatrix}.$$

This definition easily extends to larger matrices as follows.

Definition 2.2.6 (Tensor product). *Consider any $d' \times d$ matrix ρ_A and $k' \times k$ matrix ρ_B,*

$$\rho_A = \begin{pmatrix} m_{11} & m_{12} & \cdots & m_{1d} \\ m_{21} & \ddots & \ddots & m_{2d} \\ \vdots & \ddots & \ddots & \vdots \\ m_{d'1} & m_{d'2} & \cdots & m_{d'd} \end{pmatrix}, \quad \rho_B = \begin{pmatrix} n_{11} & n_{12} & \cdots & n_{1k} \\ n_{21} & \ddots & \ddots & n_{2k} \\ \vdots & \ddots & \ddots & \vdots \\ n_{k'1} & n_{k'2} & \cdots & n_{k'k} \end{pmatrix}.$$

Their tensor product is given by the $d'k' \times dk$ matrix

$$\rho_{AB} = \rho_A \otimes \rho_B = \begin{pmatrix} m_{11}\rho_B & m_{12}\rho_B & \cdots & m_{1d}\rho_B \\ m_{21}\rho_B & \ddots & \ddots & m_{2d}\rho_B \\ \vdots & \ddots & \ddots & \vdots \\ m_{d'1}\rho_B & m_{d'2}\rho_B & \cdots & m_{d'd}\rho_B \end{pmatrix}.$$

As a word of caution, beware that the tensor product, as the usual matrix product, is noncommutative.

Example 2.2.5 Consider the density matrices $\rho_A = \frac{1}{4}\begin{pmatrix} 1 & 1 & 0 \\ 1 & 2 & 1 \\ 0 & 1 & 1 \end{pmatrix}$ and $\rho_B = \frac{1}{2}\begin{pmatrix} 1 & -i \\ i & 1 \end{pmatrix}$.

Then

$$\rho_A \otimes \rho_B = \frac{1}{8}\begin{pmatrix} 1 & -i & 1 & -i & 0 & 0 \\ i & 1 & i & 1 & 0 & 0 \\ 1 & -i & 2 & -2i & 1 & -i \\ i & 1 & 2i & 2 & i & 1 \\ 0 & 0 & 1 & -i & 1 & -i \\ 0 & 0 & i & 1 & i & 1 \end{pmatrix},$$

and

$$\rho_B \otimes \rho_A = \frac{1}{8}\begin{pmatrix} 1 & 1 & 0 & -i & -i & 0 \\ 1 & 2 & 1 & -i & -2i & -i \\ 0 & 1 & 1 & 0 & -i & -i \\ i & i & 0 & 1 & 1 & 0 \\ i & 2i & i & 1 & 2 & 1 \\ 0 & i & i & 0 & 1 & 1 \end{pmatrix} \neq \rho_A \otimes \rho_B .$$

■

QUIZ 2.2.6 $\frac{1}{2}\left(\rho_A^1 + \rho_A^2\right) \otimes \rho_B = \frac{1}{2}\left(\rho_A^1 \otimes \rho_B + \rho_A^2 \otimes \rho_B\right)$ *for all* ρ_A^1, ρ_A^2 *and* ρ_B. *True or false?*

(a) *True*
(b) *False*

QUIZ 2.2.7 $\rho_A \otimes \rho_B = \rho_B \otimes \rho_A$ *for all* ρ_A *and* ρ_B. *True or false?*

QUIZ 2.2.8 *What is the tensor product* $\rho_{AB} = \rho_A \otimes \rho_B$ *of* $\rho_A = |1\rangle\langle 1|$ *and* $\rho_B = \frac{\mathbb{I}}{2}$?

(a) $\rho_{AB} = \frac{1}{4}\begin{pmatrix} 1 & 0 & 1 & 0 \\ 0 & 1 & 0 & 1 \\ 1 & 0 & 1 & 0 \\ 0 & 1 & 0 & 1 \end{pmatrix}$

(b) $\rho_{AB} = \frac{1}{2}\begin{pmatrix} 0 & 0 & 0 & 0 \\ 0 & 0 & 0 & 0 \\ 0 & 0 & 1 & 0 \\ 0 & 0 & 0 & 1 \end{pmatrix}$

(c) $\rho_{AB} = \frac{1}{2}\begin{pmatrix} 0 & 0 & 0 & 0 \\ 0 & 0 & 0 & 0 \\ 0 & 0 & 1 & 1 \\ 0 & 0 & 1 & 1 \end{pmatrix}$

2.2.6 Classical-Quantum States

In quantum cryptography we frequently find ourselves in a situation in which the "honest parties" have some classical information X about which an "adversary" – such as an eavesdropper, Eve – may hold quantum information Q. In other words, the quantum state Q is correlated with the classical information X. Since classical information is a special case of quantum information, the joint state of both X and Q can be represented by a density matrix ρ_{XQ}. What does such a density matrix look like?

Classical States

As a first step, let us pause to think about what it means for X to contain "classical information." In full generality, classical information can be modeled by a probability distribution over strings of bits x. Here x denotes the information and p_x the probability that this is the information contained in X. Suppose then that we are given a probability distribution over symbols x taken from the alphabet $\mathcal{X} = \{0, \ldots, d-1\}$, and let p_x denote the probability of symbol x. Identifying each possible value in $\mathcal{X}$

with an element of the standard basis $\{|0\rangle, \ldots, |d-1\rangle\}$, we can describe a system that is initialized in state $|x\rangle$ with probability p_x using the density matrix

$$\rho = \sum_{x=0}^{d-1} p_x |x\rangle\langle x| .$$

Note that ρ is a matrix that has the probabilities p_x on the diagonal and has all other entries equal to zero. As such, ρ is just another way to represent the distribution p_x: instead of a sequence of numbers, or a vector, we wrote the numbers on the diagonal of a matrix. Moreover, you can verify that measuring ρ in the standard basis results in outcome "x" with probability precisely p_x. In this sense, ρ is an accurate representation of the system X described above.

Definition 2.2.7 (Classical state). *Let $\{|x\rangle\}_{x=0}^{d-1}$ denote the standard basis for $\mathbb{C}^d$. A system X is said to be in a classical state, or c-state, if its density matrix ρ_X is diagonal in the standard basis, i.e. ρ_X has the form*

$$\rho = \sum_{x=0}^{d-1} p_x |x\rangle\langle x| ,$$

where $\{p_x\}_{x=0}^{d-1}$ is a probability distribution.

Thus, from now on we equate "classical state" or "classical density matrix" with "diagonal in the standard basis." The choice of the standard basis is arbitrary, as from a mathematical point of view all orthonormal bases are equivalent. Nevertheless, it is an important convention and serves as a point of connection between the classical and quantum worlds.

Classical-Quantum States

Now, let's move to states that are partially classical and partially quantum. Let's start with an example. Suppose that with probability 1/2 system X is in the classical state $|0\rangle$ and system Q is in the mixed state $\mathbb{I}/2$, and with probability 1/2 system X is in the classical state $|1\rangle$ and system Q is in the pure state $|+\rangle$. How do we write down the density matrix of the joint system XQ? In the first case, the density matrix is $|0\rangle\langle 0|_X \otimes (\mathbb{I}/2)_Q$, and in the second it is $|1\rangle\langle 1|_X \otimes |+\rangle\langle +|_Q$. Since both probabilities are equal to 1/2, overall we obtain

$$\rho_{XQ} = \frac{1}{2}|0\rangle\langle 0|_X \otimes \frac{\mathbb{I}_Q}{2} + \frac{1}{2}|1\rangle\langle 1|_X \otimes |+\rangle\langle +|_Q .$$

Check for yourself that ρ_{XQ} is a valid density matrix (remember the two conditions that need to be verified). This kind of density matrix is called a *classical-quantum state*, or cq-state for short. The reason is that the X part of the state is classical. More generally we give the following definition.

Definition 2.2.8. *A classical-quantum state, or simply* cq-state, *is a state of two subsystems, X and Q, such that its density matrix has the form*

$$\rho_{XQ} = \sum_x p_x |x\rangle\langle x|_X \otimes \rho_x^Q ,$$

where $\{p_x\}_x$ is a probability distribution and for every x, $|x\rangle$ designates the standard basis state on X and ρ_x^Q is an arbitrary density matrix on Q.

In applications to cryptography x will often represent some (partially secret) classical string that Alice creates during a quantum protocol, and ρ_x^Q some quantum information that an eavesdropper may have gathered during the protocol and which may be correlated with the string x. By convention we will usually reserve the letters X, Y, Z to denote classical registers, and use the other letters for quantum information. (More letters for quantum!)

QUIZ 2.2.9 *Which of the following states is in general a classical-quantum state?*

(a) $\rho_{AB} = \frac{1}{2} \left(\rho_A^0 \otimes \rho_B^0 + \rho_A^1 \otimes \rho_B^1 \right)$

(b) $\rho_{AB} = \frac{1}{2} \left(\rho_A^0 \otimes \rho_B^0 + |0\rangle\langle 0|_A \otimes |1\rangle\langle 1|_B \right)$

(c) $\rho_{AB} = \frac{1}{2} \left(|0\rangle\langle 0|_A \otimes \rho_B^0 + |1\rangle\langle 1|_A \otimes \rho_B^1 \right)$

QUIZ 2.2.10 *Alice prepares uniformly at random (each with probability $p_i = 1/3$) one out of three quantum states ρ_B^i, where $i \in \{0,1,2\}$, and sends this state to Bob. After preparation, the information about the state she prepared becomes encoded in a classical memory $|i\rangle\langle i|_A$ that Alice keeps. What is the correct description of the joint state that Alice and Bob share?*

(a) $\rho_{AB} = \frac{1}{3} \left(|0\rangle\langle 0|_A \otimes \rho_B^0 + |1\rangle\langle 1|_A \otimes \rho_B^1 + |2\rangle\langle 2|_A \otimes \rho_B^2 \right)$

(b) $\rho_{AB} = \frac{1}{9} \left(|0\rangle\langle 0|_A + |1\rangle\langle 1|_A + |2\rangle\langle 2|_A \right) \otimes \left(\rho_B^0 + \rho_B^1 + \rho_B^2 \right)$

(c) $\rho_{AB} = \frac{1}{2} \left(|0\rangle\langle 0|_A \otimes |1\rangle\langle 1|_A \otimes |2\rangle\langle 2|_A + \rho_B^0 \otimes \rho_B^1 \otimes \rho_B^2 \right)$

2.3 General Measurements

So far we have described how to measure a quantum state, pure or mixed, in a given orthonormal basis. Quantum mechanics allows a much more refined notion of measurement, which plays an important role both in quantum information theory and in cryptography. Indeed, in quantum information theory certain tasks, such as the task of discriminating between multiple states, can be solved more efficiently using these generalized measurements. Moreover, taking an adversarial viewpoint, in quantum cryptography it is essential to prove security for the most general kind of attack, including all measurements that an attacker could possibly make! This includes using extra qubits to make measurements, which is effectively how such generalized measurements can be realized.

2.3.1 Positive Operator-Valued Measures

If we are only interested in computing the probabilities of measurement outcomes – but do not require a complete specification of what happens to the quantum state once the measurement has been performed – then the most general kind of measurement that is allowed in quantum mechanics can be described mathematically by a positive operator-valued measure, or POVM for short.

> **Definition 2.3.1 (POVM).** *A POVM on $\mathbb{C}^d$ is a set of positive semidefinite matrices* $\{M_x\}_{x\in\mathcal{X}}$ *such that*
>
> $$\sum_x M_x = \mathbb{I}_d \,.$$
>
> *The subscript x is used as a label for the measurement outcome.*

> **QUIZ 2.3.1** *Which of the following is a valid POVM?*
>
> I. $\left\{ \begin{pmatrix} \frac{1}{2} & -\frac{1}{2} \\ -\frac{1}{2} & \frac{1}{2} \end{pmatrix}, \begin{pmatrix} \frac{1}{2} & 0 \\ 0 & \frac{1}{2} \end{pmatrix}, \begin{pmatrix} 0 & \frac{1}{2} \\ \frac{1}{2} & 0 \end{pmatrix} \right\}$
>
> II. $\left\{ \begin{pmatrix} \frac{1}{3} & 0 \\ 0 & \frac{1}{3} \end{pmatrix}, \begin{pmatrix} \frac{1}{2} & 0 \\ 0 & 0 \end{pmatrix}, \begin{pmatrix} 0 & 0 \\ 0 & \frac{1}{2} \end{pmatrix} \right\}$
>
> III. $\left\{ \begin{pmatrix} \frac{1}{2} & 0 \\ 0 & 0 \end{pmatrix}, \begin{pmatrix} 0 & 0 \\ 0 & \frac{1}{2} \end{pmatrix}, \begin{pmatrix} \frac{1}{2} & 0 \\ 0 & \frac{1}{2} \end{pmatrix} \right\}$
>
> (a) *I and II*
> (b) *I and III*
> (c) *only I*
> (d) *only III*

Having generalized our notion of measurement, we need to extend the measurement rule, i.e. the rule that specifies the probability of obtaining each possible outcome when performing the measurement on a given state with density matrix ρ.

> **Definition 2.3.2 (Generalized measurement rule).** *Let $\{M_x\}_x$ be a POVM. Then the probability p_x of observing outcome x when performing the measurement $\{M_x\}_x$ on a density matrix ρ is*
>
> $$p_x = \mathrm{tr}(M_x\rho) \,.$$
>
> *This expression is sometimes called the* Born rule.

The next two examples show that the generalized Born rule is compatible with the measurement rule we introduced before.

Example 2.3.1 Consider a probability distribution $\{p_x\}_x$ and the associated classical mixture $\rho = \sum_x p_x |x\rangle\langle x|$. If we measure ρ in the standard basis, with associated POVM $M_x = |x\rangle\langle x|$ as in Example 2.3.2, we obtain outcome x with probability

$$\mathrm{tr}(|x\rangle\langle x|\rho) = \langle x|\rho|x\rangle = p_x,$$

as expected: ρ indeed captures the classical distribution given by the probabilities p_x. ∎

Example 2.3.2 Recall that when measuring a state $|\psi\rangle = \sum_x \alpha_x |x\rangle$ in a basis such as $\{|x\rangle\}_x$, the probability of observing outcome x is given by $|\alpha_x|^2$. Let us verify that this rule is recovered as a special case of the POVM formalism. For each x let $M_x = |x\rangle\langle x|$, so that M_x is positive semidefinite (in fact, it is a projector, i.e. $M_x^2 = M_x$) and $\sum_x M_x = \mathbb{I}$ (this can be verified by using that $\{|x\rangle\}$ is a basis), as required. Let $\rho = |\psi\rangle\langle\psi|$. We can use the Born rule to compute

$$
\begin{aligned}
p_x &= \mathrm{tr}(M_x\rho)\\
&= \mathrm{tr}(|x\rangle\langle x|\rho)\\
&= \langle x|\rho|x\rangle\\
&= \sum_{x',x''} \alpha_{x'}\alpha_{x''}^* \langle x|x'\rangle\langle x''|x\rangle\\
&= |\alpha_x|^2.
\end{aligned}
$$

∎

Beyond the calculation of outcome probabilities it can be important to know what happens to a quantum state after a generalized measurement has been performed. For the case of measuring in a basis, we already know the answer: the state collapses to the basis element associated with the outcome of the measurement that is obtained.

In the case of a POVM it turns out that the information given by the operators $\{M_x\}_x$ is not sufficient to fully determine the post-measurement state. This is because such a measurement may not fully collapse the state, meaning that the post-measurement state may not be pure (this corresponds to the case where M_x has rank more than 1). Intuitively, if the measurement operator M_x does not have rank 1 there is some freedom in choosing exactly where the post-measurement state lies without affecting the outcome probabilities.

2.3.2 Generalized Measurements

The additional information needed to specify post-measurement states is a *Kraus operator representation* of the POVM.

Definition 2.3.3 (Kraus operators). *Let $M = \{M_x\}_x$ be a POVM on $\mathbb{C}^d$. A Kraus operator representation of M is a set of $d' \times d$ matrices A_x such that $M_x = A_x^\dagger A_x$ for all x.*

For any positive semidefinite matrix N, if $N = \sum_i \lambda_i |v_i\rangle\langle v_i|$ is the spectral decomposition of N, then N has a unique positive semidefinite square root which is given by $\sqrt{N} = \sum_i \sqrt{\lambda_i} |v_i\rangle\langle v_i|$. Thus, a Kraus decomposition of any POVM always exists by setting $A_x = \sqrt{M_x}$. In particular, if $M_x = |u_x\rangle\langle u_x|$ is a projector then $\sqrt{M_x} = M_x$ and we can take $A_x = M_x$. But for any unitary U_x on $\mathbb{C}^d$, $A'_x = U_x\sqrt{M_x}$ is also a valid decomposition. Hence, there is no unique Kraus representation for a given POVM. In fact, the definition even allows matrices A_x that are not square.

This means we cannot go from POVM to Kraus operators. However, given Kraus operators we can find the POVM. Thus, the most general form to write down a quantum measurement is through the full set of Kraus operators $\{A_x\}_x$. Let's see how knowledge of the Kraus operators allows us to compute post-measurement states.

Definition 2.3.4 (Post-measurement state). *Let ρ be a density matrix and $M = \{M_x\}_x$ a POVM with Kraus decomposition given by operators $\{A_x\}_x$. Suppose the measurement is performed on a density matrix ρ, and the outcome x is obtained. Then the state of the system after the measurement, conditioned on having obtained the outcome x, is*

$$\rho_{|x} = \frac{A_x \rho A_x^\dagger}{\operatorname{tr}(A_x^\dagger A_x \rho)} \ .$$

If $\operatorname{tr}(A_x^\dagger A_x \rho) = 0$ then the formula for $\rho_{|x}$ is meaningless. However, in that case the outcome x has probability 0 of occurring and so there is no need to define an associated post-measurement state.

You may want to convince yourself that when measuring a pure state $|\psi\rangle$ in an arbitrary orthonormal basis, with Kraus decomposition $A_x = \sqrt{M_x} = |x\rangle\langle x|$, the post-measurement state as defined above is precisely the basis state associated with the measurement outcome.

An important class of generalized measurements is given by the case where the M_x are projectors onto orthogonal subspaces (not necessarily of rank 1).

Definition 2.3.5. *A* projective measurement, *also called a* von Neumann measurement, *is given by a set of orthogonal projectors $M_x = \Pi_x$ such that $\sum_x \Pi_x = \mathbb{I}$. For such a measurement, unless otherwise specified, we will always use the default Kraus decomposition $A_x = \sqrt{M_x} = \Pi_x$. For such a measurement the probability q_x of observing measurement outcome x can be expressed as*

$$q_x = \operatorname{tr}(\Pi_x \rho) \ ,$$

and the post-measurement states are

$$\rho_{|x} = \frac{\Pi_x \rho \Pi_x}{\operatorname{tr}(\Pi_x \rho)} \ .$$

The following example shows how to use the formalism of generalized measurements to perform a certain task in different ways.

Example 2.3.3 Suppose we are given a two-qubit state ρ, such that we would like to measure the parity (in the standard basis) of the two qubits. A first way to do this would be to measure ρ in the standard basis, obtain two bits, and take their parity. In this case the probability of obtaining the outcome "even" would be

$$q_{\text{even}} = \langle 00|\rho|00\rangle + \langle 11|\rho|11\rangle,$$

and the post-measurement state would be the mixture of the two post-measurement states associated with outcomes $(0,0)$ and $(1,1)$, so

$$\rho_{|\text{even}} = \frac{1}{q_{\text{even}}} \left(\left(\langle 00|\rho|00\rangle \right) |00\rangle\langle 00| + \left(\langle 11|\rho|11\rangle \right) |11\rangle\langle 11| \right).$$

Now suppose that we attempt to measure the parity using a generalized measurement that directly projects onto the relevant subspaces, without measuring the qubits individually. That is, consider the projective measurement $\Pi_{\text{even}} = |00\rangle\langle 00| + |11\rangle\langle 11|$ and $\Pi_{\text{odd}} = \mathbb{I} - \Pi_{\text{even}} = |01\rangle\langle 01| + |10\rangle\langle 10|$. With this measurement the probability of obtaining the outcome "even" is

$$q'_{\text{even}} = \text{tr}(\Pi_{\text{even}}\rho) = \langle 00|\rho|00\rangle + \langle 11|\rho|11\rangle,$$

as before. However, the post-measurement state is now

$$\rho'_{|\text{even}} = \frac{1}{q'_{\text{even}}} \Pi_{\text{even}}\rho\Pi_{\text{even}}.$$

To see the difference, consider the state $\rho = |\text{EPR}\rangle\langle\text{EPR}|$ where $|\text{EPR}\rangle = \frac{1}{\sqrt{2}}(|00\rangle + |11\rangle)$. Then clearly the parity measurement should report the outcome "even" with probability 1, and you can check that this is the case for both measurements. However, the post-measurement states are different. In the first case,

$$\rho_{|\text{even}} = \frac{1}{2}|00\rangle\langle 00| + \frac{1}{2}|11\rangle\langle 11|,$$

while in the second case,

$$\rho'_{|\text{even}} = |\text{EPR}\rangle\langle\text{EPR}|$$

is unchanged! This is one of the key advantages of using generalized measurements, as opposed to basis measurements: they allow us to compute certain simple quantities on multi-qubit states (such as the parity) without fully "destroying" the state. ∎

Exercise 2.3.1 Use a projective measurement to measure the parity, *in the Hadamard basis*, of the state $|00\rangle\langle 00|$.

Exercise 2.3.2 For the same scenario as the previous exercise, compute the probabilities of obtaining measurement outcomes "even" and "odd," and the resulting post-measurement states. What would the post-measurement states have been if you had first measured the qubits individually in the Hadamard basis, and then taken the parity?

2.4   The Partial Trace

Going back to our second motivation for introducing density matrices, let us now give an answer to the following question: Given a multi-qubit state, how do we write down the "partial state" associated with a subset of the qubits? More generally, suppose ρ_{AB} is a density matrix on a tensor product space $\mathbb{C}_A^{d_A} \otimes \mathbb{C}_B^{d_B}$, and suppose that Alice holds the part of ρ corresponding to system A and Bob holds the part corresponding to system B. How do we represent the state ρ_A of Alice's system?

2.4.1   An Operational Viewpoint

The operation that takes us from ρ_{AB} to ρ_A is called the *partial trace*. It can be specified in purely mathematical terms, and we do so in the next section. Before we do that, let us try to think about the problem from an operational point of view. First, consider an easy case: if $\rho_{AB} = \rho_A \otimes \rho_B$, where ρ_A and ρ_B are both density matrices, then clearly Alice's system is defined by ρ_A. In this case, we would say that the partial trace of ρ_{AB}, when "tracing out" system B, is the density matrix ρ_A.

A slightly more complicated case is when

$$\rho_{AB} = \sum_i p_i \rho_i^A \otimes \rho_i^B \qquad (2.6)$$

is a mixture of tensor products (we will later see that this is called a "separable state"). Using the interpretation that this represents a state that is in state $\rho_i^A \otimes \rho_i^B$ with probability p_i, it would be natural to claim that Alice's share of the state is ρ_i^A with probability p_i, i.e. the partial trace of ρ_{AB}, when tracing out – i.e. ignoring – system B, is now $\rho_A = \sum_i p_i \rho_i^A$.

How about a general ρ_{AB}? Remember from Problem 1.2 that there exist some ρ that do not have a decomposition of the form (2.6), such as the EPR pair. Our idea is to "force" such a decomposition by performing the following little thought experiment. Let us *imagine* that Bob performs a complete basis measurement on his system, using an arbitrary basis $\{|u_x\rangle\}_x$. Let us introduce a POVM on the joint system of Alice and Bob that models this measurement: since Alice does nothing, we can set $M_x = \mathbb{I}_A \otimes |u_x\rangle\langle u_x|_B$, which you can check indeed defines a valid POVM. Moreover, this is a projective measurement, so we can take the Kraus operators $A_x = \sqrt{M_x} = M_x$. By definition the post-measurement states are given by

$$\rho_{|x}^{AB} = \frac{M_x \rho_{AB} M_x}{\operatorname{tr}(M_x \rho_{AB})} = \frac{\left((\mathbb{I}_A \otimes \langle u_x|)\rho_{AB}(\mathbb{I}_A \otimes |u_x\rangle)\right)_A \otimes |u_x\rangle\langle u_x|_B}{\operatorname{tr}\left((\mathbb{I}_A \otimes |u_x\rangle\langle u_x|_B)\rho_{AB}\right)}.$$

Notice how we wrote the state as a tensor product of a state on A and one on B. Make sure you understand the notation in this formula, and that it specifies a well-defined state.

The key step is to realize that, whatever the state of Alice's system A is, it shouldn't depend on any operation that Bob performs on B. After all, it may be that A is here on

Earth, and B is on Mars. Since quantum mechanics does not allow faster than light communication, as long as the two of them remain perfectly isolated, meaning that Alice doesn't get to learn the measurement that Bob performs or its outcome, then her state should remain unchanged. We can thus describe it as follows: "With probability $q_x = \text{tr}(M_x \rho_{AB})$, Alice's state is the A part of $\rho_{|x}^{AB}$." Using the rule for computing post-measurement states, we get

$$\rho_A = \sum_x q_x \frac{\left(\left(\mathbb{I} \otimes \langle u_x|\right)\rho_{AB}\left(\mathbb{I} \otimes |u_x\rangle\right)\right)_A}{\text{tr}\left(\left(\mathbb{I} \otimes |x\rangle\langle x|\right)\rho_{AB}\right)} = \sum_x \left(\mathbb{I} \otimes \langle u_x|\right)\rho_{AB}\left(\mathbb{I} \otimes |u_x\rangle\right). \tag{2.7}$$

Although we derived the above expression for Alice's state using sensible arguments, there is something you should be worried about: Doesn't it depend on the choice of basis $\{|u_x\rangle\}_x$ we made for Bob's measurement? Of course, it should not, as our entire argument is based on the idea that Alice's reduced state should not depend on any operation performed by Bob. The next exercise asks you to verify that this is indeed the case. (We emphasize that this is only the case as long as Alice doesn't learn the measurement outcome! If we fix a particular outcome x then it's a completely different story. Beware of this subtlety, it will come up repeatedly throughout the book.)

Exercise 2.4.1 Verify that the state ρ_A defined in Eq. (2.7) does not depend on the choice of basis $\{|u_x\rangle\}$. *[Hint: first argue that if two density matrices ρ, σ satisfy $\langle \phi|\rho|\phi\rangle = \langle \phi|\sigma|\phi\rangle$ for all unit vectors $|\phi\rangle$ then $\rho = \sigma$. Then compute $\langle \phi|\rho_A|\phi\rangle$, and use the POVM condition $\sum_x M_x = \mathbb{I}$ to check that you can get an expression independent of the $\{|u_x\rangle\}_x$. Conclude that ρ_A itself does not depend on $\{|u_x\rangle\}_x$.]*

Example 2.4.1 Consider the example of the EPR pair

$$|\text{EPR}\rangle_{AB} = \frac{1}{\sqrt{2}}\left(|00\rangle + |11\rangle\right). \tag{2.8}$$

Writing this as a density operator we have

$$\rho_{AB} = |\text{EPR}\rangle\langle\text{EPR}|_{AB} = \frac{1}{2}\left(|00\rangle\langle00| + |00\rangle\langle11| + |11\rangle\langle00| + |11\rangle\langle11|\right). \tag{2.9}$$

Let's measure system B in the standard basis: taking A into account we consider the POVM $M_0 = \mathbb{I}_A \otimes |0\rangle\langle0|_B$ and $M_1 = \mathbb{I}_A \otimes |1\rangle\langle1|_B$. We can then compute

$$q_0 = \text{tr}(M_0\rho)$$
$$= \frac{1}{2}\text{tr}\left((\mathbb{I} \otimes |0\rangle\langle0|)(|00\rangle\langle00| + |00\rangle\langle11| + |11\rangle\langle00| + |11\rangle\langle11|)\right)$$
$$= \frac{1}{2}(1 + 0 + 0 + 0) = \frac{1}{2},$$

and similarly $q_1 = 1/2$. The post-measurement state on A is then

$$\rho_{|0}^A = \frac{1}{2}(\mathbb{I} \otimes \langle0|)\rho_{AB}(\mathbb{I} \otimes |0\rangle) + \frac{1}{2}(\mathbb{I} \otimes \langle1|)\rho_{AB}(\mathbb{I} \otimes |1\rangle) = \frac{1}{2}|0\rangle\langle0| + \frac{1}{2}|1\rangle\langle1|.$$

Now do the same calculation using a measurement in the Hadamard basis on B, and check that you get the same result! ∎

QUIZ 2.4.1 *Suppose that Alice and Bob share the state $\frac{1}{\sqrt{2}}(|00\rangle + |11\rangle)$. Bob measures his qubit in the basis $\{|+\rangle, |-\rangle\}$ and obtains $|+\rangle$. What is the post-measurement state of Alice's qubit?*

(a) $|-\rangle$

(b) $|0\rangle$

(c) $|+\rangle$

QUIZ 2.4.2 *Suppose instead that Alice and Bob share the state $\frac{1}{\sqrt{2}}(|00\rangle - |11\rangle)$. Bob again measures his qubit in the basis $\{|+\rangle, |-\rangle\}$ and obtains $|+\rangle$. What is the post-measurement state of Alice's qubit?*

(a) $|-\rangle$

(b) $|0\rangle$

(c) $-|+\rangle$

2.4.2   A Mathematical Definition

Armed with our "operational" definition of what the partial trace *should* achieve, we now give the precise, mathematical definition of this operation.

Definition 2.4.1 (Partial trace). *Consider a general matrix*

$$M_{AB} = \sum_{ijk\ell} \gamma_{ij}^{k\ell} |i\rangle\langle j|_A \otimes |k\rangle\langle\ell|_B \,, \tag{2.10}$$

where $|i\rangle_A, |j\rangle_A$ and $|k\rangle_B, |\ell\rangle_B$ run over orthonormal bases of A and B respectively. Then the partial trace over B is defined as

$$
\begin{aligned}
M_A &= \mathrm{tr}_B(M_{AB}) \\
&= \sum_{ijk\ell} \gamma_{ij}^{k\ell} |i\rangle\langle j|_A \otimes \mathrm{tr}(|k\rangle\langle\ell|_B) \\
&= \sum_{ijk\ell} \gamma_{ij}^{k\ell} |i\rangle\langle j|_A \otimes \langle\ell|k\rangle_B \\
&= \sum_{ijk\ell} \gamma_{ij}^{k\ell} |i\rangle\langle j|_A \otimes \delta_{k\ell} \\
&= \sum_{ij} \left(\sum_k \gamma_{ij}^{kk} \right) |i\rangle\langle j|_A \,.
\end{aligned}
$$

Similarly, the partial trace over A is

$$M_B = \text{tr}_A(M_{AB}) = \sum_{ijk\ell} \gamma_{ij}^{k\ell}\, \text{tr}(|i\rangle\langle j|) \otimes |k\rangle\langle\ell| = \sum_{k\ell} \left(\sum_j \gamma_{jj}^{k\ell} \right) |k\rangle\langle\ell|\,.$$

If $M_{AB} = \rho_{AB}$ is a density matrix, meaning that it is positive semidefinite and $\text{tr}(M_{AB}) = 1$, then it is a good exercise to verify that the partial traces $\rho_A = \text{tr}_B(\rho_{AB})$ and $\rho_B = \text{tr}_A(\rho_{AB})$ are again density matrices. We refer to them as the *reduced states* of the system AB on system A and system B respectively.

The formal definition directly gives us a recipe for computing the partial trace of a state ρ_{AB}, as follows.

1. Write a decomposition of ρ_{AB} in the form (2.10). Note that you may do this for any choice of orthonormal bases that you like for systems A and B. The coefficients $\gamma_{ij}^{k\ell}$ are then simply the entries of the matrix ρ_{AB} at position $|i\rangle\langle j| \otimes |k\rangle\langle\ell|$.
2. Put a trace on the "B part" and use cyclicity of the trace to finish the computation.

In many cases, a decomposition of ρ as in (2.10) is easily found.

Example 2.4.2 Let us consider again the example of the EPR pair

$$|\text{EPR}\rangle_{AB} = \frac{1}{\sqrt{2}}(|00\rangle + |11\rangle)\,,$$

with associated density matrix

$$\rho_{AB} = |\text{EPR}\rangle\langle\text{EPR}|_{AB}$$
$$= \frac{1}{2}\left(|0\rangle\langle0|_A \otimes |0\rangle\langle0|_B + |0\rangle\langle1|_A \otimes |0\rangle\langle1|_B + |1\rangle\langle0|_A \otimes |1\rangle\langle0|_B + |1\rangle\langle1|_A \otimes |1\rangle\langle1|_B\right)\,.$$

Using the definition we can compute

$$\text{tr}_B(\rho_{AB}) = \frac{1}{2}\left(|0\rangle\langle0|_A \otimes \text{tr}(|0\rangle\langle0|_B) + |0\rangle\langle1|_A \otimes \text{tr}(|0\rangle\langle1|_B)\right.$$
$$\left. + |1\rangle\langle0|_A \otimes \text{tr}(|1\rangle\langle0|_B) + |1\rangle\langle1|_A \otimes \text{tr}(|1\rangle\langle1|_B)\right)\,.$$

Since the trace is cyclic, $\text{tr}(|0\rangle\langle1|) = \langle1|0\rangle = 0$, similarly $\text{tr}(|1\rangle\langle0|) = 0$, but $\text{tr}(|0\rangle\langle0|) = \text{tr}(|1\rangle\langle1|) = 1$ and hence

$$\text{tr}_B(\rho_{AB}) = \frac{1}{2}(|0\rangle\langle0| + |1\rangle\langle1|) = \frac{\mathbb{I}}{2}\,. \tag{2.11}$$

Convince yourself that when we take the partial trace operation over A, and hence look at the state of just Bob's qubit, we also get

$$\text{tr}_A(\rho_{AB}) = \frac{\mathbb{I}}{2}\,. \tag{2.12}$$

This is consistent with our calculations in Example 2.4.1. ∎

Exercise 2.4.2 If $\rho_{AB} = |\Phi\rangle\langle\Phi|$ is the singlet $|\Phi\rangle = (|01\rangle - |10\rangle)/\sqrt{2}$, compute ρ_A and ρ_B.

Example 2.4.3 The notion of partial trace allows us to verify that performing a unitary operation on A has no effect on the state of B, i.e. it does not change ρ_B.

$$(U_A \otimes \mathbb{I}_B)\rho_{AB}(U_A \otimes \mathbb{I}_B)^\dagger = \sum_{ijk\ell} \gamma_{ij}^{k\ell} U_A |i\rangle\langle j|_A U_A^\dagger \otimes |k\rangle\langle\ell|_B . \tag{2.13}$$

Computing again the partial trace we have

$$\begin{aligned}
\mathrm{tr}_A(U_A \otimes \mathbb{I}_B \rho_{AB} U_A^\dagger \otimes \mathbb{I}_B) &= \sum_{ijk\ell} \gamma_{ij}^{k\ell}\, \mathrm{tr}(U_A |i\rangle\langle j| U_A^\dagger) \otimes |k\rangle\langle\ell| \\
&= \sum_{ijk\ell} \gamma_{ij}^{k\ell}\, \mathrm{tr}(|i\rangle\langle j| U_A^\dagger U_A) \otimes |k\rangle\langle\ell| \\
&= \sum_{ijk\ell} \gamma_{ij}^{k\ell}\, \mathrm{tr}(|i\rangle\langle j|) \otimes |k\rangle\langle\ell| \\
&= \sum_{k\ell} \left(\sum_j \gamma_{jj}^{k\ell} \right) |k\rangle\langle\ell| = \rho_B .
\end{aligned} \tag{2.14}$$

Can you convince yourself that performing a measurement on A also has no effect on B? ∎

QUIZ 2.4.3 *What are Alice and Bob's reduced states in the joint state*

$$\rho_{AB} = \begin{pmatrix} \frac{1}{4} & 0 & 0 & \frac{1}{4} \\ 0 & \frac{1}{4} & -\frac{1}{4} & 0 \\ 0 & -\frac{1}{4} & \frac{1}{4} & 0 \\ \frac{1}{4} & 0 & 0 & \frac{1}{4} \end{pmatrix}?$$

(a) $\rho_A = \rho_B = \begin{pmatrix} \frac{1}{2} & 0 \\ 0 & \frac{1}{2} \end{pmatrix}$

(b) $\rho_A = \rho_B = \begin{pmatrix} \frac{1}{2} & -\frac{1}{2} \\ -\frac{1}{2} & \frac{1}{2} \end{pmatrix}$

(c) $\rho_A = \begin{pmatrix} \frac{3}{4} & 0 \\ 0 & \frac{1}{4} \end{pmatrix}, \rho_B = \begin{pmatrix} \frac{1}{4} & 0 \\ 0 & \frac{3}{4} \end{pmatrix}$

QUIZ 2.4.4 *Alice and Bob share a state ρ_{AB}. If Alice's reduced state is $\rho_A = |0\rangle\langle 0|$, we know that ρ_{AB} is*

(a) *pure*

(b) *mixed*

(c) *not enough information*

QUIZ 2.4.5 *Alice and Bob share a state ρ_{AB}. If Alice's reduced state is $\rho_A = \frac{1}{2}(|0\rangle\langle 0| + |1\rangle\langle 1|)$, we know that ρ_{AB} is*

(a) *pure*
(b) *mixed*
(c) *not enough information*

2.5 Secure Message Transmission

With all the math behind us we are ready to turn to our first serious cryptographic task – in fact, the most serious task of all: the secure transmission of messages. To set things up, imagine two protagonists, Alice and Bob. Alice and Bob would like to exchange classical messages between each other (e.g. they want to chat!). However, Alice and Bob are worried that the messages they exchange could be intercepted by a malicious eavesdropper, Eve. This is because, although Alice and Bob each trust that they have full control over their own secure laboratory (i.e. their bedroom), they really don't know what happens on the communication line, such as the airwaves for a cell phone conversation or the post office truck for a snail mail conversation. Alice and Bob's goal is to limit, and if possible reduce to zero, the useful information that Eve may be able to get: they want to make sure that even if Eve intercepts all the classical messages they exchange, these messages look like complete rubbish to Eve!

If you think about this setup, you will see that we are faced with a symmetry-breaking problem. This is because, if Alice wants to send a message to Bob but Eve can listen to all messages exchanged, then Eve receives everything that Bob does. So then, how can Bob understand what Alice wants to say to him, but somehow Eve has no information, even though she read the same message?

To break the symmetry we will make a crucial assumption. We will assume that Alice and Bob are in possession of a *secret key*, which is known to them but completely unknown to Eve. This key will be used to hide the messages that they exchange (Figure 2.2). Cryptosystems that make use of this assumption are called *private-key* cryptosystems.

For the time being, we will not justify our assumption about the key: this is just some secret key Alice and Bob have in common, a secret they may have agreed on a long time in the past, when they were in the same place and could whisper in each other's ears. In later chapters we will see how quantum information can be used to establish such a secret even when Alice and Bob are physically separated.

2.5.1 Shannon's Secrecy Condition and the Need for Large Keys

The mathematical framework for the description of secret communication schemes was first developed by Claude Shannon in the 1940s, well before quantum information made its apparition. According to Shannon's formalism, an encryption scheme consists of two functions. The first is the *encryption function* $\text{Enc}(m, k) = e$, which takes the key k and the message m and maps it to some encrypted message e. The original message m is often called the *plaintext*, and e the *ciphertext*. The second function is

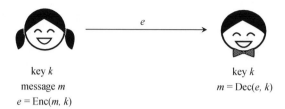

Fig. 2.2   Alice sends an encrypted message to Bob.

the *decryption function* $\text{Dec}(e,k) = m$, which takes the key k and the ciphertext e back to the plaintext (Figure 2.2).

> **Definition 2.5.1.** *An encryption scheme* (Enc, Dec) *is called* correct *if, for every key k and every plaintext m,* $\text{Dec}(\text{Enc}(k,m),k) = m$. *It is called* perfectly secure *if, for any distribution $p(\cdot)$ over the space $\mathcal{M}$ of plaintexts, the following two distributions on plaintexts are identical:*
>
> 1. *Generate a random plaintext $m \in \mathcal{M}$ with probability $p(m)$.*
> 2. *Select an arbitrary ciphertext e. Generate a uniformly random key $k \in K$. Generate a random plaintext $m \in \mathcal{M}$ with probability $p(m | \text{Enc}(m,k) = e)$.*

In the definition of perfectly secure the key k is chosen uniformly at random. This is an important condition. It expresses our assumption that Eve has no information whatsoever about the key. So from her point of view every possible key has the same a priori probability: for every k in the key space K, it holds that $p_k = 1/|K|$.

The definition may be a little hard to understand the first time that you read it. So let's paraphrase using words. We call an encryption scheme perfectly secure whenever an eavesdropper, Eve, ignorant of the key, does not gain any additional information about a plaintext message m from its encryption e. In other words, the probability $p(m)$ of the message m is the same a priori (as anyone could guess) as it is from the point of view of Eve, who has obtained e. Observe that this is a very strong notion of security: absolutely no information is gained from having access to e!

This definition is so strong that it may even seem impossible to realize: if e has "no information" about m, then how can e be decrypted to recover m? As we will soon see, there is no contradiction: it is possible that e has no information at all about m *from the point of view of an eavesdropper who does not know the secret key k*, yet e still has full information about m from the point of view of an honest party, Bob, *who does know the secret key*. This is a very subtle point: make sure you fully understand the distinction.

Note that it would be easy to come up with an encryption scheme that is "just" secret: Alice simply sends a randomly chosen e to Bob. Then, because e is random and independent of any message, of course learning e does not reveal information about the message. But clearly this scheme would not be correct: Bob cannot recover Alice's message. Similarly, it is easy to devise a scheme that is "just" correct: Alice sends $e = m$ to Bob. Clearly this is not secure since Eve also learns m. In summary, the art of encryption is to design schemes that are both correct *and* secure.

In our presentation we assumed that Alice and Bob share a secret key k, and we informally argued that such a key was needed to "break the symmetry" between Bob and the eavesdropper, Eve. Is this argument watertight – is a key really needed? As it turns out, not only is it needed but in fact the number of possible keys needs to be as large as the number of possible messages that Alice may wish to send. The following theorem, due to Shannon, proves this.

Theorem 2.5.1 *An encryption scheme* (Enc, Dec) *can only be* perfectly secure *and* correct *if the number of possible keys* $|K|$ *is at least as large as the number of possible messages* $|M|$, *that is,* $|K| \geq |M|$.

Proof   Suppose for contradiction that there exists a correct scheme using fewer keys, i.e. $|K| < |M|$. We will show that such a scheme cannot be perfectly secure. Let p be the uniform distribution over M. Consider an eavesdropper who has intercepted the ciphertext e. She can compute

$$S = \{\hat{m} \mid \exists k, \hat{m} = \text{Dec}(e, k)\} \,, \tag{2.15}$$

that is, the set of all messages $\hat{m}$ for which there exists a key k that could have resulted in the observed ciphertext e. Note that the size $|S|$ of this set is $|S| \leq |K|$, since for each possible key k we get at most one message $\hat{m}$. Since $|K| < |M|$, we thus have $|S| < |M|$. This means that there exists at least one message m such that $m \notin S$, and hence $p(m|e) = 0$. However, by definition $p(m) = 1/|M|$. This contradicts the definition of perfect security given in Definition 2.5.1. ∎

Can the bound given in the theorem be achieved: Does there exist an encryption scheme that is both correct *and* secure, and which uses precisely the minimum number of keys $|K| = |M|$? The answer is yes! We construct such a scheme in the next section.

2.5.2 The (Quantum) One-Time Pad

The *one-time pad* is arguably the simplest, yet also the most secure, encryption scheme known. We start with the "classical" version, which allows encryption of classical messages.

The Classical One-Time Pad

Imagine that Alice (the sender) wants to send a secret message m to Bob (the receiver). For simplicity, we take the message space M to be the set of all n-bit strings: $M = \{0,1\}^n$. Let us furthermore assume that Alice and Bob already share a key $k \in \{0,1\}^n$ which is just as long as the message, and is uniformly random from the point of view of the adversary, Eve. In the following definition, we use the notation $a \oplus b$ for the bitwise XOR, or equivalently addition modulo 2: for $a, b \in \{0,1\}$, $a \oplus b = a + b \bmod 2$.

Protocol 1 The classical *one-time pad* is an encryption scheme in which the encryption of a message $m \in \{0,1\}^n$ using the key $k \in \{0,1\}^n$ is given by

$$\text{Enc}(m, k) = m \oplus k = (m_1 \oplus k_1, m_2 \oplus k_2, \ldots, m_n \oplus k_n) = (e_1, \ldots, e_n) = e \,.$$

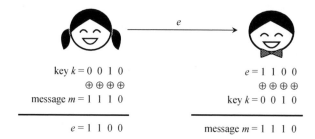

Fig. 2.3　An example of a one-time pad between Alice and Bob.

The decryption is given by

$$\mathrm{Dec}(e,k) = e \oplus k = (e_1 \oplus k_1, e_2 \oplus k_2, \ldots, e_n \oplus k_n) \ .$$

Figure 2.3 shows an example of the one-time pad. Note that since for any $j \in \{1,\ldots,n\}$, $m_j \oplus k_j \oplus k_j = m_j$, the scheme is correct. Is it secure?

To see that it satisfies Shannon's definition, consider any distribution p on M. For a uniformly random choice of key k and a fixed message m, the associated ciphertext $e = \mathrm{Enc}(m,k)$ is uniformly distributed over all n-bit strings: for any e,

$$p(\mathrm{Enc}(m,k) = e|m) = p(m \oplus k = e|m) = p(k = e \oplus m|m) = \frac{1}{2^n} \ ,$$

since k is chosen uniformly at random. Since this holds for any message m,

$$p(e) = \sum_m p(m)p(e|m) = \frac{1}{2^n} \ .$$

Applying Bayes' rule we get that

$$p(m|e) = \frac{p(m,e)}{p(e)} = \frac{p(e|m)p(m)}{p(e)} = p(m) \ ,$$

independent of m. Thus $p(m|e) = p(m)$ and the scheme is perfectly secure.

Note that our argument crucially relies on the key being uniformly distributed and independent of the eavesdropper, a condition that has to be treated with care! In Chapter 6 we will introduce a method called *privacy amplification*, which can be used to "improve" the quality of a key about which the eavesdropper may have partial information.

Remark 2.5.2 While the one-time pad is "perfectly secure" according to Shannon's definition, it does not protect against an adversary changing bits in the messages exchanged between Alice and Bob. Indeed, you can verify that for any key k, and any string x, $\mathrm{Enc}(m \oplus x, k) = \mathrm{Enc}(m,k) \oplus x$. What this means is that flipping bits of the ciphertext is equivalent to flipping bits of the plaintext, and there is no way for Bob to detect if such an operation has taken place. This would be an issue for bank transactions, since an adversary could flip the transaction amount in an arbitrary way (without learning any information about the amount itself!). For this reason, one-time pads are generally supplemented by checksums or message authentication codes which allow changes to be detected (and corrected). These are well-known classical techniques, and we will not get into them in more detail here.

QUIZ 2.5.1 *Bob received from Alice a message encoded using the one-time pad:*
e = 0010111. Bob has the key needed to decrypt the message: k = 1001011. What
is the message that Alice sent him?

(a) 1001011
(b) 0010111
(c) 1011100
(d) 1011111

There is another way to look at the classical one-time pad that brings it much closer
to the quantum version we will consider next. Consider the encryption of a single-bit
message $m \in \{0, 1\}$. Recall that we can represent this message as a pure quantum state
$|m\rangle$, or equivalently as the density matrix $|m\rangle\langle m|$. When we apply the XOR operation
the result is that the bit m is flipped whenever the key bit $k = 1$, and unchanged if $k = 0$.
That is, when $k = 1$ the state is transformed as $|m\rangle \mapsto X|m\rangle$, where recall that X is the
Pauli bit-flip matrix. Thus, in this case encryption implements the transformation
$|m\rangle\langle m| \mapsto X|m\rangle\langle m|X$ visualized in Figure 2.4.

If Alice and Bob choose a uniformly random key bit k then we can write the density
matrix for the entire system KM, where K contains the key and M the message, as

$$\rho_{KM} = \frac{1}{2}|0\rangle\langle 0|_K \otimes |m\rangle\langle m|_M + \frac{1}{2}|1\rangle\langle 1|_K \otimes X|m\rangle\langle m|_M X .$$

From the point of view of Eve, who does not have access to the system K containing
the key, the state of the message is represented by the density matrix

$$\rho_M = \frac{1}{2}|m\rangle\langle m|_M + \frac{1}{2}X|m\rangle\langle m|_M X = \frac{\mathbb{I}}{2} .$$

Note that ρ_M does *not* depend on m! Whatever m is, we get that $\rho_M = \mathbb{I}/2$. Since all
information that can be gained from receiving the encrypted message is captured in
the density matrix ρ_M, it follows that absolutely no information about m can be gained
from intercepting the encryption.

The Quantum One-Time Pad

We are finally ready for our first element of quantum cryptography: the quantum one-
time pad! Let us consider the task of encrypting a qubit, instead of a classical bit (see
Figure 2.5). As a first attempt we might try to use the classical one-time pad, and see if
it works for qubits as well. Does it? Well, the qubits $|0\rangle$ and $|1\rangle$ are encrypted just like
classical messages are. But what about the qubit in state $|+\rangle$? Because $X|+\rangle = |+\rangle$, in
this case whatever key Alice and Bob share, the qubit is "encrypted" to itself. This is
certainly not secure!

The difficulty is that a quantum encryption scheme should hide information in
all possible bases the qubit could be encoded in. In the classical case, applying the
bit-flip operator X allowed us to encrypt any bit expressed in the standard basis. If
we are allowed other bases, we should also encrypt a bit encoded in the Hadamard

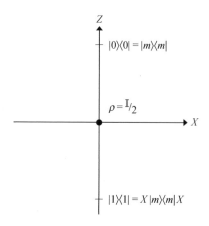

Fig. 2.4 Classical one-time pad in the XZ-plane of the Bloch sphere for $m = 0$.

Fig. 2.5 General form of a quantum one-time pad. Alice encrypts the message qubit ρ with key k by applying unitary $U(k)$. Bob decrypts by undoing the unitary according to the key k.

basis. This could be done by applying a Z instead of an X, because $Z|+\rangle = |-\rangle$ and vice versa.

But what about other bases? What operation do we need to apply to encrypt information encoded in them? And how do we combine all these operations so that the same encryption scheme works for *all* qubits?

At this point it may seem miraculous that quantum encryption is at all possible using only a finite amount of key! But it is possible, and in fact all we need are *two* bits of key, for every qubit.

Amazingly, it is enough to handle both the standard and the Hadamard bases, and all other bases will follow. Let's see how this works. To flip in both bases, we apply the unitary operator $X^{k_1} Z^{k_2}$, where $k_1, k_2 \in \{0, 1\}$ are two key bits chosen uniformly at random. With this choice of encryption operation, an arbitrary single qubit ρ is transformed as

$$\rho \mapsto \frac{1}{4} \sum_{k_1, k_2 \in \{0,1\}} X^{k_1} Z^{k_2} \rho Z^{k_2} X^{k_1} . \tag{2.16}$$

Now let's verify that this securely encrypts *any* single-qubit density matrix ρ. For this, remember the Bloch sphere representation of ρ (Figure 2.6). Remember also the fact that the Pauli matrices pairwise anti-commute. Using this we can make a small calculation,

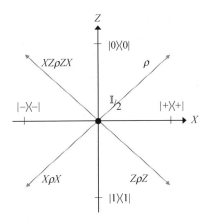

Fig. 2.6   Quantum one-time pad in the XZ-plane of the Bloch sphere. A qubit ρ is encoded by two key bits: the operations $\mathbb{I}, X, Z, XZ$ are performed on the qubit with equal probability. The resulting mixture of states is the maximally mixed state (represented by the origin of the diagram).

$$\frac{1}{4}\left(X + XXX + ZXZ + XZXZX\right) = \frac{1}{4}\left(X + X - ZZX - XZZXX\right)$$
$$= \frac{1}{4}\left(X + X - X - X\right)$$
$$= 0 ,$$

where we used the fact that the Pauli matrices are observables (i.e. they are Hermitian and square to identity), and $\{X, Z\} = XZ + ZX = 0$. The interpretation of this calculation is that if we apply either $\mathbb{I}$, X, Z or XZ with equal probability to the Pauli matrix X then we obtain 0. Moreover, the same calculation can be done on the matrices Y and Z, and we obtain the same result, 0.

Exercise 2.5.1   Show that similarly, for any $M \in \{X, Y, Z\}$, we have

$$\frac{1}{4}\sum_{k_1, k_2 \in \{0,1\}} X^{k_1} Z^{k_2} M Z^{k_2} X^{k_1} = 0 . \qquad (2.17)$$

Now let's use that any single-qubit state can be written as

$$\rho = \frac{1}{2}\left(\mathbb{I} + v_x X + v_y Y + v_z Z\right) .$$

By linearity and the calculation in Exercise 2.5.1 we then get that for any ρ,

$$\frac{1}{4}\sum_{k_1, k_2 \in \{0,1\}} X^{k_1} Z^{k_2} \rho Z^{k_2} X^{k_1} = \frac{\mathbb{I}}{2} . \qquad (2.18)$$

What this equation means is precisely that from the point of view of anyone who does not know k_1, k_2 the bit- and phase-flipped state is completely independent of the input ρ, which means that all information contained in ρ is hidden from the eavesdropper,

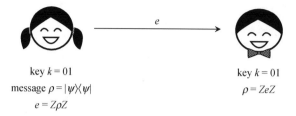

key $k = 01$
message $\rho = |\psi\rangle\langle\psi|$
$e = Z\rho Z$

key $k = 01$
$\rho = ZeZ$

Fig. 2.7   An example of a one-time pad using unitary operations.

who only "sees" $\frac{\mathbb{I}}{2}$ independently of ρ. This leads to the following quantum encryption scheme.

Protocol 2 The quantum one-time pad is an encryption scheme for qubits. The key $k = (k_1, k_2)$ is chosen uniformly at random in $K = \{0,1\}^2$. To encrypt a qubit in state ρ, Alice applies the unitary operation $X^{k_1} Z^{k_2}$ to ρ. To decrypt, Bob applies the inverse operation $(X^{k_1} Z^{k_2})^\dagger = Z^{k_2} X^{k_1}$.

The fact that the scheme is correct follows by definition, since the decryption operation is the inverse of the encryption operation. See Figure 2.7 for an example. For security, we haven't given a complete formal definition for the quantum case. Doing so would take us too far so early in the book; if you are impatient you may jump ahead to Chapter 12. Intuitively, the scheme is perfectly secure because, just as for the classical case, when we compute the reduced density matrix of an encrypted qubit, having traced out the key, we obtain something that is completely independent of the message itself.

To conclude, we observe that the quantum one-time pad can easily be extended to n qubits by applying either $\mathbb{I}, X, Z$ or XZ on each qubit, depending on two key bits associated with that qubit. This means that to encrypt n qubits, we use $2n$ bits of classical key. In Chapter 12 we will show that this is optimal: quantum information requires twice as many bits of key as classical information for perfectly secure encryption.

Exercise 2.5.2 Show that the collection of all (normalized) tensor products of Pauli matrices

$$P^s = \frac{1}{2^n} X^{s_1} Z^{s_2} \otimes X^{s_3} Z^{s_4} \otimes \ldots \otimes X^{s_{2n-1}} Z^{s_{2n}}$$

with $s \in \{0,1\}^{2n}$ form an orthogonal basis for the vector space of all $2^n \times 2^n$ matrices, i.e. for all $s, t \in \{0,1\}^{2n}$, $\mathrm{tr}((P^s)^\dagger P^t) = \delta_{st}$. In particular, any density matrix ρ on n qubits has a unique decomposition of the form

$$\rho = \frac{1}{2^n}\left(\mathbb{I}^{\otimes n} + \sum_{s \neq 0} v_s P^s\right), \tag{2.19}$$

for some complex coefficients v_s.

QUIZ 2.5.2 *Alice encodes the qubit $|\psi\rangle$ using the quantum one-time pad. Eve is ignorant about the key bits k_1 and k_2. What is the state of the encoded qubit as seen by Eve?*

(a) $\rho = \frac{1}{2}(|\psi\rangle\langle\psi| + XZ|\psi\rangle\langle\psi|ZX)$
(b) $\rho = \frac{\mathbb{I}}{2}$
(c) $\rho = X^{k_1}Z^{k_2}|\psi\rangle\langle\psi|Z^{k_2}X^{k_1}$

QUIZ 2.5.3 *What is the state of the encoded qubit as seen by Bob, who does know the key bits k_1 and k_2?*

(a) $\rho = \frac{1}{2}(|\psi\rangle\langle\psi| + XZ|\psi\rangle\langle\psi|ZX)$
(b) $\rho = \frac{\mathbb{I}}{2}$
(c) $\rho = X^{k_1}Z^{k_2}|\psi\rangle\langle\psi|Z^{k_2}X^{k_1}$

CHAPTER NOTES

For additional background on probability theory you may consult any one of the many textbooks available, such as D. G. Kelly, *Introduction to Probability* (Macmillan, 1994) or S. M. Ross, *A First Course in Probability* (Pearson Prentice Hall, 2010). For the new elements of the quantum formalism introduced in this chapter, we recommend the textbook by A. Nielsen and I. L. Chuang, *Quantum Computation and Quantum Information* (Cambridge University Press, 2000). For a more advanced discussion focused on quantum information theory, the book by M. M. Wilde, *Quantum Information Theory* (Cambridge University Press, 2013), provides a wealth of information.

Shannon in his 1949 paper (Communication theory of secrecy systems. *Bell System Technical Journal*, **28**(4):656–715, 1949) formally introduced the notion of perfect secrecy for classical communication and showed that the one-time pad achieves perfect secrecy. The task of encrypting quantum information is first considered by A. Ambainis, et al. (Private quantum channels. In *Proceedings of FOCS*, 2000. arXiv:quant-ph/0003101) and P. O. Boykin and V. Roychowdhury (Optimal encryption of quantum bits. quant-ph/0003059, 2000), who introduce the quantum one-time pad and show that it achieves perfect secrecy for quantum encryption. We will return to the topic of quantum encryption in Chapter 12.

PROBLEMS

2.1 Classical one-time pad

We meet up with our favorite protagonists, Alice and Bob. As you know by now, Alice and Bob often encounter an adversary named Eve who is intent on listening in on their conversations. In order to protect themselves Alice and Bob have, during the the last quantum cryptography conference, exchanged a large amount of classical key which they can use to encrypt messages. Alice knows that a safe way to encrypt is to use a classical one-time pad, but she feels like this uses a large amount of key. She comes up with the following encoding scheme, which she claims is also secure but uses less key. Alice's scheme goes as follows. For i ranging from 1 to n, the total number of bits in her message, Alice does the following:

Step 1. Alice flips a coin.
Step 2. If the result is tails she uses a shared key bit to encode the i-th message bit, via addition modulo 2 as in the one-time pad.
Step 3. If the result is heads, she uses a fresh random bit r, generated on the fly, to encode the i-th message bit, again via addition modulo 2.

Alice claims that this procedure uses less key, but is this really true?

1. How many bits of key will Alice use on average for an n-bit message?
2. This gain in key length probably comes at a price. Which of the following statements about the protocol is true?

   I. The protocol is secure and correct (Bob can decode the message but Eve cannot).

II. The protocol is not secure but correct (Bob and Eve can decode the message).

III. The protocol is secure but not correct (neither Bob nor Eve can decode the message).

IV. The protocol is not secure and not correct (Eve can decode the message but Bob cannot).

2.2 Density matrices

Suppose that Alice and Bob share a device which they can use to send qubits to each other. In this problem we investigate what happens when the device sends states that are noisy.

Let us imagine that Alice wants to send the state $|0\rangle$ to Bob. However, 50% of the time the quantum device outputs the state $|1\rangle$ instead.

1. Give an expression for ρ, the density matrix describing the state that Bob receives.

2. Imagine that Bob receives two identical, independent copies of this density matrix. He chooses to measure one of them in the standard basis and the other in the Hadamard basis. What are the distributions of the outcomes $0, 1, +, -$?

3. Now suppose that the machine on Alice's side is not noisy but simply wrong and consistently prepares qubits in the state $|+\rangle$. If Bob again has two states and measures one of them in the standard basis and one of them in the Hadamard basis, what is the distribution of outcomes?

2.3 Classical-quantum states

Consider again Alice's faulty qubit transmission device. Imagine that, as in the previous problem, when asked to produce a $|0\rangle$ state the device returns either $|0\rangle$ or $|+\rangle$ with 50% probability. But now suppose further that the device also returns a classical flag that indicates which state was produced, 0 for $|0\rangle$ and 1 for $|+\rangle$. The joint state of the flag and the qubit produced can be described by the classical-quantum state

$$\rho_{XA} = \frac{1}{2}|0\rangle\langle0|_X \otimes |0\rangle\langle0|_A + |1\rangle\langle1|_X \otimes |+\rangle\langle+|_A,$$

where X is used to designate the flag bit, and A the qubit. Now imagine Alice generates a state using her machine and sends it to Bob, while keeping the classical flag to herself. From the point of view of Bob, this situation is exactly analogous to the one described in the previous problem. In particular, Bob receives a qubit that is in the mixed state

$$\rho_B = \frac{1}{2}|0\rangle\langle0| + \frac{1}{2}|+\rangle\langle+|.$$

Now Bob can choose to measure either in the standard basis or in the Hadamard basis.

1. Which one of Bob's possible measurement settings, standard or Hadamard, will give him the highest probability of getting outcome 0 (+ in the Hadamard basis)?

Now imagine that Alice also sends the flag X to Bob. Thus Bob receives two qubits in the joint state ρ_{XA}.

2. Which of the following strategies allows Bob to recover Alice's intended qubit, $|0\rangle$, with certainty?

    I. If the flag value is 0 Bob measures in the standard basis, and in the Hadamard basis otherwise.

   II. If the flag value is 0 Bob measures in the Hadamard basis, and in the standard basis otherwise.

  III. The flag value does not affect Bob's chances of getting the right result (outcome 0 in the standard basis, outcome + in the Hadamard basis).

2.4 Quantum one-time pad

In the chapter we saw that two classical bits of key are needed to encrypt one qubit: one for choosing whether to apply a Z, and another for choosing whether to apply an X. This was necessary because the X operation has no effect on the $|+\rangle$ state and the Z operation has no effect on the $|0\rangle$ state. Now, Alice has come up with a clever idea for a protocol that uses only one bit of key per qubit. Instead of an X or a Z operation she will apply a Hadamard H, which is the unitary transformation such that $H|0\rangle = |+\rangle$ and $H|+\rangle = |0\rangle$. This allows her to avoid the problem of leaving either standard basis states or Hadamard basis states unchanged by the encryption, while only using one bit of key (for deciding whether or not to apply H) per qubit. Before Alice rushes to publish her discovery it might be worthwhile to check if her scheme is truly secure.

1. Specify a decryption operation such that Alice's scheme is a correct encryption scheme.

2. Is Alice's scheme secure?

2.5 Unambiguous quantum state discrimination

In this problem we will explore a practical advantage to performing a general POVM rather than a projective measurement. Consider the following scenario: Bob sends Alice a qubit prepared in one of the two nonorthogonal states $|0\rangle$ and $|+\rangle$, each with probability 1/2. Alice wants to determine which state Bob has prepared. To this end she performs a measurement on Bob's qubit whose measurement outcome identifies it as either $|0\rangle$ or $|1\rangle$. Alice's goal is to minimize the probability of mis-identifying $|0\rangle$ as $|+\rangle$, or vice versa. Let us first restrict her to projective measurements.

1. Suppose that Alice measures in the basis $\{|0\rangle, |1\rangle\}$. She identifies the state as $|0\rangle$ if she gets the outcome $|0\rangle$ and as $|+\rangle$ if she gets the outcome $|1\rangle$. What are p, her probability of incorrectly identifying $|0\rangle$, and q, her probability of incorrectly identifying $|+\rangle$?

2. Suppose instead Alice measures in the basis $\{|+\rangle, |-\rangle\}$. She identifies the state as $|+\rangle$ if she gets the outcome $|+\rangle$ and as $|0\rangle$ if she gets the outcome $|-\rangle$. What are p, her probability of incorrectly identifying $|0\rangle$, and q, her probability of incorrectly identifying $|+\rangle$?

One can show (you may try!) that Alice cannot do better than the above with any projective measurement. That is, no projective measurement gives her a smaller average probability of mis-identification $(p+q)/2$. Now suppose that we allow Alice to perform a general measurement. In particular, consider the following POVM with three elements:

$$E_1 = \frac{\sqrt{2}}{1+\sqrt{2}}|1\rangle\langle1|,$$

$$E_2 = \frac{\sqrt{2}}{1+\sqrt{2}}\frac{(|0\rangle - |1\rangle)(\langle0| - \langle1|)}{2},$$

$$E_3 = \mathbb{I} - E_1 - E_2.$$

3. Alice identifies the state as $|+\rangle$ if she gets outcome 1, as $|0\rangle$ if she gets outcome 2, and makes no identification if she gets outcome 3. What are her probabilities of mis-identifying $|0\rangle$ and $|+\rangle$ as the other, respectively?

2.6 Robustness of GHZ and W states

In this problem we explore two classes of N-qubit states that are especially useful for cryptography and communication, but behave very differently under tracing out a single qubit. Let's first define them for $N = 3$:

$$\text{GHZ state:}\quad |\text{GHZ}_3\rangle = \frac{1}{\sqrt{2}}(|000\rangle + |111\rangle),$$

$$\text{W state:}\quad |W_3\rangle = \frac{1}{\sqrt{3}}(|100\rangle + |010\rangle + |001\rangle).$$

Note that both states are symmetric under permutation of the three qubits, so without loss of generality we may trace out the last one, tr_3. Also, we have analogously $|\text{GHZ}_2\rangle = \frac{1}{\sqrt{2}}(|00\rangle + |11\rangle)$ and $|W_2\rangle = \frac{1}{\sqrt{2}}(|10\rangle + |01\rangle)$.

In the following we consider the *overlap* between N-qubit GHZ and W states with one qubit discarded (i.e. traced out) and their $(N-1)$-qubit counterparts. The overlap of density matrices ρ and σ is defined as $\text{tr}\rho\sigma$, a measure of "closeness" that generalizes the expression $|\langle\phi|\psi\rangle|^2$ for pure states.

1. Calculate the overlap between $|\text{GHZ}_2\rangle\langle\text{GHZ}_2|$ and $\text{tr}_3|\text{GHZ}_3\rangle\langle\text{GHZ}_3|$.
2. Calculate the overlap between $|W_2\rangle\langle W_2|$ and $\text{tr}_3|W_3\rangle\langle W_3|$.

Now we generalize to the N-qubit case. As you might expect, $|\text{GHZ}_N\rangle = \frac{1}{\sqrt{2}}(|0\rangle^{\otimes N} + |1\rangle^{\otimes N})$ and $|W_N\rangle$ is an equal superposition of all N-bit strings with exactly one 1 and $N - 1$ 0's.

3. What is the overlap $\text{tr}(|\text{GHZ}_{N-1}\rangle\langle\text{GHZ}_{N-1}| \text{tr}_N |\text{GHZ}_N\rangle\langle\text{GHZ}_N|)$ in the limit $N \to \infty$?
4. What is the overlap $\text{tr}(|W_{N-1}\rangle\langle W_{N-1}| \text{tr}_N |W_N\rangle\langle W_N|)$ in the limit $N \to \infty$?

The interpretation of these results is that W states are more "robust" against loss of a single qubit than GHZ states.

CHEAT SHEET

Trace

Given a matrix M, its trace is given by $\text{tr}(M) = \sum_i M_{ii}$, i.e. the sum of its diagonal elements. The trace operation is cyclic, i.e. for any two matrices M, N, $\text{tr}(MN) = \text{tr}(NM)$.

Density matrices

If we prepare a quantum system in the state ρ_x with probability p_x, then the state of the system is given by the density matrix

$$\rho = \sum_x p_x \rho_x .$$

Bloch representation of density matrices: any qubit density matrix can be written as

$$\rho = \frac{1}{2} \left(\mathbb{I} + v_x X + v_z Z + v_y Y \right) ,$$

and the Bloch vector $\vec{v} = (v_x, v_y, v_z)$ satisfies $\|\vec{v}\|_2 \leq 1$, with equality if and only if ρ is pure.

Probability of measurement outcomes on a density matrix

If a quantum state with density matrix ρ is measured in the basis $\{|w_j\rangle\}_j$, then the probability of obtaining each outcome $|w_j\rangle$ is given by

$$p_{w_j} = \langle w_j | \rho | w_j \rangle = \text{tr}(\rho |w_j\rangle\langle w_j|) .$$

Combining density matrices

For density matrices $\rho_A = \begin{pmatrix} a_{11} & a_{12} \\ a_{21} & a_{22} \end{pmatrix}$ and $\rho_B = \begin{pmatrix} b_{11} & b_{12} \\ b_{21} & b_{22} \end{pmatrix}$ representing qubits A and B, the joint density matrix is given by

$$\rho_{AB} = \rho_A \otimes \rho_B := \begin{pmatrix} a_{11}\rho_B & a_{12}\rho_B \\ a_{21}\rho_B & a_{22}\rho_B \end{pmatrix} = \begin{pmatrix} a_{11}b_{11} & a_{11}b_{12} & a_{12}b_{11} & a_{12}b_{12} \\ a_{11}b_{21} & a_{11}b_{22} & a_{12}b_{21} & a_{12}b_{22} \\ a_{21}b_{11} & a_{21}b_{12} & a_{22}b_{11} & a_{22}b_{12} \\ a_{21}b_{21} & a_{21}b_{22} & a_{22}b_{21} & a_{22}b_{22} \end{pmatrix} .$$

Partial trace

Given a bipartite matrix ρ_{AB} which has a decomposition of the form

$$\rho_{AB} = \sum_{ijk\ell} \gamma_{ij}^{k\ell} |i\rangle\langle j|_A \otimes |k\rangle\langle\ell|_B ,$$

where $\{|i\rangle_A\}_i$ and $\{|k\rangle_B\}_k$ are orthonormal bases of A and B respectively, the partial trace over system A yields the reduced state ρ_B:

$$\rho_B = \text{tr}_A(\rho_{AB}) = \sum_{ijk\ell} \gamma_{ij}^{k\ell} \text{tr}(|i\rangle\langle j|) \otimes |k\rangle\langle\ell|_B = \sum_{k\ell} \left(\sum_j \gamma_{jj}^{k\ell} \right) |k\rangle\langle\ell|_B .$$

Properties of the Pauli matrices X, Y, Z

For any $S_1, S_2 \in \{X, Y, Z\}$, $\{S_1, S_2\} = 2\delta_{S_1 S_2}\mathbb{I}$ where the anti-commutator is $\{A, B\} = AB + BA$. This implies the following properties.

- Zero trace: $\text{tr}(S_1) = 0$.
- Orthogonality: $\text{tr}(S_1^\dagger S_2) = 0$.
- Unitary: $S_1^\dagger S_1 = S_1 S_1^\dagger = \mathbb{I}$.
- Square to identity: $S_1^2 = \mathbb{I}$.

3

Quantum Money

In this chapter we put our freshly acquired formalism of qubits and measurements to good use by exploring a rather ancient cryptographic task: money! While traditional coins and bills can always be copied, the idea for "uncloneable" quantum money was discovered in the first paper ever written on quantum information, by Stephen Wiesner in the 1970s. Wiesner's key observation was that the possibility to encode information in different bases, such as the standard basis and the Hadamard basis, provides a natural mechanism for copy-protection. In this chapter we explain Wiesner's idea and take the opportunity to deepen our understanding of quantum states and measurements.

So what is money? Generally, a bill has two components. First, there is a physical object, such as a piece of paper or metal. Second, there is some identifier associated with the physical object, such as a serial number. The serial number is created on the day that the bill is minted, and it is used to specify all kinds of information about the bill, such as its value, its provenance, the date on which it was minted, etc. This information is kept by the bank as a means to keep track of all valid money in circulation.

The main security guarantee that one wants of money is that it cannot be duplicated. This is what the "paper" part of the bill is meant for: if the bill only consisted of a serial number, this number could be easily copied and the amount of real currency associated with it spent twice. A piece of paper is technically a little harder to duplicate than a mere number ... but not impossible!

Remember the *no-cloning principle* from Chapter 1. Informally, this principle states that there is no quantum operation that can perfectly copy an arbitrary qubit. In other words, *qubits cannot be duplicated*. You can see where this is going, right? Let us first explore a very simple (but flawed) idea for a quantum money scheme.

3.1 A (Too) Simple Quantum Money Scheme

Let us give a preliminary definition of a quantum bill as a physical object that consists of two parts: first, a classical "serial number" $ that uniquely identifies the state and a copy of which is stored by the bank, and second, a quantum state $|\psi_\$\rangle$ that constitutes the "uncloneable" component of the bill. Think of the serial number as a string of bits that is publicly known and records general information about the bill, such as when it was created, how much money it is worth (such as its gold equivalent), etc. Generally we will take the serial number to be $\$ \in \{0,1\}^n$ for some integer n that is sufficiently large that we never run out of serial numbers, such as $n = 1024$.

Let's see the simplest quantum money scheme you might think of. To create a quantum bill, first generate a serial number $\$ \in \{0,1\}^n$ uniformly at random. Then create an n-qubit quantum state such that the i-th qubit is initialized in the standard basis state equal to the i-th bit of $\$$. In other words, create the quantum state $|\psi_\$\rangle = |\$\rangle$. The quantum bill is the pair $(\$, |\psi_\$\rangle)$ of the serial number and the state associated with it. Since qubits (and a fortiori n qubits) cannot be cloned, the scheme is secure, right?

Of course not! This scheme has no secret information. Given a quantum bill $(\$, |\psi_\$\rangle)$ it is very easy to create an unlimited number of identical copies of it, simply by using the serial number to prepare the state $|\psi_\$\rangle$. This does not violate the no-cloning principle, because we are given a classical description of the state: it is the standard basis state associated with the n-bit string $\$$. Given this classical description, and a quantum computer, it is straightforward to create as many copies of $|\psi_\$\rangle$ as desired. In fact, even if we didn't have access to the classical serial number, the scheme would be entirely broken, as an attacker could first measure $|\psi_\$\rangle$ in the standard basis to obtain $\$$, and then create as many copies of it as desired.

In case you're not sure why the no-cloning principle does not apply, remember that the impossible task is to design a quantum machine that has the ability to clone *every* state. But there still can be machines that clone specific families of states, such as all standard basis states. An interesting money scheme will necessarily involve states that are more complicated than simple standard basis states!

3.2 Wiesner's Quantum Money

In Wiesner's money scheme, each serial number $\$$ is made of two strings $x_\$, \theta_\$ \in \{0,1\}^n$, where n is an integer that parametrizes the security of the scheme (for example, $n = 1024$). In the following, we drop the subscript $\$$ and simply write x, θ for $x_\$, \theta_\$$. It is important that the pair (x, θ) is kept secret: it is generated at random by the bank on the day when the quantum bill is minted, but it is never revealed, even to the honest holder of the bill.

For $x_i, \theta_i \in \{0, 1\}$ we introduce the notation $|x_i\rangle_{\theta_i} = H^{\theta_i}|x_i\rangle$.[1] Then the money state associated with $\$$ is the n-qubit state $|\psi_\$\rangle$ whose i-th qubit is in state $|x_i\rangle_{\theta_i}$:

$$|\psi_\$\rangle = |x_1\rangle_{\theta_1} \otimes \cdots \otimes |x_n\rangle_{\theta_n}.$$

QUIZ 3.2.1 *Suppose that $n = 2$, and consider a serial number $\$$ such that $x_\$ = 01$ and $\theta_\$ = 10$. Then the associated quantum money state is*

(a) $|\psi_\$\rangle = |0\rangle|1\rangle$
(b) $|\psi_\$\rangle = |0\rangle|+\rangle$
(c) $|\psi_\$\rangle = |+\rangle \otimes |1\rangle$

---

1 The four possible states are thus $|0\rangle, |1\rangle$ (for $\theta = 0$) and $|+\rangle, |-\rangle$ (for $\theta = 1$). These states are often referred to as "BB'84 states" – we will see why in Chapter 8.

Now that we've described a scheme, can you break it? That is, can you forge multiple copies of a quantum bill, given a single copy as input? If not, then why?

Based on the intuition that quantum information cannot in general be cloned, intuitively we shouldn't be able to copy quantum bills. However, as we saw, the no-cloning theorem requires care in its application: in particular, "classical" states of the form $|000\rangle, |001\rangle$, etc. certainly can be copied! So does the no-cloning theorem apply in our setting, or not? To answer this we need to go back to the proof of the theorem, given in Section 1.6.3. If you examine the proof closely you will notice that the theorem already applies in the case when the only states considered are $|0\rangle, |1\rangle, |+\rangle, |-\rangle$. This seems to rule out a perfect cloning machine for our quantum bills. However, let's be careful! If you measure each qubit of $|\psi_\$\rangle$ in either the standard or the Hadamard basis, without knowing which is the correct basis, you expect to get the right answer for approximately half the qubits. So you "learn" half the state in this way. What if you could learn more? What if you could recover 99% of the qubits? Or all the qubits, 99% of the time? This is not obviously ruled out by the "qualitative" no-cloning theorem that we have seen. If this were the case, would we still want to consider the scheme to be secure, even though perfect cloning is impossible?

To answer this question we have to go through one of the most important exercises in cryptography: introducing a security definition! Until now we have been arguing about security at a very intuitive level; to make progress we need to establish firm foundations to support our investigation.

3.2.1 Definition of Quantum Money

To specify a quantum money scheme we need to answer the following questions: How (and by whom) is a quantum bill generated? And what is the procedure for determining the validity of a (candidate) bill? More formally, we define a quantum money scheme to consist of two procedures, each meant to answer one of these two questions:

- A *state generation procedure* $\text{GEN}(1^n)$: This is the procedure applied by the bank to mint money. It takes as input an integer n specified in unary called the "security parameter" (intuitively, the larger n is, the more secure is the scheme).[2] The procedure returns a triple $(\$, |\psi_\$\rangle, k_\$)$ of a classical serial number $\$$, a quantum state $|\psi_\$\rangle$, and a classical "private key" $k_\$$ that specifies secret information about the bill that is kept by the bank.
- A *bill verification procedure* $\text{VER}(\$, |\psi\rangle, k)$: This is the procedure executed by the bank to verify a quantum state. It takes as input a pair $(\$, |\psi\rangle)$ of a serial number and a quantum state, as well as a key k, and returns either "accept" or "reject."

Note that the state generation procedure GEN does not explicitly specify a denomination for the quantum bills. The simplest implementation of the scheme will associate an identical value to each money state, such as 1€. It is also possible to associate

---

2 The reason that n is specified as a string of 1's of length n, as opposed to using its binary representation, is a convention motivated by the standard requirement that GEN runs in time bounded by a polynomial in the length of its input. Here we want to allow GEN to run in time polynomial in n, not just polynomial in $\log n$. Since we do not focus on algorithmic efficiency, you can ignore this requirement.

different values to the bills: in this case, we can imagine that a bill's value is specified as a (classical) integer accompanying the state, and it is also kept together with the serial number in the bank's records (so that a user cannot arbitrarily change the value of their money state). Here, we stay with the simpler definition of assuming that all money states have the same value.

Note also that our specification of the verification procedure implicitly destroys the money state: VER takes as input $, $|\psi\rangle$ and k and only returns "accept" or "reject." In general, it may seem desirable that valid money states are returned to the user, so that it is possible to verify a state without being forced to destroy it. This, however, creates security risks which we will explore in Section 3.5 below. For now we stick with the definition and consider that verification always entirely destroys the money state (if you find that unreasonable, imagine that the bank generates a fresh bill to compensate the user).

Let's see how the abstract formalism looks for the case of Wiesner's scheme:

- The state generation procedure $\text{GEN}_W(1^n)$ first selects a serial number \$ in an arbitrary way (for example, the serial numbers can be chosen sequentially, or they can be sampled at random and contain a time stamp, etc.). Then it selects two strings $x, \theta \in \{0,1\}^n$ uniformly at random. The bank records the information $k_\$ = (x, \theta)$ and it creates the state

$$|\psi_\$\rangle = \bigotimes_{i=1}^{n} \left(H^{\theta_i} |x_i\rangle \right) .$$

  Finally, GEN_W returns the triple $(\$, |\psi_\$\rangle, k_\$)$.
- The state verification procedure $\text{VER}_W(\$, |\psi\rangle, k)$ proceeds as follows. First, it interprets the key k as a pair of n-bit strings (x, θ). (If k is not formatted in this way then the procedure rejects.) Then, it applies a Hadamard gate to each of the n qubits of $|\psi\rangle$, for $i = 1, \ldots, n$, such that $\theta_i = 1$. Finally, it measures all qubits in the standard basis to obtain a string $y \in \{0,1\}^n$. It returns "accept" if $y = x$, and "reject" otherwise.[3]

Since all the operations that we described can be implemented on a quantum computer, the pair $(\text{GEN}_W, \text{VER}_W)$ is a well-defined quantum money scheme. But this doesn't make it an *interesting* quantum money scheme! Indeed, the (too) simple scheme from the previous section is also "well-defined." To make it interesting, we need to introduce *correctness* and *security* requirements.

3.2.2 Correctness and Security of Quantum Money

The first property that a quantum money scheme should satisfy is called the *correctness* property: valid money states should always be accepted by the verification procedure. Formally,

$$\forall n \geq 1, \quad \text{VER}\big(\text{GEN}(1^n)\big) = \text{``accept''} . \tag{3.1}$$

---

3 In this description the serial number \$ is not used. In a "real" interaction, the user would give the (claimed) quantum bill $|\psi\rangle$ as well as the classical serial number to the bank, who would then use \$ to look up the shared secret key k and then perform $\text{VER}_W(\$, |\psi\rangle, k)$.

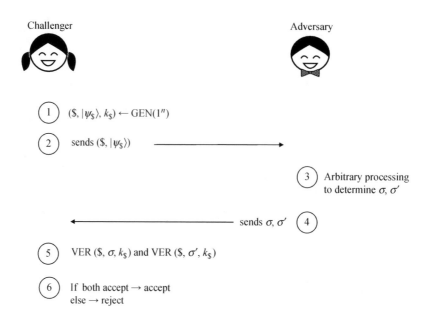

Challenger   Adversary

1   ($, |$\psi_\$$⟩, $k_\$$) ← GEN(1^n)

2   sends ($, |$\psi_\$$⟩)   ───────────────→

                                          3   Arbitrary processing
                                              to determine σ, σ'

    ←───────────────   sends σ, σ' 4

5   VER ($, σ, $k_\$$) and VER ($, σ', $k_\$$)

6   If both accept → accept
    else → reject

Fig. 3.1   The CLONE game. The challenger implements the actions described, whereas the adversary may use an arbitrary (quantum) strategy to perform their actions.

This seems like an absolute minimum requirement, as otherwise the bank would not accept its own correctly minted bills. Note, however, that it doesn't prevent the verification procedure from accepting *all* states! This would still be a correct money scheme according to our definition. However, intuitively it would be far from secure, since any user could create bills out of nothing and still pass verification. To prevent this and other more subtle attacks, we need to introduce a *security condition* for quantum money.

How should we define security? Informally, we would like it to be impossible to "duplicate" a quantum money state: given a valid quantum bill, a user should be able to spend it once (this is guaranteed by (3.1)), but not twice. In cryptography we often formalize a security notion through a "game" that expresses precisely the kind of situation that the scheme should prevent from happening. Here, the forbidden situation is that an adversary to the scheme manages to copy a quantum bill. Let's formulate this requirement as the following game, played between an "adversary" and a "challenger." The idea is that the challenger is trusted (they play their part in the game as described), but the adversary may employ any kind of malicious behavior to maximize their chances of winning in the game.

- The challenger executes the procedure ($, |$\psi_\$$⟩, $k_\$$) ← GEN(1^n).[4] They keep $k_\$$ to themselves and give ($, |$\psi_\$$⟩) to the adversary.

---

4 The notation X ← PROC(Y) means that we use the variable X to denote the outcome of running the procedure PROC on input Y.

- The adversary returns *two* quantum states σ and σ', each of the same number of qubits as $|\psi_\$\rangle$. (It is up to the adversary how these states are obtained.)
- The challenger executes $\text{VER}(\$, \sigma, k_\$)$ and $\text{VER}(\$, \sigma', k_\$)$. They accept if and only if both verification attempts accept.

We call this game CLONE since in the game the adversary is challenged to create two copies σ and σ' of the quantum bill (see Figure 3.1). The game captures the intuitive security notion described above: a strategy for the adversary that wins in the game is equivalent to an attack on the quantum money scheme.

Note that in the game we do not assume that the adversary returns two pure single-qubit states. The reason is that it is physically impossible to tell if a state is pure or mixed: indeed, a mixed state is nothing but a distribution over pure states, so allowing mixed states is similar to allowing the adversary to apply a randomized strategy to implement its attack. In particular, σ and σ' could each consist of a certain number of qubits taken out of a bigger entangled state ρ (for example, they could be the two halves of an EPR pair). The procedure $\text{VER}(\$, \sigma, k_\$)$ is still well-defined: it applies whatever measurement VER would apply on a pure state $|\psi\rangle$ to the mixed state σ.

> **Definition 3.2.1.** *For any $\varepsilon \geq 0$ we say that a quantum money scheme is ε-secure if the maximum probability with which any adversary can succeed in the game CLONE is at most ε, where the probability is taken over the randomness in GEN, VER, and any randomness used by the adversary.*

It is important to realize that the above is a *definition*. The definition captures a specific type of attack, which is modeled in the security game. Any such attack corresponds to some quantum actions chosen by the adversary. Such actions can take as input $(\$, |\psi_\$\rangle)$, but also any extra qubits that the adversary wants to use. If a scheme is shown to be secure it exactly means that such attacks are impossible: it means no more and no less. Indeed, we have to be careful with words, and the definition does not capture *all* attacks! For example, if the adversary is allowed to break into the bank and steal the database of secret keys $k_\$$, then all guarantees are off. The way that this is prevented in the security game is when we write that "[The challenger] keeps $k_\$$ to itself" in the first step. If the challenger (the bank) does *not* "keep $k_\$$ to itself" (e.g. the adversary manages to steal it), then this scenario falls outside of the rules of the game, and the security definition no longer provides any guarantees.

In the remainder of this chapter we study the security of Wiesner's quantum money scheme $(\text{GEN}_W, \text{VER}_W)$. Before we do so, we need to introduce the most general possible kind of transformation that the adversary could employ; this is the theory of *quantum channels*.

3.3 Quantum Channels

What is the most general operation that a quantum attacker, or more abstractly a quantum device (or computer), can make?

We have already seen in Section 1.6 that quantum operations are unitary. However, in the context of cloning, we are looking at a map that increases the number of qubits. A unitary does not change the dimension of the state it acts on! How is this possible? This is because there are two other operations that are considered valid quantum operations:

- Preparation of an extra qubit, often called an "ancilla." This is the operation that, given a quantum state $|\psi\rangle$, appends to it a qubit in state $|0\rangle$: $|\psi\rangle \mapsto |\psi\rangle |0\rangle$. Note that this operation increases the dimension of space by a factor 2, because $\dim(\mathbb{C}^d \otimes \mathbb{C}^2) = 2d$. The operation can be repeated any number of times to add more qubits.
- Removing qubits, or, mathematically, *tracing out*: This is the operation that, given a density matrix $\rho_{1\cdots(n+1)}$ on $n+1$ qubits, for any $n \geq 0$, takes the partial trace over the $(n+1)$-st qubit: $\rho_{1\cdots(n+1)} \mapsto \mathrm{tr}_{(n+1)}(\rho_{1\cdots(n+1)}) = \rho_{1\cdots n}$. This operation reduces the dimension by a factor 2. It can be repeated any number of times to remove more qubits, and can be applied on any qubit that one desires (not just the last one).

> **Definition 3.3.1 (Quantum channel).** *For any $n, m \geq 0$ a quantum channel from n qubits to m qubits is any operation $\mathcal{N}(\cdot)$ from $(\mathbb{C}^2)^{\otimes n}$ to $(\mathbb{C}^2)^{\otimes m}$ that can be expressed as a sequence of (i) extra qubit preparation, (ii) unitary operation, and (iii) tracing out.*

Although the definition requires that all the extra qubit preparation comes first, and the tracing out comes last, Problem 3.1 shows that allowing arbitrary orders does not make the definition more general.

Because each of the three operations described in the definition takes density matrices to density matrices, a quantum channel always takes density matrices to density matrices. Because they can represent any operation that can be performed by a quantum entity, aside from their use in modeling adversaries in cryptography, quantum channels play an important role throughout quantum information, for example to characterize noise.

A well-known example of a quantum channel is the bit-flip channel, from one qubit in system A to one qubit in system A, given in its Kraus decomposition (see Box 3.1) by

$$\mathcal{N}_{\text{bitflip}}(\rho_A) = (1-p)\rho_A + pX\rho_A X , \qquad (3.2)$$

where $p \in [0,1]$ can be understood as the probability that a bit-flip X is applied in the standard basis.

The expression in (3.2) defines $\mathcal{N}_{\text{bitflip}}$ through its Kraus operators $N_1 = \sqrt{1-p}\mathbb{I}$ and $N_2 = \sqrt{p}X$ which satisfy

$$N_1 N_1^\dagger + N_2 N_2^\dagger = (1-p)\mathbb{I} + p\mathbb{I} = \mathbb{I} ,$$

as required. To practice with the definition, let us see how to express $\mathcal{N}_{\text{bitflip}}$ as a sequence of qubit addition, unitary, and tracing-out operations.

BOX 3.1   Kraus Decomposition of a Quantum Channel

Any quantum channel $\mathcal{N}$ can be expressed in terms of Kraus operators $\{N_i\}_i$ as

$$\mathcal{N}(\rho) = \sum_i N_i \rho N_i^\dagger \ ,$$

where $\sum_i N_i^\dagger N_i = \mathbb{I}$. Examples of quantum channels include the following: the identity channel $\mathcal{N}(\rho) = \rho$ whose only Kraus operator is $N_1 = \mathbb{I}$; unitary channels $\mathcal{N}_U(\rho) = U\rho U^\dagger$ for a unitary U, with a single Kraus operator $N_1 = U$, any POVM $\{M_i\}$, with Kraus operators $N_i = U_i\sqrt{M_i}$ for any unitary U_i; the qubit trivial channel $\mathcal{N}(\rho) = \mathbb{I}/2$ for all ρ, whose Kraus operators are $(1/2)\mathbb{I}$, $(1/2)X$, $(1/2)Y$, and $(1/2)Z$; and many more!

First of all, it is clear that we will need to use all three operations. This is because, although $\mathcal{N}_{\text{bitflip}}$ sends one qubit to one qubit, unless $p = 0$ or $p = 1$ it is not unitary: for example, the pure state $|0\rangle$ is sent to a mixed state, which wouldn't happen with a unitary. Now if we could add a qubit in system E, and find U_{AE} such that for any $|\psi\rangle_A$,

$$U_{AE} |\psi\rangle_A \otimes |0\rangle_E = \sqrt{1-p}\,|\psi\rangle_A \otimes |0\rangle_E + \sqrt{p}X\,|\psi\rangle_A \otimes |1\rangle_E \ ,$$

then we would be done. This is because tracing out E on the right-hand side above would create the desired noisy mixture of $|\psi\rangle_A$ with probability $1 - p$, and $X\,|\psi\rangle_A$ with probability p. To define such a unitary, we should also specify what happens if E is initially in the state $|1\rangle_E$. Let's take

$$U_{AE} |\psi\rangle_A \otimes |1\rangle_E = \sqrt{p}\,|\psi\rangle_A \otimes |0\rangle_E - \sqrt{1-p}X\,|\psi\rangle_A \otimes |1\rangle_E \ ,$$

which would give us the definition

$$U_{AE} = \sqrt{1-p}\,\mathbb{I}_A \otimes |0\rangle\langle 0|_E + \sqrt{p}Z_A \otimes |1\rangle\langle 0|_E$$
$$+ \sqrt{p}\,\mathbb{I}_A \otimes |0\rangle\langle 1|_E - \sqrt{1-p}Z_A \otimes |1\rangle\langle 1|_E \ .$$

This is a well-defined matrix, and it achieves what we want in terms of the channel $\mathcal{N}_{\text{bitflip}}$, but we still need to verify that it is a valid unitary transformation. The next exercise asks you to check this.

Exercise 3.3.1 Verify that $U_{AE}U_{AE}^\dagger = \mathbb{I}_{AE}$.

Similar to the bit-flip channel, we can define a *phase-flip* channel by

$$\mathcal{N}_{\text{phaseflip}}(\rho_A) = (1-p)\rho_A + pZ\rho_A Z \ .$$

It is a good exercise to find a decomposition of this channel as a sequence of qubit addition, unitary, and tracing-out operations.

Exercise 3.3.2   The map $\mathcal{N}_{\text{depolarizing}}(\rho_A) = (1-p)\rho_A + \frac{p}{2}\mathbb{I}_A$ is called *depolarizing noise* with strength p. Find the Kraus operators for this channel, and then find a decomposition of it as a sequence of qubit addition, unitary, and tracing-out operations. *[Hint: remember the identity (2.18) showing security of the quantum one-time pad!]*

BOX 3.2   POVM Measurements as Projective Measurements

We record a useful consequence of our discussion regarding measurements. Suppose we are given a general POVM with Kraus elements $\{M_i\}$. Then the map

$$\mathcal{N} : \rho \mapsto \sum_i |i\rangle\langle i| \otimes M_i \rho M_i^\dagger$$

is a quantum channel, with Kraus operators $N_i = |i\rangle \otimes M_i$. Since it is a quantum channel, it can be implemented as a sequence of ancilla preparation, unitary operation, and tracing out. Now, by definition of $\mathcal{N}(\cdot)$ it is clear that performing the POVM M on ρ, or computing $\mathcal{N}(\rho)$ and then measuring the first register in the standard basis, lead to outcomes that have the same distribution. However, the second operation takes the form of a *projective* measurement, applied on a larger system (after adding the ancillas). This is because the "tracing out" part can always be ignored, as it has no effect on the outcome. We have thus discovered the *Naimark dilation principle*. This principle states that for any POVM there is a *projective* measurement on a larger system that produces exactly the same measurement outcome distributions, for any input state. This principle is very useful in cryptography, where any simplification of an adversary's "attack" that we can get our hands on is welcome help!

3.4   Attacks on Wiesner's Scheme

As a warm-up before proving security, let us explore some simple "attacks" on Wiesner's quantum money scheme. Indeed, it is always instructive to try to *break* a cryptographic scheme, by designing the best possible attack on it. By trying hard enough, you can have surprises! Many schemes proposed since the birth of cryptography could have been broken by their inventor, had they tried hard enough, and this would have avoided some embarrassing situations!

In general, an "attack" on a cryptographic system is a procedure that breaks the security definition. In the case of quantum money, our goal is to define actions for the "adversary" in the game CLONE such that the adversary's success probability is as high as possible.

For simplicity, let us first consider the case of the single-qubit version of Wiesner's money. This corresponds to choosing $n = 1$ in CLONE. In this case the adversary is given by the challenger a single-qubit state $|\psi_\$\rangle \in \{|0\rangle, |1\rangle, |+\rangle, |-\rangle\}$. It is also given the serial number, but since this is chosen by GEN independently of anything else, it does not provide any useful information. The adversary does not know which of the four-qubit states has been chosen, and it has to return two single-qubit density matrices σ and σ' (or equivalently, a two-qubit density matrix ρ) such that the probability of both qubits of ρ passing verification is maximized.

Since we know exactly what the verification procedure in Wiesner's scheme does, we can write out explicitly the adversary's maximum success probability as a function of the state ρ that it returns. Let ρ_0, ρ_1, ρ_+, and ρ_- be the two-qubit density matrix returned by the adversary on challenge $|0\rangle$, $|1\rangle$, $|+\rangle$, and $|-\rangle$ respectively. Then the probability that ρ_0 is accepted is the probability that a measurement of its two qubits in the standard basis yields the outcome $|00\rangle$, which is just $\langle 00|\rho_0|00\rangle$. Similarly, the probability that ρ_1 is accepted is $\langle 11|\rho_1|11\rangle$. For ρ_+ and ρ_-, it is $\langle ++|\rho_+|++\rangle$ and $\langle --|\rho_-|--\rangle$ respectively. Since in CLONE the quantum bill is chosen uniformly at random among the four possibilities, the adversary's success probability is the average of these four quantities, i.e.

$$p_{\text{succ}} = \frac{1}{4} \left(\langle 0|\langle 0|\rho_0|0\rangle|0\rangle + \langle 1|\langle 1|\rho_1|1\rangle|1\rangle \right.$$
$$\left. + \langle +|\langle +|\rho_+|+\rangle|+\rangle + \langle -|\langle -|\rho_-|-\rangle|-\rangle \right). \tag{3.3}$$

The adversary's goal is to define a quantum operation, i.e. a quantum channel as described in the previous section, that sends single-qubit states $|\psi\rangle \in \{|0\rangle, |1\rangle, |+\rangle, |-\rangle\}$ to two-qubit density matrices ρ_0, ρ_1, ρ_+, ρ_- in a way that (3.3) is maximized. Intuitively, what blocks us from doing this outright is the *no-cloning principle* (see Section 1.6.3). If you remember, this principle states that no two nonorthogonal states can be perfectly copied. Since, for example, $|0\rangle$ and $|+\rangle$ are not orthogonal, there is no quantum operation that can copy both!

Of course, this impossibility result does not say that we cannot try to clone *approximately*, and see how well we do! So let us give it a try.

3.4.1 Measure-and-Prepare Attacks

A first class of attacks for the adversary is to choose a basis in which to measure $|\psi_\$\rangle$, and then attempt to prepare two copies of it based on the information obtained from the measurement outcome.

As a warm-up, consider the case where the adversary measures in the standard basis. Let's say that if they obtain outcome 0, they return $\rho = |0\rangle\langle 0| \otimes |0\rangle\langle 0|$, and if they get outcome 1, they return $\rho = |1\rangle\langle 1| \otimes |1\rangle\langle 1|$. What is the success probability of this attack? If $\psi_\$ \in \{0, 1\}$ then the state is perfectly cloned. But if $\psi_\$ \in \{+, -\}$ then the outcome from the measurement is a uniformly random bit, so $\rho_+ = \rho_- = \frac{1}{4}\mathbb{I} \otimes \mathbb{I}$. Overall,

$$p_{\text{succ}} = \frac{1}{4}\left(1 + 1 + \frac{1}{4} + \frac{1}{4}\right) = \frac{5}{8}.$$

QUIZ 3.4.1 *Suppose the attacker measures* $|\psi\rangle$ *in the Hadamard basis to obtain an outcome* $x \in \{+, -\}$, *and returns the Hadamard basis state* $\rho = |x\rangle\langle x| \otimes |x\rangle\langle x|$. *What is the success probability of this attack?*

(a) $\frac{1}{2}$

(b) $\frac{5}{8}$

(c) $\frac{3}{4}$

(d) 1

What about measurements in other bases? In the problems you will analyze all attacks that directly measure the qubit, and then prepare a new two-qubit state depending on the measurement outcome obtained. But we could consider even more general attacks. For example, there is no reason to limit the adversary to a basis measurement: making use of extra qubits it can also implement a more general POVM, with more than two outcomes. How can we classify such a broad class of attacks? This points you to the difficulty of proving security of a cryptographic scheme in general: there is no limit to the ingenuity of adversaries! Let's progress slowly by considering a second class of attacks.

3.4.2 Cloning Attacks

Instead of measuring the money state before creating two new ones, the adversary could attempt to "clone" the state right away. This means that the adversary is directly applying a quantum channel from one qubit to two qubits. The simplest case of such a channel adds a single ancilla qubit to the quantum money state, and then applies a unitary U to the two qubits to obtain the two "clones." Starting from (3.3) and using that here $\rho_x = U(|x\rangle\langle x| \otimes |0\rangle\langle 0|)U^\dagger$ for $x \in \{0, 1, +, -\}$ we can rewrite the success probability of such a "simple cloning attack" as

$$p_s = \frac{1}{4}\left(\left|\langle 0|\langle 0|U|0\rangle|0\rangle\right|^2 + \left|\langle 1|\langle 0|U|1\rangle|0\rangle\right|^2\right.$$
$$\left. + \left|\langle +|\langle 0|U|+\rangle|0\rangle\right|^2 + \left|\langle -|\langle 0|U|-\rangle|0\rangle\right|^2\right). \tag{3.4}$$

Before we even try to analyze this, we should be conscious that there can be even more general attacks! Why would the adversary limit themselves to preparing a *pure* two-qubit state: in general, they may prepare a mixed state, by using more than one ancilla qubit and then tracing out unneeded qubits in the end. This could be helpful because the adversary could, for example, use some of the additional ancilla qubits to flip some random bits and, depending on the outcomes, prepare a different state. To see what such an attack could look like, let's consider a specific example.

Example 3.4.1 Consider the following map T from single-qubit quantum money states to two-qubit density matrices, where we write $|\psi_{10}\rangle = \frac{1}{\sqrt{2}}(|01\rangle + |10\rangle)$:

$$|0\rangle\langle 0| \mapsto \rho_0 = \frac{2}{3}|00\rangle\langle 00| + \frac{1}{3}|\psi_{10}\rangle\langle\psi_{10}|\,,$$

$$|1\rangle\langle 1| \mapsto \rho_1 = \frac{2}{3}|11\rangle\langle 11| + \frac{1}{3}|\psi_{10}\rangle\langle\psi_{10}|\,,$$

$$|+\rangle\langle +| \mapsto \rho_+ = \frac{1}{12}\left(2|00\rangle + \sqrt{2}|\psi_{10}\rangle\right)\left(2\langle 00| + \sqrt{2}\langle\psi_{10}|\right)$$
$$+ \frac{1}{12}\left(2|11\rangle + \sqrt{2}|\psi_{10}\rangle\right)\left(2\langle 11| + \sqrt{2}\langle\psi_{10}|\right)\,,$$

$$|-\rangle\langle -| \mapsto \rho_- = \frac{1}{12}\left(2|00\rangle - \sqrt{2}|\psi_{10}\rangle\right)\left(2\langle 00| - \sqrt{2}\langle\psi_{10}|\right)$$
$$+ \frac{1}{12}\left(2|11\rangle - \sqrt{2}|\psi_{10}\rangle\right)\left(2\langle 11| - \sqrt{2}\langle\psi_{10}|\right)\,. \qquad\blacksquare$$

Problem 3.3 guides you through the verification that the map defined in Example 3.4.1 is a valid quantum operation, by decomposing it into a sequence of (i) ancilla preparation, (ii) unitary transformation, and (iii) tracing out. Assuming this has been verified, let's see how well an adversary using it to make a cloning attempt will succeed. For this, we just need to evaluate its success probability using (3.4). For the case of the density matrices from Example 3.4.1, this is straightforward. Working through the calculation, we find

$$p_s = \frac{1}{4}\left(\frac{2}{3} + \frac{2}{3} + \frac{2}{3} + \frac{2}{3}\right) = \frac{2}{3}\,.$$

(Make sure that you are able to verify that each of the four $2/3 \approx 0.667$ is correct!) As you can see, this attack is only very marginally better than the simple prepare-and-measure attack we considered earlier; that achieves a success probability of $5/8 = 0.625$. After so many calculations, this may come as a disappointment. However, we studied the map T for a good reason. Indeed, it is possible to verify by direct calculation that T has the property that for *any* pure single-qubit state $|\psi\rangle$ it holds that

$$\langle\psi|\langle\psi|T(|\psi\rangle\langle\psi|)|\psi\rangle|\psi\rangle = \frac{2}{3}\,.$$

In words, the quantum map T has the ability to "clone" any single-qubit state with success probability $2/3$, not only the four BB'84 states that appear in Wiesner's quantum money scheme. Moreover, it is possible to show that $2/3$ is the optimal success probability of such a "universal cloning map" for all single-qubit states.

In Problem 3.4 you will show that there is an even better attack on Wiesner's scheme, which succeeds in CLONE with probability $3/4$. Is it possible to design a more secure scheme, for which the best attack would succeed with probability $2/3 < 3/4$? Indeed, it is possible! The natural idea is to increase the number of single-qubit states used for the bills. By considering a six-state scheme, where the two additional states are

$$\frac{1}{\sqrt{2}}\left(|0\rangle + i|1\rangle\right)\,, \qquad \frac{1}{\sqrt{2}}\left(|0\rangle - i|1\rangle\right)\,,$$

it is possible to get a better scheme, such that the optimal cloning attack only has success probability $2/3$. Our calculation using the map T shows that this is optimal: adding more states than this will not help, because however many states the scheme uses, there is always a cloning attack, given by the map T, that succeeds with probability $2/3$. The only way to do better is to move away from single-qubit states: in the next section we explore improvements based on considering bills made of multiple qubits.

Finally, let us give a little intuition for the definition of T. It turns out that, from a mathematical point of view, the map can be expressed as follows:

$$T(\rho) = \Pi_s \left(\rho \otimes \frac{1}{2}\mathbb{I} \right) \Pi_s \,, \tag{3.5}$$

where Π_s is the orthogonal projection onto the symmetric subspace of the two-qubit space $\mathbb{C}^4$, i.e. the 3-dimensional subspace spanned by the vectors $|00\rangle$, $|11\rangle$, and $|\psi_{10}\rangle$. At first it is not obvious that this is a valid quantum map, but it is, and you can verify that it is identical to the map T defined earlier.

Exercise 3.4.1   Show that the definition of T in (3.5) is a valid quantum channel by finding its Kraus decomposition. *[Hint: write $\mathbb{I} = |0\rangle\langle0| + |1\rangle\langle1|$ and expand $\Pi_s(\rho \otimes \frac{1}{2}\mathbb{I})\Pi_s$ as a sum of two terms by linearity.]*

Intuitively, what T does is that it "maximally symmetrizes" its input state by adding a qubit initialized in the totally mixed state and then projecting both qubits, the quantum money qubit and the extra workspace qubit, into the symmetric subspace. This has the effect of "smearing out" the quantum information present in $|\psi_\$\rangle\langle\psi_\$|$ across both qubits and results in the optimal way to approximately clone an arbitrary qubit.

3.4.3 n-Qubit Cloning Attacks

So far we have considered cloning attacks on the single-qubit version of Wiesner's scheme, and claimed that the single-qubit scheme is $3/4$-secure according to Definition 3.2.1. In general, a security of $\varepsilon = 3/4$ is not very satisfactory, as it means that an adversary still has a rather large chance of succeeding in the security game, i.e. to create a forgery. We would like this probability to be as small as possible, without increasing the complexity of the scheme by too much.

A possibility to achieve this is to consider higher-dimensional states. The simplest way to create a high-dimensional state is to put many qubits together. If it is "3/4-hard" to clone a quantum bill made of a single qubit, how hard is it to clone a bill made of n such qubits, where n is a large integer, say $n = 1024$?

Let's consider the corresponding security game. In the game, the adversary receives a single copy of an n-qubit quantum money state $|\psi_\$\rangle$, such that $|\psi_\$\rangle$ is a tensor product of single-qubit states $|x_1\rangle_{\theta_1}, \ldots, |x_n\rangle_{\theta_n}$ and x, θ are uniformly random n-bit strings chosen by the challenger. Suppose further that the adversary attempts to clone the qubits one by one, applying a map T_1 on the first qubit, T_2 on the second, etc., such that each map sends one qubit to two qubits. This produces the state

$$T(|\$\rangle\langle\$|) = T_1 \otimes \cdots \otimes T_n(|\$\rangle\langle\$|) = T_1(|x_1\rangle\langle x_1|_{\theta_1}) \otimes \cdots \otimes T_n(|x_n\rangle\langle x_n|_{\theta_n}). \qquad (3.6)$$

Averaging over all possible random choices of the challenger, the success probability of this map is

$$\frac{1}{2^{2n}} \sum_{\$=(x.\theta)} \langle\$|\langle\$| T(|\$\rangle\langle\$|) |\$\rangle |\$\rangle$$

$$= \left(\frac{1}{4} \sum_{x_1.\theta_1} \langle x_1|_{\theta_1} \langle x_1|_{\theta_1} T(|x_1\rangle\langle x_1|_{\theta_1}) |x_1\rangle_{\theta_1} |x_1\rangle_{\theta_1} \right)$$

$$\cdots \left(\frac{1}{4} \sum_{x_n.\theta_n} \langle x_n|_{\theta_n} \langle x_n|_{\theta_n} T(|x_n\rangle\langle x_n|_{\theta_n}) |x_n\rangle_{\theta_n} |x_n\rangle_{\theta_n} \right),$$

where to get the second expression we used that $|\psi_\$\rangle$ is a product state, (3.6), the distributive identity $(A \otimes B) \cdot (C \otimes D) = (AC \otimes BD)$ for any (not necessarily square) matrices, and we re-ordered the $2n$ qubits $(1 \cdots n)(1' \cdots n')$ as $(11') \cdots (nn')$. Since we know that each term on the right-hand side is at most $3/4$, we immediately get that the success probability of this type of attack is at most $(3/4)^n$. This number goes to 0 very fast (*exponentially fast*) as $n \to \infty$, so it is a security level we are happy with: for example, choosing $n = 1024$ already brings the success probability down to a number that is so small that even if the adversary was able to try an attack every nanosecond, it would take them far more than the age of the universe (which is of order 10^{25} nanoseconds) to successfully break the scheme.

Unfortunately, this analysis only considers a very specific type of attack: attacks that attempt to clone the qubits one by one, independently of each other. Could there be a better attack, one that simultaneously takes all the qubits into account? In general, this would be modeled by an arbitrary quantum transformation T from n to $2n$ qubits. As you can imagine, analyzing such a general attack can take a lot of work! The idea is to find an argument that shows that no such attack can succeed with substantially higher probability than the independent attacks considered above. In the case of Wiesner's scheme this can be done, and it is known that no attack on the n-qubit scheme can have higher success probability than the one that consists in applying the optimal single-qubit attack independently on each qubit.

3.5   The Elitzur–Vaidman Bomb Tester

In the previous section we considered "cloning attacks" on Wiesner's quantum money scheme, and argued that for the n-qubit scheme the best such attack has success probability at most $(3/4)^n$ in the security game. This should make us confident in the security of the scheme. However, we have to be careful. A security definition in cryptography only covers the kinds of attacks that are captured by the underlying security game ... real life can be more complex!

In particular, note that the definition of quantum money requires one to specify a verification procedure, but it does not say what should happen with a bill once it has been verified. So what should the bank do with the user's bill, once it has executed the verification procedure on it?

It is natural to consider that, if the bill is accepted, the bank returns it to the user, while if it is rejected, then the bank destroys the bill (and even fines the malicious user, or puts her in prison). We'll soon see that this a priori reasonable assumption turns out to *break* security: if the bank returns valid bills to their owner, even if it only does so when the bill has successfully passed verification, a malicious user can take advantage of this fact to break the scheme and clone any valid bill!

But wait, didn't we *prove* security? We did, but in our model the adversary does not have access to what is sometimes called a "verification oracle." In our security game, the adversary receives one bill and she has to produce two; to do this her only resource is whatever quantum operation she can apply in her laboratory – there is no interaction with the "bank" or the challenger in-between the two phases. Let's now see how the adversary has a cloning attack if in addition she is allowed to submit "candidate bills" to the bank (or, in the security game, to the challenger) for verification and the bank returns bills that were declared valid.

3.5.1   A Cloning Attack Against Permissive Banks

Let's first consider a simpler scenario where we assume that verification *always* returns a bill to the user, even if the bill failed verification. At first it may seem like this is not really a problem: if the bill is invalid, it will remain invalid, so what is the adversary going to do with it anyways? We'll call a bank that returns invalid bills to the user a *permissive bank*.

Let's play the security game CLONE in this context. Suppose that the adversary (let's call her Eve) got her hands on a valid quantum bill $(\$, |\psi_\$\rangle)$. Recall that in Wiesner's scheme the bill is made of n qubits, in state $|x_1\rangle_{\theta_1}, \ldots, |x_n\rangle_{\theta_n}$. If Eve sends this state for verification to the bank, the bank accepts it. Now suppose Eve sets the first qubit $|x_1\rangle_{\theta_1}$ aside, replaces it with a $|0\rangle$, and sends the modified bill for verification. If the bill is rejected, she tries with $|1\rangle, |+\rangle, |-\rangle$, one after the other. At least one of these must be accepted. As soon as her attempt is accepted, she keeps the qubit, and proceeds in the same way with the second qubit, etc. Once she is done with all n qubits, she has two copies of the valid bill: the one made of the qubits set aside at each step, and the one made of the "guessed" qubits. Moreover, this attack only required going through at most $4n$ verifications!

Observe that this attack relies on the fact that, once a qubit is accepted, it will be accepted in all future attempts. For example, if the correct first qubit is $|+\rangle$, but Eve submits a $|0\rangle$, she has probability $1/2$ of being accepted, since $|\langle 0|+\rangle|^2 = 1/2$. But if she is accepted, then her qubit is projected to the post-measurement state $|+\rangle$, so she will succeed with probability 1 in any subsequent verification attempt.

This suggests a slightly more streamlined attack. Instead of going through the four possibilities $|0\rangle, |1\rangle, |+\rangle, |-\rangle$ in sequence, Eve can guess a uniformly random state for the qubit, effectively replacing it by the density matrix $\rho = \mathbb{I}/2$. Then, verification will succeed with probability $1/2$ in all cases, since

$$\langle 0|\frac{\mathbb{I}}{2}|0\rangle = \langle 1|\frac{\mathbb{I}}{2}|1\rangle = \langle +|\frac{\mathbb{I}}{2}|+\rangle = \langle -|\frac{\mathbb{I}}{2}|-\rangle = \frac{1}{2}.$$

Moreover, whenever verification succeeds, the state is automatically projected onto the correct one. On expectation, this attack requires only $2n$ verification attempts to succeed. However, note that, since Eve will fail n times on expectation, for the attack to be effective we need the bank to return invalid bills without putting Eve in prison. What if it doesn't?

3.5.2 Nonpermissive Banks

In retrospect you may think that allowing the bank to return invalid bills is an obvious mistake: clearly, if Eve submits an invalid bill, she should be sent to prison right away! In this situation we can think of verification as a "bomb" that explodes any time it is given the wrong state. Observe, however, that if Eve submits a state that is "slightly invalid," such as a state obtained by applying a small unitary rotation to a valid quantum bill, then her chances of succeeding in verification remain relatively high, depending on the angle of the rotation: this observation will be crucial to her attack!

Eve's goal is, given one copy of the "magic state" that is the only input on which the "bomb" does *not* explode, to find two such inputs ... while staying alive (i.e. out of prison). Let's see how it is possible to do this. Informally, the idea is to make use of a phenomenon called the "quantum Zeno effect." Starting from her valid quantum bill, Eve is going to make very small modifications of it, and "test" these modifications against the verification procedure. Since the modifications are small, they are very likely to be accepted. Yet Eve will make them in such a way that she still learns information about her quantum bill.

Let's see how to do this one qubit at a time. Eve does the following.

1. Eve initializes an extra qubit to state $|0\rangle$. Including the quantum money qubit, her entire state is $|x\rangle_\theta |0\rangle$, for unknown $x, \theta \in \{0, 1\}$.
2. Eve chooses a small rotation parameter $\delta \in (0, \pi)$ and applies the unitary rotation $R_\delta = \begin{pmatrix} \cos\delta & -\sin\delta \\ \sin\delta & \cos\delta \end{pmatrix}$ to her second qubit (the ancilla qubit).
3. Eve applies a control-X operation from the second qubit to the first. That is, if the extra qubit is in state $|1\rangle$ then she applies an X-flip on the money qubit, and otherwise she does nothing.
4. Finally, Eve submits the money qubit for verification to the bank.

Intuitively, for small δ this procedure modifies the money state only a little bit (since $R_\delta |0\rangle = \cos\delta |0\rangle + \sin\delta |1\rangle$ only has a small amplitude $\sin\delta \approx \delta$ on $|1\rangle$), so that Eve's chances of succeeding in verification should be high (in particular, if $\delta = 0$ then the money state is unchanged and she succeeds with probability 1).

Let's examine what happens to the money state in all possible cases.

QUIZ 3.5.1

Suppose that $\theta = 1$, i.e. $|\psi_\$\rangle = |+\rangle$ or $|\psi_\$\rangle = |-\rangle$. After one step of the procedure described above, if $|\psi\rangle = |+\rangle$ then the state is

(a) $|+\rangle \otimes (R_\delta |0\rangle)$
(b) $|+\rangle \otimes (R_{-\delta} |0\rangle)$

and if $|\psi\rangle = |-\rangle$ then the state is

(a) $|-\rangle \otimes (R_\delta |0\rangle)$
(b) $|-\rangle \otimes (R_{-\delta} |0\rangle)$

Suppose that Eve repeats the procedure an even number 2N of times, with an angle $\delta = \frac{\pi}{4N}$. What is the final state of the control qubit in the case $|\psi_\$\rangle = |+\rangle$?

(a) $|+\rangle \otimes |0\rangle$
(b) $-|+\rangle \otimes |0\rangle$
(c) $|+\rangle \otimes |1\rangle$
(d) $-|+\rangle \otimes |1\rangle$

And in the case $|\psi_\$\rangle = |-\rangle$?

(a) $|-\rangle \otimes |0\rangle$
(b) $-|-\rangle \otimes |0\rangle$
(c) $|-\rangle \otimes |1\rangle$
(d) $-|-\rangle \otimes |1\rangle$

Now suppose that $\theta = 0$, so $|\psi\rangle = |x\rangle$, for some unknown $x \in \{0, 1\}$. In this case, right before the verification attempt, Eve's state is

$$(\cos \delta) |x\rangle \otimes |0\rangle + (\sin \delta)(X |x\rangle) \otimes |1\rangle \ .$$

Verification measures in the standard basis and accepts if and only if the outcome is $|x\rangle$. Hence in our setting it accepts with probability $\cos^2 \delta$, in which case the state gets projected to $|x\rangle |0\rangle$, i.e. the same state as originally. With probability $\sin^2 \delta \approx \delta^2$ (for small δ), verification fails and Eve goes to jail.

Now, suppose δ is very small, say $\delta = \pi/4N$ for some large integer N, as in Quiz 3.5.1. Suppose Eve executes the procedure described above $(1/\delta)$ times. If $|\psi\rangle$ is $|0\rangle$ or $|1\rangle$, then (unless Eve has been caught) the ancilla qubit is $|0\rangle$. Moreover, the chance that she is caught is at most $(1/\delta) \sin^2 \delta \leq \pi/4N$, which can be made arbitrarily small by choosing the number of iterations large enough.

Next, if the state is $|\psi\rangle = |-\rangle$, since the number of repetitions is even, the ancilla qubit is in state $|0\rangle$. Finally, if it is $|\psi\rangle = |+\rangle$, the final state of the ancilla qubit is $R_{\pi/2\delta \cdot \delta} |0\rangle = R_{\pi/2} |0\rangle = |1\rangle$.

Overall, Eve observes a $|1\rangle$ when measuring the ancilla qubit if and only if the state is $|\psi\rangle = |+\rangle$: in this case, she has perfectly identified it! Replacing the controlled operation in step 2 by a control $(-X)$, Z, or $(-Z)$, she can similarly perfectly identify the cases where $|\psi\rangle$ is $|-\rangle$, $|0\rangle$, or $|1\rangle$ respectively. Moreover, since in each case the chance that she is caught is less than $1/400n$, she manages to succeed in identifying the first qubit perfectly, while getting caught only with probability $1/100n$. Repeating this procedure independently for all n qubits yields a perfect classical description of the money state, with only a 1% chance of going to jail. Moreover, clearly this 1% can be made much lower if Eve is willing to be just a little more careful: by generalizing our analysis you can easily show that, if she wishes to be caught with probability at most p, then $\sim n/p$ total verification attempts will be enough.

CHAPTER NOTES

The idea of quantum money is one of the oldest ideas in quantum information, and the oldest idea in quantum cryptography: it originates in S. Wiesner's paper (Conjugate coding. *SIGACT News*, **15**:78–88, 1983), which, although only published in 1983, had been in circulation since the mid-1970s. In particular, the idea for quantum key distribution, which we will describe in detail in Chapter 8, builds on Wiesner's idea for quantum money. Even though Wiesner introduced the scheme described in this chapter, the idea that its security could be shown based on the no-cloning principle only appears almost two decades later in the book by H.-K. Lo, T. Spiller, and S. Popescu, *Introduction to Quantum Computation and Information* (World Scientific, 1998), and a formal proof of the security bound $\varepsilon = 3/4$ from Section 3.4 is given by A. Molina, T. Vidick, and J. Watrous (Optimal counterfeiting attacks and generalizations for Wiesner's quantum money. In *Conference on Quantum Computation, Communication, and Cryptography*, pp. 45–64. Springer, 2012). The idea for the "bomb tester" attack described in Section 3.5 is due to A. Brodutch, et al. (An adaptive attack on Wiesner's quantum money. *Quantum Information and Computing*, **16**(11&12):1048–1070, 2016). The n-qubit Wiesner scheme and the six-qubit scheme are analyzed in Molina et al., where the optimal attack bounds claimed in this chapter are shown formally using a semidefinite programming formulation.

One of the main drawbacks of Wiesner's scheme is that it requires *private verification*, i.e. only the bank has the capacity to verify a quantum bill. Different methods to mitigate this limitation have been considered. To alleviate the verification task it is possible to devise variants of Wiesner's scheme that allow verification to be achieved through classical communication with the bank only, as shown, for example, by F. Pastawski, et al. (Unforgeable noise-tolerant quantum tokens. *Proceedings of the National Academy of Sciences*, **109**(40):16079–16082, 2012) and D. Gavinsky (Quantum money with classical verification. In *IEEE 27th Conference on Computational Complexity*, pp. 42–52. IEEE, 2012). Going further, one may imagine money schemes where verification is *public*, i.e. it can be performed by anyone. The notion of public-key quantum money is introduced by S. Aaronson (Quantum copy-protection and quantum money. In *24th Annual IEEE Conference on Computational Complexity*, pp. 229–242. IEEE, 2009), and one of the most popular schemes is the one by S. Aaronson and P. Christiano (Quantum money from hidden subspaces. In *Proceedings of the Forty-Fourth Annual ACM Symposium on Theory of Computing*, pp. 41–60, 2012), which is based on "subspace states," a generalization of Wiesner's money states. At the time of writing, the only public-key money schemes known rely on advanced cryptographic assumptions, such as the random oracle model, for their security; it is an open question to obtain a scheme based on standard assumptions.

PROBLEMS

3.1 Composing quantum maps

Show carefully that the composition of two quantum maps according to Definition 3.3.1 is still a quantum map, i.e. that a sequence of (i), (ii), (iii) as in the definition repeated twice can be "re-ordered" so that all the ancilla preparation comes first and all the tracing out comes last, without changing how the map operates on any quantum state.

3.2 Measurement attacks

Consider all cloning attacks that take the following form. The adversary decides on an arbitrary orthonormal basis $(|u_0\rangle, |u_1\rangle)$ for the single-qubit space $\mathbb{C}^2$. It then measures the challenger's state $|\psi_S\rangle$ in the basis $(|u_0\rangle, |u_1\rangle)$ to obtain an outcome $b \in \{0, 1\}$. Finally, the adversary returns the density matrix $\rho = |u_b\rangle\langle u_b| \otimes |u_b\rangle\langle u_b|$.

1. Express the success probability of this attack as a function of the coefficients α, β of $|u_0\rangle = \alpha|0\rangle + \beta|1\rangle$. (Since $|u_1\rangle$ is orthogonal to $|u_0\rangle$, without loss of generality $|u_1\rangle = \beta|0\rangle - \alpha|1\rangle$.)

2. Find the choice of α, β that maximizes the success probability (don't forget about complex numbers!).

3. Did you find an attack that is better than the ones considered in Section 3.4?

3.3 A cloning map

In this problem we verify that the map T defined in Example 3.4.1 is a valid quantum map.

1. Start with a simple "sanity check": verify that each of the four matrices ρ_0, ρ_1, ρ_+, and ρ_- is a valid density matrix.

However, this is not enough: for example, these conditions are satisfied by the "optimal cloning map" $|\psi_S\rangle\langle\psi_S| \mapsto |\psi_S\rangle\langle\psi_S| \otimes |\psi_S\rangle\langle\psi_S|$, but we know that there exists no such map! To see that T is a well-defined map, we verify that it can be implemented by (i) adding two ancilla qubits in state $|00\rangle_{BC}$, (ii) a unitary transformation, and (iii) a tracing-out operation.

2. Consider the following map V, where $|\psi_{11}\rangle_{AB} = \frac{1}{\sqrt{2}}(|10\rangle_{AB} - |01\rangle_{AB})$.

$$|0\rangle_A |00\rangle_{BC} \mapsto \frac{2}{\sqrt{6}} |00\rangle_{AB}|0\rangle_C + \frac{1}{\sqrt{3}} |\psi_{11}\rangle_{AB}|1\rangle_C \,,$$

$$|1\rangle_A |00\rangle_{BC} \mapsto \frac{1}{\sqrt{3}} |\psi_{11}\rangle_{AB}|0\rangle_C + \frac{2}{\sqrt{6}} |11\rangle_{AB}|1\rangle_C \,.$$

Check that the two states on the right-hand side, call them $|v_0\rangle$ and $|v_1\rangle$, are orthonormal. Using Remark 1.6.3 this shows that it is possible to extend V to a valid unitary operation on the entire three-qubit space $\mathbb{C}^8$.

3. Show that the map T is identical to the composition of adding two ancilla qubits in state $|00\rangle_{BC}$, applying V, and tracing out the third qubit. That is, for all states $|\psi\rangle$,

$$T(|\psi\rangle\langle\psi|_A) = \text{tr}_C \left(V(|\psi\rangle\langle\psi|_A \otimes |00\rangle\langle00|_{BC})V^\dagger \right) .$$

This justifies that T is a valid quantum map, because it can be written as a sequence of three valid operations.

3.4 An optimal attack
Let

$$N_1 = \frac{1}{\sqrt{12}} \begin{pmatrix} 3 & 0 \\ 0 & 1 \\ 0 & 1 \\ 1 & 0 \end{pmatrix} \quad \text{and} \quad N_2 = \frac{1}{\sqrt{12}} \begin{pmatrix} 0 & 1 \\ 1 & 0 \\ 1 & 0 \\ 0 & 3 \end{pmatrix}.$$

1. Show that (N_1, N_2) are valid Kraus operators in the definition of a quantum channel $\mathcal{N}(\rho) = N_1 \rho N_1^\dagger + N_2 \rho N_2^\dagger$ mapping one qubit to two qubits.
2. Show that using $\mathcal{N}$, a quantum adversary succeeds in the game CLONE for Wiesner's quantum money scheme with probability 3/4.

4

The Power of Entanglement

The EPR pair $|\text{EPR}\rangle = \frac{1}{\sqrt{2}}(|00\rangle + |11\rangle)$ is the most special two-qubit state that we have encountered so far. What makes it special is that it cannot be written as a tensor product $|\psi_1\rangle \otimes |\psi_2\rangle$. In this chapter we give a name to this property: any such state will be called *entangled*. We will define entanglement formally and explore some of the reasons that make it such a fascinating topic in quantum information! To whet your appetite, let it already be said that in later chapters we will see that entanglement allows us to guarantee security using the laws of nature. Beyond cryptography, entanglement is a necessary ingredient in quantum communication, in the most impressive quantum algorithms, such as Shor's algorithm for factoring, and in the design of quantum error-correcting codes.

4.1 Entanglement

If we combine two qubits A and B, each of which is in a pure state, the joint state of the two qubits is given by

$$|\psi\rangle_{AB} = |\psi_1\rangle_A \otimes |\psi_2\rangle_B . \tag{4.1}$$

Any two-qubit state that is either directly of this form, or is a probabilistic mixture of states of this form, is called *separable*. Entangled states are states that are *not* separable. In other words, a pure state $|\psi\rangle_{AB}$ is entangled if and only if

$$|\psi\rangle_{AB} \neq |\psi_1\rangle_A \otimes |\psi_2\rangle_B , \tag{4.2}$$

for any possible choice of $|\psi_1\rangle$ and $|\psi_2\rangle$. A mixed state ρ is entangled if and only if it cannot be written as a convex combination of pure product states of the form in Eq. (4.1). Let's make this into a definition.

> **Definition 4.1.1 (Entanglement).** *Consider two quantum systems A and B. The joint state ρ_{AB} is* separable *if there exists a probability distribution $\{p_i\}_i$, and sets of density matrices $\{\rho_i^A\}_i, \{\rho_i^B\}_i$ such that*
>
> $$\rho_{AB} = \sum_i p_i \rho_i^A \otimes \rho_i^B . \tag{4.3}$$
>
> *If there exists no such decomposition then ρ_{AB} is called* entangled.

Remark 4.1.1 An important remark is that if $\rho_{AB} = |\psi\rangle\langle\psi|_{AB}$ is a pure state, then $|\psi\rangle_{AB}$ is separable if and only if there exists $|\psi\rangle_A, |\psi\rangle_B$ such that

$$|\psi\rangle_{AB} = |\psi\rangle_A \otimes |\psi\rangle_B \, .$$

In other words, a pure state is separable if and only if it can be written in the form (4.3) where in addition there is a single term in the convex decomposition. This property is not obvious to see; we will show it once we discover a more general method for writing pure states of two systems, the Schmidt decomposition, later on in the chapter.

We emphasize that entanglement is a property that is always defined with respect to a fixed bipartition AB of the system. If we consider, for example, a three-qubit state $|\psi\rangle_{ABC}$ then it makes sense to ask if the state is entangled between A and BC, between AB and C, etc., and in general the answer to these questions can be different. It is a good exercise to find examples of this – though we recommend that you progress through the chapter first, as deciding entanglement is hard! We return to the question of multipartite entanglement, i.e. entanglement between more than two systems, in Section 4.5.

So what are examples of entangled or separable states, and how do we demonstrate the difference between them? Let's go over some examples to start building intuition about these complex questions. A simple case is the case of pure states, for which a decomposition of the form (4.1) can often be found simply by staring.

Example 4.1.1 Consider the state

$$|\psi\rangle_{AB} = \frac{1}{\sqrt{2}}\left(|01\rangle_{AB} + |11\rangle_{AB}\right) \, .$$

This state is not entangled, since $|\psi\rangle_{AB} = \frac{1}{\sqrt{2}}\left(|0\rangle_A + |1\rangle_A\right) \otimes |1\rangle_B = |+\rangle_A \otimes |1\rangle_B.$ ∎

QUIZ 4.1.1 *Is the state* $|\psi\rangle = \frac{1}{\sqrt{2+\sqrt{2}}}\left(|00\rangle + |+0\rangle\right)$ *entangled?*

(a) *Yes*

(b) *No*

Things become far more delicate when we look at nonpure states. Sometimes, a separable decomposition is obvious.

Example 4.1.2 Consider the density matrix

$$\rho_{AB} = \frac{1}{2}|0\rangle\langle0|_A \otimes |1\rangle\langle1|_B + \frac{1}{2}|+\rangle\langle+|_A \otimes |-\rangle\langle-|_B \, . \tag{4.4}$$

Such a state is in the form of (4.3), so it is not entangled: it is separable. Similarly, any cq-state, i.e. a state of the form $\rho_{XQ} = \sum_x p_x |x\rangle\langle x|_X \otimes \rho_x^Q$, is separable. ∎

The two examples above illustrate an important point. When we first discover the notion of entanglement, it is very tempting to say that "a state is entangled if the two subsystems are correlated, and it is separable if they are independent." However, for both the examples we can see that this is not correct. For example, for ρ_{AB} in (4.4) we see that a measurement of both qubits in the standard basis yields the outcome $(0, 1)$ with probability $5/8$, and all other pairs of outcomes with probability $1/8$; the two outcome bits are not independent. In this case we say that the two subsystems are "classically correlated." Let's look at the EPR pair for an example of correlation that is stronger than a classical correlation.

Example 4.1.3 Consider the EPR pair

$$|\text{EPR}\rangle_{AB} = \frac{1}{\sqrt{2}} \left(|00\rangle_{AB} + |11\rangle_{AB} \right) . \tag{4.5}$$

This state is entangled, because it is pure but cannot be written in the form (4.1). In this chapter we will see multiple proofs of this. In fact, we will see that this state is, in a precise sense, the "most entangled" state of two qubits and is thus often referred to as a *maximally entangled* state.

As we already saw in Example 1.4.4, the EPR pair has the special property that it can be written in many symmetric ways. For example, in the Hadamard basis

$$|\text{EPR}\rangle_{AB} = \frac{1}{\sqrt{2}} \left(|++\rangle_{AB} + |--\rangle_{AB} \right) . \tag{4.6}$$

Thus, measurements of both qubits of $|\text{EPR}\rangle_{AB}$ in the standard basis, or the Hadamard basis, always produce the same outcome. ∎

The example demonstrates a state such that measurement outcomes on both subsystems are correlated, in multiple different bases. In fact, it turns out that this property can even be used to *characterize* the EPR pair: it is the only two-qubit state having this property; we will show this in Chapter 8 and use it to prove security of quantum key distribution. For now let's show that this property does make the EPR pair "special," in the sense of not being separable.

Exercise 4.1.1 Suppose that ρ_{AB} is a two-qubit separable state. Show that if a measurement of both qubits of ρ_{AB} in the standard basis always yields the same outcome, then a measurement of both qubits in the Hadamard basis necessarily has nonzero probability of giving different outcomes. Deduce a proof that the EPR pair (4.5) is not a separable state.

Here is another way to see that the correlations in the EPR pair are stronger than those present in a separable state, which nevertheless looks similar to it. Remember that it is always very important to make the distinction between a superposition and a mixture. We saw in Example 2.2.4 a single-qubit example of the difference between these two notions. Similarly, it is important to make the distinction between the two states

$$\rho_{AB} = \frac{1}{2}|0\rangle\langle0|_A \otimes |0\rangle\langle0|_B + \frac{1}{2}|1\rangle\langle1|_A \otimes |1\rangle\langle1|_B$$

and

$$\sigma_{AB} = |\text{EPR}\rangle\langle\text{EPR}|_{AB} .$$

For the state ρ_{AB}, if A is measured in the standard basis then whenever $|0\rangle_A$ is observed the state on B is $|0\rangle_B$; likewise when $|1\rangle_A$ is observed, the state on B is $|1\rangle_B$. This is also true for σ_{AB}. However, consider measuring system A of ρ_{AB} in the Hadamard basis. The corresponding measurement operators are $|+\rangle\langle+|_A \otimes \mathbb{I}_B, |-\rangle\langle-|_A \otimes \mathbb{I}_B$. The post-measurement state conditioned on obtaining the outcome $|+\rangle_A$ is then

$$\begin{aligned}
\rho_{|+_A}^{AB} &= \frac{(|+\rangle\langle+|_A \otimes \mathbb{I}_B)\rho_{AB}(|+\rangle\langle+|_A \otimes \mathbb{I}_B)}{\text{tr}((|+\rangle\langle+|_A \otimes \mathbb{I}_B)\rho_{AB})} \\
&= 2 \cdot \left(\frac{1}{2}\frac{1}{2}|+\rangle\langle+|_A \otimes |0\rangle\langle0|_B + \frac{1}{2}\frac{1}{2}|+\rangle\langle+|_A \otimes |1\rangle\langle1|_B\right) \\
&= |+\rangle\langle+|_A \otimes \frac{\mathbb{I}_B}{2} ,
\end{aligned}$$

and we see that the reduced state on B, $\rho_{|+_A}^B = \frac{\mathbb{I}_B}{2} = \rho_B$, is maximally mixed. In contrast, using that the state $|\text{EPR}\rangle_{AB}$ can be rewritten as

$$|\text{EPR}\rangle_{AB} = \frac{1}{\sqrt{2}}(|00\rangle_{AB} + |11\rangle_{AB}) = \frac{1}{\sqrt{2}}(|++\rangle_{AB} + |--\rangle_{AB}) ,$$

when σ_{AB} is measured with respect to the Hadamard basis on system A, conditioned on the outcome $|+\rangle_A$, the reduced state on B is $\sigma_{|+_A}^B = |+\rangle\langle+|_B$. In particular, this state is pure: it is very different from the totally mixed state we obtained by performing the same experiment on ρ_{AB}. This is a sense in which the correlations in σ_{AB} are stronger than those in ρ_{AB}. Here, we have that for both bases, knowing the measurement outcome on A allows us to perfectly predict the outcome on B.

4.2   Purifications

Let's approach the problem of determining entanglement in a more systematic way. For this we need to learn about different ways of representing bipartite states, as well as methods for going from pure to mixed states and vice versa. In Chapter 2 we saw such a method, which sometimes creates a mixed state from a pure state: this is the partial trace operation. Even if the state of the larger system is pure, the reduced state can sometimes be mixed. In fact, according to Remark 4.1.1 this is a signature of entanglement in the larger state. This is because if a pure bipartite state is not entangled, it can be written as a product (4.1), and in this case tracing out one of the systems leaves a pure state on the other system.

So we know that mixed states sometimes arise from pure states, by "forgetting" (tracing out) some of the information. Is it possible to reverse this process? Suppose we are given a density matrix ρ_A describing a quantum state on system A. Is it always possible to find a pure state $\rho_{AB} = |\psi\rangle\langle\psi|_{AB}$ such that $\text{tr}_B(\rho_{AB}) = \rho_A$? As we will see, the answer is yes, and any such state is called a *purification* of ρ_A.

> **Definition 4.2.1 (Purification).** *Given a density matrix ρ_A, a pure state $|\psi_{AB}\rangle$ is a purification of ρ_A if $\text{tr}_B(|\psi\rangle\langle\psi|_{AB}) = \rho_A$.*

Let's see how an arbitrary density matrix ρ_A can be purified. As a first step, diagonalize ρ_A, expressing it as a mixture

$$\rho_A = \sum_{j=1}^{d_A} \lambda_j |\phi_j\rangle\langle\phi_j| \,,$$

where λ_j are the eigenvalues of ρ_A and $|\phi_j\rangle$ the eigenstates. Since ρ_A is a density matrix the λ_j are non-negative and sum to 1. We've seen this interpretation of density matrices before: ρ_A can be thought of as the description of a quantum system that is in a probabilistic mixture of being in state $|\phi_j\rangle$ with probability λ_j. But who "controls" which part of the mixture A is in?

Let's introduce an imaginary system B which achieves just this. Let $\{|j\rangle_B\}_{j \in \{1,\dots,d_B\}}$ be the standard basis for a system B of dimension $d_B = d_A$, and consider the pure state

$$|\psi\rangle_{AB} = \sum_{j=1}^{d_A} \sqrt{\lambda_j} |\phi_j\rangle_A \otimes |j\rangle_B \,, \tag{4.7}$$

where $\{|j\rangle_B\}_j$ is the standard basis on system B. Suppose we measure the B system of $|\psi\rangle_{AB}$ in the standard basis. We know what will happen: we will obtain outcome j with probability $\langle\psi|_{AB} M_j |\psi\rangle_{AB}$, where $M_j = \mathbb{I}_A \otimes |j\rangle\langle j|_B$. A short calculation will convince you that this equals λ_j. Since we're using a projective measurement, we can describe the post-measurement state easily as being proportional to $M_j |\psi\rangle\langle\psi|_{AB} M_j$, and looking at the A system only we find that it is $|\phi_j\rangle\langle\phi_j|_A$.

To summarize, a measurement of system B of the state $|\psi\rangle_{AB}$ gives outcome j with probability λ_j, and the post-measurement state on A is precisely $|\phi_j\rangle\langle\phi_j|$. This implies that $\text{tr}_B(|\psi\rangle\langle\psi|_{AB}) = \rho_A$.

 Exercise 4.2.1   Verify this computation of $\text{tr}_B(|\psi\rangle\langle\psi|_{AB})$ using the mathematical definition of the partial trace, Definition 2.4.1.

 QUIZ 4.2.1 *Which of the following two-qubit pure states is a purification of the single-qubit density matrix $\frac{1}{2}(|0\rangle\langle 0| + |1\rangle\langle 1|)$?*

   I. $\frac{1}{\sqrt{2}}(|00\rangle + |11\rangle)$

   II. $\frac{1}{\sqrt{2}}(|+-\rangle - |-+\rangle)$

   III. $\frac{1}{2}(|00\rangle + |01\rangle + |10\rangle + |11\rangle)$

(a) *only I*
(b) *I and II*
(c) *II and III*
(d) *I, II, and III*

Now that we know that purifications exist, and we know how to compute them, we can ask if purifications are unique. You'll notice that in the construction above we made the choice of the standard basis for system B, but any other basis would have worked just as well. So it seems like we at least have a choice of basis on system B: there is a "unitary degree of freedom." To see that this is the only freedom that we have in choosing a purification, we first need to learn about a very convenient representation of bipartite pure states, the *Schmidt decomposition*.

4.2.1   The Schmidt Decomposition

The purification that we constructed in (4.7) has a special form: it is expressed as a sum, with non-negative coefficients whose squares sum to 1, of tensor products of basis states for the A and B systems respectively. As we saw, this particular form is convenient because it lets us compute the reduced states in A and B very easily. Unfortunately not every state is always given in this way: for example, if we write $|\psi\rangle_{AB} = \frac{1}{\sqrt{2}}(|0\rangle_A|0\rangle_B + |+\rangle_A|1\rangle_B)$ then this is a valid quantum state but the two states $|0\rangle_A, |+\rangle_A$ on A are not orthogonal. But maybe the same state can be written in a more convenient form? The answer is yes, and it is given by the Schmidt decomposition.

Theorem 4.2.1 (Schmidt decomposition) *Consider quantum systems A and B with dimensions d_A, d_B respectively, and let $d = \min(d_A, d_B)$. Any pure bipartite state $|\psi\rangle_{AB}$ has a Schmidt decomposition*

$$|\psi\rangle_{AB} = \sum_{i=1}^{d} \sqrt{\lambda_i}|u_i\rangle_A|v_i\rangle_B , \qquad (4.8)$$

where $\lambda_i \geq 0$ and $\{|u_i\rangle_A\}_i, \{|v_i\rangle_B\}_i$ are collections of orthonormal vectors. The coefficients $\sqrt{\lambda_i}$ are called Schmidt coefficients *and $|u_i\rangle_A, |v_i\rangle_B$* Schmidt vectors.

We won't give a detailed proof of the theorem here. The main idea is to start by expressing $|\psi\rangle_{AB} = \sum_{j,k} \alpha_{j,k}|j\rangle_A|k\rangle_B$ using the standard bases of A and B, and then write the singular value decomposition of the $d_A \times d_B$ matrix with coefficients $\alpha_{j,k}$ to recover the $\sqrt{\lambda_i}$ (the singular values), the $|u_i\rangle_A$ (the left eigenvectors), and the $|v_i\rangle_B$ (the right eigenvectors).

The Schmidt decomposition has many interesting consequences. A first consequence is that it provides a simple recipe for computing reduced density matrices: given a state of the form (4.8) we immediately get $\rho_A = \sum_i \lambda_i|u_i\rangle\langle u_i|_A$ and $\rho_B = \sum_i \lambda_i|v_i\rangle\langle v_i|_B$. An important observation is that ρ_A and ρ_B have the same eigenvalues, which are precisely the squares of the Schmidt coefficients. So given any two density matrices ρ_A and ρ_B there exists a pure bipartite state $|\psi\rangle_{AB}$ such that $\rho_A = \text{tr}_B(|\psi\rangle\langle\psi|_{AB})$ and $\rho_B = \text{tr}_A(|\psi\rangle\langle\psi|_{AB})$ if and only if ρ_A and ρ_B have the same eigenvalues! Without the Schmidt decomposition this is not at all an obvious fact to prove.

The same observation also implies that the Schmidt coefficients are uniquely defined: they are the square roots of the eigenvalues of the reduced density matrix. The Schmidt vectors are also unique, up to degeneracy and choice of phase: if an eigenvalue has an associated eigenspace of dimension 1 only, then the associated Schmidt vector must be the corresponding eigenvector. If the eigenspace has dimension more than 1, we can choose as Schmidt vectors any basis for the subspace. And note that in (4.8) we can always multiply $|u_i\rangle$ by $e^{i\theta_i}$, and $|v_i\rangle$ by $e^{-i\theta_i}$, so there is a phase degree of freedom.

Another important consequence of the Schmidt decomposition is that it provides us with a way to measure entanglement between the A and B systems in a pure state $|\psi_{AB}\rangle$. A first, rather rough but convenient such measure is given by the number of nonzero coefficients $\sqrt{\lambda_j}$. This measure is called the *Schmidt rank*. Since the Schmidt rank is uniquely defined for any given state, it allows us to tell if a state is entangled or not: if the Schmidt rank is 1 then the state is a product state, and if it is strictly larger than 1 then the state is entangled.

Definition 4.2.2 (Schmidt rank). *For any bipartite pure state with Schmidt decomposition* $|\psi\rangle_{AB} = \sum_{i=1}^{d} \sqrt{\lambda_i} |a_i\rangle_A |b_i\rangle_B$ *the* Schmidt rank *is defined as the number of nonzero coefficients* $\sqrt{\lambda_i}$. *It is also equal to* $\mathrm{rank}(\rho_A)$ *and* $\mathrm{rank}(\rho_B)$.

QUIZ 4.2.2 *True or false? For any density matrix ρ_{AB}, the reduced states ρ_A and ρ_B have equal rank.*

QUIZ 4.2.3 *Let $|+\rangle = \frac{1}{\sqrt{2}}(|0\rangle + |1\rangle)$ and $|-\rangle = \frac{1}{\sqrt{2}}(|0\rangle - |1\rangle)$. Which of the following is a Schmidt decomposition of the state $\frac{1}{\sqrt{2}}|02\rangle + \frac{1}{\sqrt{8}}(|10\rangle + |11\rangle + |20\rangle - |21\rangle)$?*

I. $\frac{1}{\sqrt{2}}|02\rangle + \frac{1}{2}(|1+\rangle + |2-\rangle)$

II. $\frac{1}{\sqrt{2}}|02\rangle + \frac{1}{4}(|1-\rangle - |2+\rangle)$

III. $\frac{1}{\sqrt{2}}|02\rangle + \frac{1}{\sqrt{8}}(|1\rangle + |2\rangle)|0\rangle + \frac{1}{\sqrt{8}}(|1\rangle - |2\rangle)|1\rangle$

(a) *only I*

(b) *I and II*

(c) *I and III*

(d) *I, II, and III*

QUIZ 4.2.4 *What is the Schmidt rank of the state found in the previous quiz?*

(a) 2

(b) 3

(c) 4

(d) 5

BOX 4.1   Measuring Entanglement

The Schmidt coefficients provide a finer way to measure entanglement than the Schmidt rank. A natural measure, called "entropy of entanglement," consists in taking the entropy of the distribution specified by the squares of the coefficients. We will learn about entropy in detail in the next chapter. For now, think of it as a quantitative measure of randomness, or spread, in the Schmidt coefficients. If the entropy is 0 then there is only a single coefficient, equal to 1, and the state is not entangled. But as soon as the entropy is positive the state is entangled.

Using entropy allows us to make finer distinctions than the Schmidt rank. When there are only two Schmidt coefficients, λ_1 and $\lambda_2 = 1 - \lambda_1$, then the entropy of entanglement is simply $H(\lambda_1)$, where the function H is defined as $H(x) = -x\log(x) - (1-x)\log(1-x)$ for $x \in [0,1]$. This measure distinguishes the entanglement in the two states

$$|\psi\rangle = \frac{1}{\sqrt{2}}|00\rangle + \frac{1}{\sqrt{2}}|11\rangle \quad \text{and} \quad |\phi\rangle = \sqrt{1-\varepsilon}|00\rangle + \sqrt{\varepsilon}|11\rangle.$$

For small $0 < \varepsilon < 1/2$ both states have the same Schmidt rank, but the first one has entanglement entropy 1 whereas the second has entanglement entropy $H(\varepsilon)$. Since $H(\varepsilon)$ is monotonically increasing for $0 < \varepsilon \leq 1/2$, we find that according to this new measure the state $|\phi\rangle$ is less entangled than the state $|\psi\rangle$. The reason that we call the EPR pair "maximally entangled" is that its entanglement entropy is maximal among all two-qubit states.

4.2.2 Uhlmann's Theorem

Let us return to the topic of the freedom in choosing purifications of a density matrix. We saw that there is at least a unitary degree of freedom by choosing a basis on the purifying system B. Uhlmann's theorem states that this is the only freedom available.

Theorem 4.2.2 (Uhlmann's theorem) *Consider a density matrix ρ_A and a purification of ρ_A given by $|\psi\rangle_{AB}$. Then another state $|\phi\rangle_{AB}$ is also a purification of ρ_A if and only if there exists a unitary U_B such that*

$$|\phi\rangle_{AB} = \mathbb{I}_A \otimes U_B |\psi\rangle_{AB} \ .$$

We already saw a proof of the "if" part of the theorem. To show the converse, i.e. that two purifications must always be related by a unitary, consider the Schmidt decompositions:

$$|\phi\rangle_{AB} = \sum_i \sqrt{\lambda_i} |u_i\rangle_A |v_i\rangle_B \ ,$$

$$|\psi\rangle_{AB} = \sum_i \sqrt{\mu_i}\, |w_i\rangle_A\, |z_i\rangle_B\ .$$

As we know, the λ_i are uniquely defined: they are the eigenvalues of ρ_A. So if $|\phi\rangle_{AB}$ and $|\psi\rangle_{AB}$ are both purifications of the same ρ_A, we must have $\lambda_i = \mu_i$. Now suppose for simplicity that all eigenvalues are nondegenerate. Then the $|u_i\rangle_A$ are also uniquely determined: they are the eigenvectors of ρ_A associated to the λ_i. Therefore $|u_i\rangle_A = |w_i\rangle_A$ as well! Thus we see that the only choice we have left is the $|v_i\rangle_B$ or $|z_i\rangle_B$: since the density matrix ρ_B of the purification is not specified a priori, we may choose any orthonormal basis of the B system. Since any two orthonormal bases of the same space are related by a unitary matrix, this choice of basis is precisely the degree of freedom that is guaranteed by Uhlmann's theorem. We will make use of Uhlmann's theorem later to show that certain protocols *cannot* be secure!

QUIZ 4.2.5 *True or false? Uhlmann's theorem guarantees that there exists a unitary on Bob's B qubit alone that takes the two-qubit density matrix*

$$\rho_{AB} = \begin{pmatrix} \frac{1}{2} & 0 & 0 & \frac{1}{2} \\ 0 & 0 & 0 & 0 \\ 0 & 0 & 0 & 0 \\ \frac{1}{2} & 0 & 0 & \frac{1}{2} \end{pmatrix} \quad to \quad \rho'_{AB} = \begin{pmatrix} 0 & 0 & 0 & 0 \\ 0 & \frac{1}{2} & -\frac{1}{2} & 0 \\ 0 & -\frac{1}{2} & \frac{1}{2} & 0 \\ 0 & 0 & 0 & 0 \end{pmatrix}$$

QUIZ 4.2.6 *True or false? Uhlmann's theorem guarantees that some unitary on Bob's B qubit alone takes*

$$\rho_{AB} = \begin{pmatrix} \frac{1}{4} & 0 & 0 & \frac{1}{4} \\ 0 & \frac{1}{4} & -\frac{1}{4} & 0 \\ 0 & -\frac{1}{4} & \frac{1}{4} & 0 \\ \frac{1}{4} & 0 & 0 & \frac{1}{4} \end{pmatrix} \quad to \quad \rho'_{AB} = \begin{pmatrix} \frac{1}{4} & 0 & 0 & \frac{1}{4} \\ 0 & \frac{1}{4} & \frac{1}{4} & 0 \\ 0 & \frac{1}{4} & \frac{1}{4} & 0 \\ \frac{1}{4} & 0 & 0 & \frac{1}{4} \end{pmatrix}$$

4.3 Two Applications

Let us discuss two cryptographic applications of the notions we just introduced. Both applications are based on the same four states:

$$\begin{aligned} |\psi_{00}\rangle_{AB} &= \tfrac{1}{\sqrt{2}}(|00\rangle_{AB} + |11\rangle_{AB})\,, & |\psi_{01}\rangle_{AB} &= \tfrac{1}{\sqrt{2}}(|00\rangle_{AB} - |11\rangle_{AB})\,, \\ |\psi_{10}\rangle_{AB} &= \tfrac{1}{\sqrt{2}}(|10\rangle_{AB} + |01\rangle_{AB})\,, & |\psi_{11}\rangle_{AB} &= \tfrac{1}{\sqrt{2}}(|10\rangle_{AB} - |01\rangle_{AB})\,. \end{aligned} \tag{4.9}$$

These states are called the *Bell states*. Observe that they are orthonormal and thus form a basis of $\mathbb{C}^2 \otimes \mathbb{C}^2$, the space of two qubits A and B. In Example 2.4.1 we calculated the reduced density matrix of Alice's system A for one of those states, the EPR pair $|\psi_{00}\rangle_{AB}$:

$$\rho_{00}^{A} = \text{tr}_{B}(|\psi_{00}\rangle\langle\psi_{00}|_{AB})$$
$$= \frac{1}{2}\Big(|0\rangle\langle0|_{A}\,\text{tr}_{B}(|0\rangle\langle0|_{B}) + |0\rangle\langle1|_{A}\,\text{tr}_{B}(|0\rangle\langle1|_{B})$$
$$+ |1\rangle\langle0|_{A}\,\text{tr}_{B}(|1\rangle\langle0|_{B}) + |1\rangle\langle1|_{A}\,\text{tr}_{B}(|1\rangle\langle1|_{B})\Big)$$
$$= \frac{1}{2}\big(|0\rangle\langle0|_{A} + |1\rangle\langle1|_{A}\big) = \frac{\mathbb{I}_{A}}{2}\,.$$

Calculating the reduced states on either A or B for each of the states in (4.9) always gives the same result,

$$\rho_{00}^{A} = \rho_{01}^{A} = \rho_{10}^{A} = \rho_{11}^{A} = \frac{\mathbb{I}}{2}\,,$$
$$\rho_{00}^{B} = \rho_{01}^{B} = \rho_{10}^{B} = \rho_{11}^{B} = \frac{\mathbb{I}}{2}\,.$$

The four Bell states in (4.9) are perfectly distinguishable given both qubits, because they are orthogonal and thus we can measure in a basis that contains them to identify them without ambiguity (see Box 1.2). Yet they all have the same reduced density matrices, which means that any measurement of system A or system B alone will yield the same outcome distribution on any of the states (because the measurement rule only depends on the density matrix). This means that the states are globally distinguishable, but locally indistinguishable. These two facts are key to the following applications.

4.3.1 Secret Sharing

Our first application is called *secret sharing*. Imagine that a country owns nuclear weapons, but wants to make sure that both the queen (Alice) and king (Bob) have to come together to activate them. One solution would be to give half of the launch codes $s = (s_1, \ldots, s_\ell) \in \{0,1\}^\ell$ to Alice, and the other half to Bob, thereby making sure that they both need to reveal their share of the information in order for the weapons to be activated. A drawback of this scheme is that each of the royal parties does have significant information about the launch codes, namely half of the bits. And what if there is only one bit? (Although that wouldn't be very secure, would it …)

The goal of a secret sharing scheme is to divide the information s into shares in such a way that any unauthorized set of parties (in the example, Alice or Bob alone) cannot learn *anything* at all about the secret. Remembering the idea behind the one-time pad, a much better scheme would be to choose a random string $r \in \{0,1\}^\ell$ and give r to Alice and $r \oplus s$ to Bob. In this case neither Alice nor Bob individually has any information about s; their respective secrets appear uniformly random. Yet when they come together they can easily recover s!

From this example we see that given a random classical bit one can construct a secret sharing scheme between Alice and Bob that shares a single secret bit s. However, they can do better if they are each given a qubit instead. Consider the case that Alice and Bob are given one of the four Bell states in (4.9). Since the reduced state of each of those states on each subsystem is maximally mixed, neither Alice nor Bob can gain any information on which of the states $|\psi_{00}\rangle_{AB}, |\psi_{01}\rangle_{AB}, |\psi_{10}\rangle_{AB}, |\psi_{11}\rangle_{AB}$ they have one qubit of. However, due to the fact that these states together form a basis, when Alice and Bob come together they can perform a measurement in that basis that perfectly distinguishes which state $|\psi_{ab}\rangle$ they have, yielding two bits of information, a and b.

Exercise 4.3.1   Suppose there are now three parties, Alice, Bob, and Charlie (the prime minister is also given a share of the nuclear codes!). Give a secret sharing scheme, based on a tripartite entangled state, such that no individual party has any information about the secret, but the three of them together are able to recover the secret. Better still, can you give a scheme such that no two of them together have any information about the secret?

QUIZ 4.3.1   *True or false? Any scheme for sharing a classical secret among n parties requires at least n/2 shares to recover the secret.*

QUIZ 4.3.2   *Charlie wants to share a four-bit classical secret between Alice and Bob in such a way that neither can recover it alone. What is the minimum number of qubits either Alice or Bob must hold?*

(a) 1
(b) 2
(c) 3
(d) 4

4.3.2 Superdense Coding

A second application of entanglement is called *superdense coding*. The task in dense coding consists in sending classical bits of information from Alice to Bob by encoding them in a quantum state that is as small as possible. Let's see how using entanglement we can send two classical bits using a single qubit.

Suppose that Alice and Bob share the state $|\psi_{00}\rangle_{AB} = \frac{1}{\sqrt{2}}(|00\rangle_{AB} + |11\rangle_{AB})$, and that Alice performs a unitary on her qubit as indicated in Table 4.1, depending on which bits $ab \in \{00, 01, 10, 11\}$ she wants to send to Bob.

As we already saw, the four states on the right-hand side in Table 4.1 form the Bell basis, and in particular they are perfectly distinguishable (see Box 1.2). Hence, if Alice sends her qubit over to Bob, he can perform a measurement in the Bell basis and recover both of Alice's classical bits.

While we won't show it here, it is possible to show that this is impossible without entanglement: without entanglement, a qubit cannot encode more than one bit of information. This fact is known as *Holevo's theorem* in quantum information theory.

Table 4.1 Unitary operation performed by Alice in order to encode her two classical bits $ab \in \{0,1\}^2$.		
Classical information a,b	Unitary $X_A^a Z_A^b$	Final joint state
00	$\mathbb{I}_A$	$\frac{1}{\sqrt{2}}(|00\rangle_{AB} + |11\rangle_{AB})$
01	Z_A	$\frac{1}{\sqrt{2}}(|00\rangle_{AB} - |11\rangle_{AB})$
10	X_A	$\frac{1}{\sqrt{2}}(|10\rangle_{AB} + |01\rangle_{AB})$
11	$X_A Z_A$	$\frac{1}{\sqrt{2}}(|10\rangle_{AB} - |01\rangle_{AB})$

4.4   Bell Nonlocality

Entanglement has many counterintuitive properties. A very important one is that it allows correlations between two particles – two qubits – that cannot be replicated classically. The very first example of such correlations was identified by Bell in the 1960s: Bell showed that the predictions of quantum theory are incompatible with those of any classical theory satisfying a natural notion of *locality*. How does one show anything like this?

The modern way to understand Bell nonlocality is by means of so-called *nonlocal games*. Let us imagine that we play a game with two players, which we will – surprise – call Alice and Bob, as depicted in Figure 4.1. Alice has a system A, and Bob has a system B. In the game we will ask Alice and Bob questions and collect answers from each of them. Let us denote the possible questions to Alice and Bob x and y, and label their answers a and b respectively. We will play this game many times, and in each round choose the questions to ask with some probability $p(x,y)$. As you might expect, our game has some rules. We denote these rules using a predicate $V(a,b|x,y)$, which takes the value "1" if a and b are winning answers for questions x and y, and "0" otherwise. To be fair, Alice and Bob know the rules of the game given by $V(a,b|x,y)$, and also the distribution $p(x,y)$. They can agree on any strategy before the game starts. However, once we start asking questions they are no longer allowed to communicate. Of interest to us will be the probability that Alice and Bob win the game, maximized over all possible strategies. That is,

$$p_{\text{win}} = \max_{\text{strategy}} \sum_{x,y} p(x,y) \sum_{a,b} V(a,b|x,y)\, p(a,b|x,y)\,,$$

where $p(a,b|x,y)$ is the probability that Alice and Bob produce answers a and b given x and y according to their chosen strategy.

What strategies are allowed? In a classical world, Alice and Bob can only have a classical strategy. A deterministic classical strategy is given by functions $f_A(x) = a$ and $f_B(y) = b$ that take the questions x and y to answers a and b. We then have $p(a,b|x,y) = 1$ whenever $a = f_A(x)$ and $b = f_B(y)$, and $p(a,b|x,y) = 0$ otherwise. Possibly, Alice and Bob also use shared randomness. That is, they have another string, which takes the value r with probability $p(r)$. In physics, r is also referred to as a hidden variable, but we will take the more operational viewpoint of shared randomness. In a strategy using shared randomness r, classical Alice and Bob can still only apply functions, except that now the function can also depend on the shared randomness r: $a = f_A(x, r)$ and $b = f_B(y, r)$. In terms of the probabilities we then have $p(a, b | x, y, r) = 1$ if $a = f_A(x, r)$ and $b = f_B(y, r)$, and $p(a, b | x, y, r) = 0$ otherwise. This gives

$$p(a, b | x, y) = \sum_r p(r) p(a, b | x, y, r) .$$

Does shared randomness help Alice and Bob? For a classical strategy based on shared randomness we have

$$
\begin{aligned}
p_{\text{win}} &= \max_{\text{class. strat.}} \sum_{x, y} p(x, y) \sum_{a, b} V(a, b | x, y) \sum_r p(r) p(a, b | x, y, r) \\
&= \max_{\text{class. strat.}} \sum_r p(r) \left(\sum_{x, y} p(x, y) \sum_{a, b} V(a, b | x, y) \, p(a, b | x, y, r) \right) \quad (4.10)
\end{aligned}
$$

Note that the quantity in brackets is largest for some particular value(s) of r. Since Alice and Bob want to maximize their winning probability they can fix the best possible r. Doing this gives them a deterministic strategy $a = f_A(x, r)$ and $b = f_A(y, r)$ where r is now fixed. We have thus shown that optimal classical strategies are always deterministic. This is a very useful fact when trying to find the best classical strategy, and we will soon use it on an example game.

Why would we care about this at all? It turns out that for many games a *quantum* strategy can achieve a higher winning probability. This is of fundamental importance for our understanding of nature, as well as quantum cryptography! Specifically, a *quantum strategy* means that Alice and Bob can pick a state ρ_{AB} to share, and agree on measurements to perform depending on their respective questions. That is, x and y will label a choice of measurement, and a and b are the outcomes of that measurement.

What's more, observing a higher winning probability is a signature of entanglement: quantumly, Alice and Bob can achieve a higher winning probability *only* if they are entangled, making such games into *tests* for entanglement. This is because if ρ_{AB} is separable, then the maximum winning probability (maximized over all possible measurements by Alice and Bob) can again be written in the form (4.10), where now the summation is over all possible product states in the decomposition of ρ_{AB} as a separable mixture.

Testing whether the state shared by Alice and Bob is entangled forms a crucial element in quantum key distribution, as we will see in later chapters.

4.4.1 Example of a Nonlocal Game: CHSH

Let us have a look at a simple example of a nonlocal game. The game is based on an inequality discovered by four physicists in the late 1960s, Clauser, Horne, Shimony, and Holt. For this reason it is called the "CHSH game." This game will turn out to be extremely useful for quantum cryptography. At the start of the game, the referee sends two bits x and y to Alice and Bob respectively, where we choose x with uniform probabilities $p(x = 0) = p(x = 1) = 1/2$ and independently y with probabilities

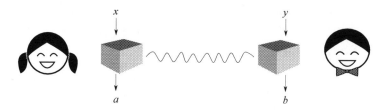

Fig. 4.1   A nonlocal game. Alice and Bob are given questions x and y, and must return answers a and b. If Alice and Bob are quantum, then x and y label measurement settings and a and b are measurement outcomes.

$p(y = 0) = p(y = 1) = 1/2$. Alice and Bob are asked to return answer bits a and b (Figure 4.1). Alice and Bob win the game if and only if

$$x \cdot y = a + b \quad \mathrm{mod}\ 2 \ .$$

In terms of the predicate $V(a, b|x, y)$ this means that $V(a, b|x, y) = 1$ if $x \cdot y = a + b$ mod 2 and $V(a, b|x, y) = 0$ otherwise. We are interested in the probability that Alice and Bob win the game. This probability can be written as

$$p_{\mathrm{win}}^{\mathrm{CHSH}} = \frac{1}{4} \sum_{x,y \in \{0,1\}} \sum_{\substack{a,b \\ a+b \ \mathrm{mod}\ 2 = x \cdot y}} p(a, b|x, y) \ ,$$

where $p(a, b|x, y)$ is the probability that Alice and Bob answer a and b given questions x and y. What can Alice and Bob do to win this game?

Classical Winning Probability

If Alice and Bob are entirely classical beings, then each of their answers must be a direct function of their question. For example, if $x = 0$, then Alice and Bob could agree as part of their strategy that Alice will then always answer $a = 0$. We see that as long as $x = 0$ or $y = 0$, then $x \cdot y = 0$. In this case, Alice and Bob want to achieve $a + b$ mod 2 = 0. However, if $x = y = 1$ then they would like to give answers such that $a + b$ mod 2 = 1. What makes this difficult for Alice and Bob is that they cannot communicate during the game.

It it not difficult to see (you may wish to check!), by trying out all possible deterministic strategies for Alice and Bob,[1] that classically the maximum winning probability that can be achieved is

$$p_{\mathrm{win}}^{\mathrm{CHSH}} = \frac{3}{4} \ .$$

Alice and Bob can achieve this winning probability with the strategy of answering $a = b = 0$ always, which means $a + b$ mod 2 = 0, which is correct in three out of the four possible cases. Only when $x = y = 1$ will Alice and Bob make a mistake.

---

1 Remember that we showed that even though shared randomness is allowed in principle, it never helps improve upon the best deterministic strategy.

BOX 4.2   A Quantum Strategy in the CHSH Game

In this strategy Alice and Bob each have one qubit of an EPR pair. We label the qubit held by Alice A and the one held by Bob B, so their shared state is

$$|\psi\rangle_{AB} = \frac{1}{\sqrt{2}} (|0\rangle_A |0\rangle_B + |1\rangle_A |1\rangle_B) .$$

Alice's measurements are as follows. When $x = 0$, Alice measures her qubit in the basis $\{|0\rangle, |1\rangle\}$. Otherwise, when $x = 1$, she measures in the basis $\{|+\rangle, |-\rangle\}$. Bob does something slightly different: when his question is $y = 0$, Bob measures his qubit in the basis $\{|v_1\rangle, |v_2\rangle\}$, where $|v_1\rangle = \cos(\pi/8)|0\rangle + \sin(\pi/8)|1\rangle$ and $|v_2\rangle = -\sin(\pi/8)|0\rangle + \cos(\pi/8)|1\rangle$. When $y = 1$, he measures in the basis $\{|w_1\rangle, |w_2\rangle\}$, where $|w_1\rangle = \cos(\pi/8)|0\rangle - \sin(\pi/8)|1\rangle$ and $|w_2\rangle = \sin(\pi/8)|0\rangle + \cos(\pi/8)|1\rangle$.

Quantum Winning Probability

It turns out that Alice and Bob can do significantly better with a quantum strategy using shared entanglement. Indeed, suppose that Alice and Bob use the strategy described in Box 4.2. How good is this strategy?

Consider first the case where $x = 0, y = 0$. This means that Alice measures in the basis $\{|0\rangle, |1\rangle\}$ and Bob in the basis $\{|v_1\rangle, |v_2\rangle\}$. The probability of winning, conditioned on $x = 0, y = 0$, is given by

$$
\begin{aligned}
p_{\text{win}|x=0,y=0} &= p(a = 0, b = 0 | x = 0, y = 0) + p(a = 1, b = 1 | x = 0, y = 0) \\
&= |\langle 0|_A \langle v_1|_B \cdot |\psi\rangle_{AB}|^2 + |\langle 1|_A \langle v_2|_B \cdot |\psi\rangle_{AB}|^2 \\
&= 2 \left| \frac{1}{\sqrt{2}} \cos \frac{\pi}{8} \right|^2 = \cos^2 \frac{\pi}{8} .
\end{aligned}
$$

Next the probability of winning, conditioned on $x = 0, y = 1$, is given by a similar expression:

$$
\begin{aligned}
p_{\text{win}|x=1,y=0} &= p(a = 0, b = 0 | x = 0, y = 1) + p(a = 1, b = 1 | x = 0, y = 1) \\
&= |\langle 0|_A \langle w_1|_B \cdot |\psi\rangle_{AB}|^2 + |\langle 1|_A \langle w_2|_B \cdot |\psi\rangle_{AB}|^2 \\
&= 2 \left| \frac{1}{\sqrt{2}} \cos \frac{\pi}{8} \right|^2 = \cos^2 \frac{\pi}{8} .
\end{aligned}
$$

Exercise 4.4.1   Show similarly that

$$p_{\text{win}|x=0,y=1} = p_{\text{win}|x=1,y=1} = \cos^2 \frac{\pi}{8} .$$

As a result, the quantum strategy for Alice and Bob from Box 4.2 succeeds with overall probability $\cos^2 \frac{\pi}{8} \approx 0.85$, which is strictly larger than the best classical strategy! This is the power of entanglement.[2]

QUIZ 4.4.1 *Suppose Alice and Bob hold a joint state* $|\psi\rangle = \frac{1}{\sqrt{2}}(|0\rangle_A |\phi\rangle_B + |1\rangle_A |\eta\rangle_B)$ *for some orthogonal pure states* $|\phi\rangle$ *and* $|\eta\rangle$. *Can they win the CHSH game with probability exactly* $\cos^2 \frac{\pi}{8}$?

(a) *Yes*
(b) *No*

QUIZ 4.4.2 *Suppose Alice and Bob hold a joint state* $\rho_{AB} = \mathrm{tr}_E |\psi_{ABE}\rangle\langle\psi_{ABE}|$ *which is not pure. Can they win the CHSH game with probability exactly* $\cos^2 \frac{\pi}{8}$?

(a) *Yes*
(b) *No*

4.4.2 Implications

The counterintuitive effects of entanglement have far-reaching consequences. The first is of a conceptual nature. You may have been wondering what *actually* happens when we measure a quantum particle. Sure, there is a probabilistic rule that we learned about. But "in reality," shouldn't it be the case that the outcome is a predetermined property of the particle – it's just that our "formalism" doesn't really allow us to say it, but only gives us access to probabilities? Maybe every particle has a classical "cheat sheet" attached to it, so that the cheat sheet can be used to specify the outcome for any possible measurement that we can make on it?

Now observe that such a cheat sheet, when computed for the two particles in an EPR pair, could be used to construct a classical strategy in the CHSH game: for every x, we'd look up Alice's answer a in the "cheat sheet" associated with her qubit. In physics, such cheat sheets are also called local hidden variables.

The fact that quantum strategies can beat classical strategies in the CHSH game implies that nature does not work that way! There are no classical cheat sheets, and nature is inherently quantum. Many experiments of ever-increasing accuracy have been performed that verify that Alice and Bob can indeed achieve a higher winning probability in the CHSH game than the classical world would allow. This tells us that the world is not classical, but we need more sophisticated tools to describe it – such as quantum mechanics. It also means that, when trying to build the ultimate computing and communication devices, we should make full use of what nature allows and "go quantum."

---

2 Later on, in Chapter 9, we will see that $\cos^2 \frac{\pi}{8}$ is the optimal winning probability for quantum players in the CHSH game.

We will see in the following chapters how to use this simple game to verify the presence of entanglement, test unknown quantum devices, and even create secure encryption keys.

4.5   The Monogamy of Entanglement

We know that two systems A and B can be in a joint pure state that is entangled, such as the "maximally entangled" EPR pair $|EPR\rangle_{AB}$. All our examples, however, had to do with entanglement between two systems A and B. What about a third system, call it C for Charlie? Of course, we could always consider three EPR pairs, $|EPR\rangle_{AB}$, $|EPR\rangle_{BC}$, and $|EPR\rangle_{AC}$. If this is the state of the three systems, however, we can't really talk about tripartite entanglement, because the correlations are always between any two of the three parties. Is it possible to create a joint state $|\psi\rangle_{ABC}$ in which the strong correlations of the EPR pair are shared simultaneously between all three systems?

Let us first argue that, if we require that A and B are strictly in an EPR pair, then it is impossible for C to share any correlation with the qubits that form the EPR pair.

Example 4.5.1 Let $\rho_{AB} = |\psi\rangle\langle\psi|_{AB}$ be an arbitrary pure state. Since ρ_{AB} is pure, its only nonzero eigenvalue is $\lambda_1 = 1$. Thus, by Uhlmann's theorem any purification of ρ_{AB} must have the form $\rho_{ABC} = |\psi\rangle\langle\psi|_{AB} \otimes |\phi\rangle\langle\phi|_C$ for an arbitrary state $|\phi\rangle_C$ of system C. But this is a pure state, whose Schmidt rank across the partition $AB : C$ is equal to 1: it is not entangled! If furthermore we take $|\psi\rangle_{AB}$ to be an EPR pair, then you can further compute that $\rho_{AC} = \frac{\mathbb{I}}{2} \otimes \rho_C$, meaning that not only is C uncorrelated with A, but from the point of view of C, A looks maximally mixed, i.e. it is completely random. The same holds for ρ_{BC}.   ∎

Qualitatively, the *monogamy* of entanglement captures precisely this phenomenon: monogamy postulates that if two systems are maximally entangled with each other then they cannot have any entanglement with any other system. In particular, these two systems must be in tensor product with the remainder of the universe. Monogamy is important in cryptography because it places a limit on the strength of quantum correlations: they can be stronger than classical, certainly, but they are not arbitrarily strong either! Let's see different ways to make this phenomenon more quantitative.

4.5.1   Quantifying Monogamy

Example 4.5.1 demonstrates monogamy of the maximally entangled EPR pair. What about more general states? Could they exhibit entanglement across three different parties? This is possible to some extent. For example, consider the three-qubit *GHZ* state

$$|GHZ\rangle_{ABC} = \frac{1}{\sqrt{2}}(|000\rangle + |111\rangle) .$$

Then you can verify that this state is entangled across any of the three possible partitions of the three qubits, $A : BC$, $AB : C$, or $AC : B$. However, the following exercise shows that this entanglement is *not* maximal as soon as one of the qubits is "dropped."

Exercise 4.5.1 Compute the reduced density matrix ρ_{AB} of the state $|\text{GHZ}\rangle_{ABC}$ on the first two qubits. Show that this reduced density is separable, by computing an explicit decomposition $\rho_{AB} = \sum_i p_i \rho_A^{(i)} \otimes \rho_B^{(i)}$.

As the exercise shows, the correlations in the GHZ state, when considering any given pair of qubits, are weaker than those of a maximally entangled state (indeed, they are not even entangled at all). To quantify entanglement in mixed states more finely than the "entangled/not entangled" distinction, one possibility is to use so-called *entanglement measures* $E(A:B)$. An entanglement measure is any function of bipartite density matrices that satisfies certain desirable properties. We already saw two such measures, the Schmidt rank and the entanglement entropy (Box 4.1); however, they only apply to pure bipartite states. For states that are not pure the situation is much more complicated, and there is no standard entanglement measure that satisfies all the properties that we would like. Among these properties, there is one that expresses monogamy as follows: for any tripartite density matrix ρ_{ABC} it requires that

$$E(A:B) + E(A:C) \leq E(A:BC). \tag{4.11}$$

One way to interpret this inequality is that, whatever the *total* entanglement that A has with B and C (right-hand side), this entanglement must split additively between entanglement with B and with C (left-hand side). You may think this is obvious – but in fact very few entanglement measures are known to satisfy the monogamy inequality (4.11)! Finding good measures of entanglement is an active area of research.

> **QUIZ 4.5.1** *Consider a GHZ state* $|\text{GHZ}\rangle_{ABC} = \frac{1}{\sqrt{2}}(|000\rangle_{ABC} + |111\rangle_{ABC})$. *What is* $\rho_A = \text{tr}_{BC}(\rho_{ABC})$?
>
> **(a)** $\rho_A = |+\rangle\langle+|$
> **(b)** $\rho_A = \frac{\mathbb{I}}{2}$
> **(c)** $\rho_A = |0\rangle\langle0|$

> **QUIZ 4.5.2** *Alice, Bob, and Charlie share the GHZ state* $|\text{GHZ}\rangle_{ABC} = \frac{1}{\sqrt{2}}(|000\rangle_{ABC} + |111\rangle_{ABC})$. *Which pairs of qubits have nonzero entanglement between them, when the third qubit is traced out?*
>
> **(a)** *AB, BC, CA*
> **(b)** *AB, BC*
> **(c)** *CA*
> **(d)** *none*

> **QUIZ 4.5.3** *Now suppose that Alice, Bob, and Charlie share the three-qubit state* $\frac{1}{\sqrt{3}}(|100\rangle_{ABC} + |010\rangle_{ABC} + |001\rangle_{ABC})$. *Which pairs of qubits have nonzero entanglement?*
>
> **(a)** *AB, BC, CA*
> **(b)** *AB, BC*
> **(c)** *CA*
> **(d)** *none*

4.5.2 A Three-Player CHSH Game

Another, more intuitive way of measuring monogamy is through the use of nonlocal games, such as the CHSH game that we discussed in Section 4.4.1. To see monogamy at play we introduce a three-player variant of this game, in which Alice is required to successfully play the CHSH game simultaneously with two different partners, Bob and Charlie. That is, Alice is sent a random x, Bob a random y, and Charlie a random z; they have to provide answers a, b, and c respectively, such that

$$x \cdot y = a + b \mod 2 \quad \text{and} \quad x \cdot z = a + c \mod 2.$$

Can they do it? The fact, discussed in Example 4.5.1, that the EPR pair has no entangled extension to three parties should give you a hint that things are going to be difficult for our three protagonists!

It is possible to make an even stronger statement. Consider a different three-player variant of the CHSH game, played as follows.

- The referee, who is setting up the game, selects two of the three players at random, and sends each of them the message "You've been selected!" together with a label that indicates which player in the CHSH game they are supposed to "impersonate."
- The referee plays the CHSH game with the selected players, sending each of them a random question and checking their answers for the CHSH condition. The third player is completely ignored.

Now, what do you think is the players' maximum success probability in this game? For the case of classical players the answer should be clear: $3/4$. Indeed, there is nothing more or less that they can do in this variant than in the original two-player CHSH. (Make sure that you are convinced of this fact. What is an optimal strategy for the three players?)

What about quantum players? Can they win with probability $\cos^2(\pi/8)$? Why not? Let's think of a possible extension of the two-player strategy we saw in Section 4.4.1 (Box 4.2). First of all we need the three players, Alice, Bob, and Charlie, to decide on an entangled state to share. Given that they know two of them are going to be asked to play CHSH, it is natural to set things up with three EPR pairs, one between Alice and Bob, another between Bob and Charlie, and the third between Alice and Charlie.

Now the game starts, and two players are told they are to play the game. However, the sticky point is that each of the selected players is not told with whom they are to play the game! So, for instance, Alice will know she has been selected, but will not be told who is her partner – Bob or Charlie. Which EPR pair is she going to use to implement her strategy?

It turns out there is no answer to this question: Alice is stuck! But maybe we chose the wrong entangled state for them to share. What if they share a GHZ state instead? Wait, no, this will not work, because in Exercise 4.5.1 we showed that the reduced density matrix of a GHZ state on two systems is always separable. So, if the three players share a GHZ state, whichever two players get selected to play the CHSH game will in fact share no entanglement at all. So maybe there is a better generalization of the EPR pair we should use?

In fact, there does not exist any such state! Although we won't do it here, it is possible to show that the optimal winning probability in the three-player CHSH game described above, for quantum players, is no larger than the classical optimum: $3/4$. This is a powerful demonstration of monogamy of entanglement, showing in particular that there is no nice extension of the EPR pair to a tripartite state – at least not one that allows any two of them to win the CHSH game. We will return to a similar manifestation of monogamy by analyzing a "tripartite guessing game" in the next chapter.

CHAPTER NOTES

The Schmidt decomposition, and additional properties of it, is covered in Section 2.5 of A. Nielsen and I. L. Chuang, *Quantum Computation and Quantum Information* (Cambridge University Press, 2000). We recommend working through as many possible examples and exercises around this decomposition as you can find, as it is a very important tool in quantum information.

The CHSH game is named after its inventors, Clauser, Horne, Shimony, and Holt (Proposed experiment to test local hidden-variable theories. *Physical Review Letters*, **23**(15):880, 1969), who formulated it as an inequality which refined the pioneering work of John Bell on quantum nonlocality. Although they already knew that by using entanglement it is possible to succeed with probability $\cos^2 \frac{\pi}{8}$ in the game, the optimality of this value was only shown much later, by Boris Tsirelson (Quantum generalizations of Bell's inequality. *Letters in Mathematical Physics*, **4**(2):93–100, 1980). In reference to these pioneering works, an upper bound on the classical success probability in a nonlocal game is often referred to as a "Bell inequality," and a bound on the quantum success probability as a "Tsirelson inequality" (*Reviews of Modern Physics*, **86**:419, 2014).

The idea that entanglement is a monogamous resource is discussed in a paper by B. M. Terhal (Is entanglement monogamous? *IBM Journal of Research and Development*, **48**(1):71–78, 2004), who attributes it to a talk by Bennett. The three-player CHSH game discussed in Section 4.5.2 is analyzed in a paper by B. Toner (Monogamy of non-local quantum correlations. *Proceedings of the Royal Society of London A: Mathematical, Physical and Engineering Sciences*, **465**(2101):59–69, 2009).

PROBLEMS

4.1 The CHSH game, first take

Alice and Bob would like to play the CHSH game. Sadly they do not possess a machine that can generate entanglement at will. Instead they have a machine that can generate the following bipartite quantum states:

$$\rho_1 = \frac{3}{4} |\text{EPR}\rangle \langle \text{EPR}| + \frac{1}{16} \mathbb{I} \,,$$

$$\rho_2 = |00\rangle \langle 00| \,,$$

with $|\text{EPR}\rangle$ being the EPR pair. Alice and Bob would like to identify the state that can produce the highest CHSH value.

1. Which one of these would generate the highest CHSH value (using any possible measurements)?

Now imagine that Alice and Bob try to build a better machine, one that produces the EPR pair (which they know will give them the highest possible CHSH value). Sadly their machine doesn't quite produce the EPR pair. Instead it produces the state $|\psi_{ab}\rangle$ with probability $p_{ab} = 0.25$, where for $a, b \in \{0, 1\}$ we have

$$|\psi_{00}\rangle = |\text{EPR}\rangle = \frac{1}{\sqrt{2}}(|00\rangle + |11\rangle) \,,$$

$$|\psi_{01}\rangle = \frac{1}{\sqrt{2}}(|00\rangle - |11\rangle) \,,$$

$$|\psi_{10}\rangle = \frac{1}{\sqrt{2}}(|01\rangle + |10\rangle) \,,$$

$$|\psi_{11}\rangle = \frac{1}{\sqrt{2}}(|01\rangle - |10\rangle) \,.$$

The machine also tells Alice and Bob which state it produces. Alice and Bob are quite happy with their efforts because all of these states will give them the maximal CHSH value (they are maximally entangled). The easiest way to see that this is true is by noting that all of these states can be transformed to an EPR pair by Bob, applying an operation to his qubit based on the number $a, b \in \{0, 1\}$ he gets from the machine.

2. For each of the states $|\psi_{ab}\rangle$, $a, b \in \{0, 1\}$, which operation should Bob apply in order to change the outputted state to the EPR pair?

Now imagine that the part of the machine that tells Alice and Bob which state $|\psi_{ab}\rangle$ it produces breaks! This means that they don't know which state the machine outputs.

3. What is their probability of winning the CHSH game if they apply the strategy that is optimal for the EPR pair? *[Hint: write down the density matrix they receive from the machine.]*

4.2 The CHSH game, second take

In the following questions we will consider different strategies for the CHSH game. We will look at how the winning probability depends on which measurements Alice and Bob perform. We will say that Alice and Bob play the CHSH game by measuring in the bases $\{|b^0_{Ax}\rangle, |b^1_{Ax}\rangle\}$ and $\{|b^0_{Bx}\rangle, |b^1_{Bx}\rangle\}$. This means that if, for example, Alice receives $x = 0$, she will measure in the basis $\{|b^0_{A0}\rangle, |b^1_{A0}\rangle\}$ and output the measurement outcome, and similarly for $x = 1$ and $y = 0$, $y = 1$ for Bob.

1. Imagine Alice and Bob both randomly output 0 or 1 with probability $p_0 = p_1 = 1/2$ independently of the input bits. What is their winning probability?

2. Now Alice and Bob share a maximally entangled state,

$$|\text{EPR}\rangle_{AB} = \frac{1}{\sqrt{2}}(|0\rangle_A |0\rangle_B + |1\rangle_A |1\rangle_B) \,,$$

which they measure in the standard basis, that is $|b^0_{Ax}\rangle = |b^0_{Bx}\rangle = |0\rangle$ and $|b^1_{Ax}\rangle = |b^1_{Bx}\rangle = |1\rangle$ for both $x = 0$ and $x = 1$. What will their corresponding winning probability be in such a scenario?

3. Now they share a state

$$\rho_{AB} = \frac{1}{2}(|0\rangle\langle 0|_A \otimes |0\rangle\langle 0|_B + |1\rangle\langle 1|_A \otimes |1\rangle\langle 1|_B) \,,$$

which they measure in the standard basis, that is $|b^0_{Ax}\rangle = |b^0_{Bx}\rangle = |0\rangle$ and $|b^1_{Ax}\rangle = |b^1_{Bx}\rangle = |1\rangle$ for both $x = 0$ and $x = 1$. What will their corresponding winning probability be in such a scenario?

4. Now Alice and Bob share a maximally entangled state $|\psi\rangle_{AB} = \frac{1}{\sqrt{2}}(|01\rangle_{AB} + |10\rangle_{AB})$, which is orthogonal to the state $|\text{EPR}\rangle_{AB} = \frac{1}{\sqrt{2}}(|00\rangle_{AB} + |11\rangle_{AB})$ that Alice and Bob shared in the optimal strategy described in the chapter. Do they need to use different measurements for state $|\psi\rangle_{AB}$ than for state $|\text{EPR}\rangle_{AB}$ to obtain the optimal winning probability in this case?

5. Consider again the same setting as in the previous question, where Alice and Bob play the CHSH game with the optimal bases as for the state $|\text{EPR}\rangle_{AB}$ but using the state $|\psi\rangle_{AB}$. Can they obtain the optimal winning probability by performing some classical processing on either the inputs x, y or their outputs?

4.3 Dimension of a purifying system

Alice and Bob share a pure state divided between them as follows: Alice holds a d-dimensional *qudit*, i.e. a system with basis states labeled $\{|0\rangle, |1\rangle, \ldots, |d-1\rangle\}$ for some d. Bob, on the other hand, holds some number $m \geq 0$ of qubits.

1. Suppose Alice's qudit is in the state $\frac{1}{2}(|0\rangle\langle 0| + |3\rangle\langle 3|)$. What is the minimum number of qubits Bob can have, given that the joint state is pure?

2. Suppose Alice's qudit is in the state $\frac{1}{4}(|0\rangle\langle 0| + |1\rangle\langle 1| + |2\rangle\langle 2| + |3\rangle\langle 3|)$. What is the minimum number of qubits Bob can have, given that the joint state is pure?

3. Suppose Alice's qudit is in the state

$$\frac{1}{4}(|1\rangle\langle 1| + |2\rangle\langle 2| + |3\rangle\langle 3|) + \frac{1}{8}(|4\rangle\langle 4| + |4\rangle\langle 5| + |5\rangle\langle 4| + |5\rangle\langle 5|) .$$

What is the minimum number of qubits Bob can have, given that the joint state is pure?

4.4 Robustness of GHZ and W states, Part 2

We return to the multi-qubit GHZ and W states introduced in Problem 2.6. As a reminder,

$$|\text{GHZ}_N\rangle = \frac{1}{\sqrt{2}}(|0\rangle^{\otimes N} + |1\rangle^{\otimes N}) ,$$

$$|W_N\rangle = \frac{1}{\sqrt{N}}(|10\cdots 0\rangle + |01\cdots 0\rangle + \cdots + |00\cdots 1\rangle) .$$

In this chapter we learned to distinguish product states from (pure) entangled states by calculating the Schmidt rank of $|\psi\rangle_{AB}$, i.e. the rank of the reduced state $\rho_A = \text{tr}_B |\psi\rangle\langle\psi|$. In particular ρ is pure if and only if $|\psi\rangle$ has Schmidt rank 1. In the following, we denote by tr_N the operation of tracing out only the last of N qubits.

1. What are the ranks r_{GHZ} of $\text{tr}_N |\text{GHZ}_N\rangle\langle\text{GHZ}_N|$ and r_W of $\text{tr}_N |W_N\rangle\langle W_N|$, respectively? (Note that these are the Schmidt ranks of $|\text{GHZ}_N\rangle$ and $|W_N\rangle$ if we partition each of them between the first $N-1$ qubits and the last qubit.)

Let us now introduce a more discriminating (in fact, continuous) measure of the entanglement of a state $|\psi\rangle_{AB}$: namely, the *purity* of the reduced state ρ_A given by $\text{tr}\,\rho_A^2$. First let's see how this works in practice with the extreme cases in d dimensions:

2. What are the purities $\text{tr}\,(\rho^2)$ for $\rho = |0\rangle\langle 0|$ and the "maximally mixed" state $\rho = \frac{1}{d}\mathbb{I}_d$, respectively?

3. Is the purity of ρ_A higher or lower for more entangled states $|\psi\rangle_{AB}$? Can you explain this in terms of the definition $\mathrm{tr}\left(\rho_A^2\right)$?

Now consider again the behavior of the N-qubit GHZ and W states with one qubit discarded (i.e. traced out):

4. What is the purity of $\mathrm{tr}_N |\mathrm{GHZ}_N\rangle\langle\mathrm{GHZ}_N|$ in the limit $N \to \infty$?
5. What is the purity of $\mathrm{tr}_N |W_N\rangle\langle W_N|$ in the limit $N \to \infty$?

Discuss the implications for the "robustness" of multipartite entanglement under loss of one qubit in GHZ versus W states. What can we say about losses of more than one qubit?

4.5 A secret shared among three people

In this chapter you learned about sharing a classical secret between two people using an entangled state. Here we will create a scheme that shares a classical secret among three people: Alice, Bob, and Charlie. We will do that by giving Alice, Bob, and Charlie a GHZ-like state of the form

$$|\psi_b\rangle = \frac{1}{\sqrt{2}}\left(|0_A\rangle |0_B\rangle |0_C\rangle + (-1)^b |1_A\rangle |1_B\rangle |1_C\rangle\right)$$

with $b \in \{0, 1\}$ being the "secret" we want to share.

1. Imagine that Alice wants to perform a local measurement on her qubit to find the secret bit. What is her reduced density matrix?

Convince yourself that this means that Alice cannot find the secret on her own.

2. Now imagine that Alice and Bob would like to discover the secret bit without Charlie being involved. What would their reduced state look like?

Convince yourself that the same holds for the combinations BC and AC and that this implies that they cannot find the secret!

Now let's imagine that a terrible snowstorm keeps Alice, Bob, and Charlie confined to their houses. However, they have the ability to apply operations to their own qubits as well as measure them. Finally, they each possess a radio through which they can communicate classical information. They would like to find out the secret bit. However, they want to also do it in a way that guarantees that they succeed, i.e. they want to perform a protocol that finds b with probability 1. Alice, Bob, and Charlie propose to each other the following measurement schemes.

Alice's proposal

Step 1. Alice, Bob measure in the standard basis and Charlie in the Hadamard basis
Step 2. Alice, Bob send their result to Charlie
Step 3. Charlie adds the measurements results modulo 2 to obtain a bit x
Step 4. Charlie obtains b as $b \cdot x = 1$ modulo 2

Bob's proposal

Step 1. Alice, Bob, Charlie apply a Hadamard operation to their qubit
Step 2. Alice, Bob, Charlie measure in the standard basis
Step 3. Alice, Bob send their result to Charlie
Step 4. Charlie adds the measurements results modulo 2 to obtain a bit x
Step 5. Charlie obtains b as $b + x = 0$ modulo 2

Charlie's proposal

Step 1. Alice, Bob, Charlie measure in the standard basis
Step 2. Alice, Bob send their result to Charlie
Step 3. Charlie adds the measurements results modulo 2 to obtain a bit x
Step 4. Charlie obtains b as $b + x = 0$ modulo 2

3. Which scheme will correctly (with probability 1) produce the bit b with Charlie?

QUIZ SOLUTIONS

Quiz 4.1 (b); Quiz 4.2.1 (a); Quiz 4.2.2 True; Quiz 4.2.3 (c); Quiz 4.2.4 (b); Quiz 4.2.5 True; Quiz 4.2.6 False; Quiz 4.3.1 False; Quiz 4.3.2 (b); Quiz 4.4.1 (a); Quiz 4.4.2 (b); Quiz 4.5.1 (b); Quiz 4.5.2 (d); Quiz 4.5.3 (a)

CHEAT SHEET

Purification of states

Given any density matrix diagonalized as $\rho_A = \sum_i \lambda_i |\phi_i\rangle\langle\phi_i|_A$, a purification of ρ_A is

$$|\psi\rangle_{AB} = \sum_i \sqrt{\lambda_i} |\phi_i\rangle_A |w_i\rangle_B,$$

for any set of orthonormal vectors $\{|w_i\rangle_B\}_i$.

Schmidt decomposition of bipartite pure states

Any bipartite pure state $|\psi\rangle_{AB}$ can be written in the form

$$|\psi\rangle_{AB} = \sum_{i=1}^{d} \sqrt{\lambda_i} |u_i\rangle_A |v_i\rangle_B,$$

where $\{|u_i\rangle_A\}_i, \{|v_i\rangle_B\}_i$ are orthonormal vector sets, and $\sum_{i=1}^{d} \lambda_i = 1$.

CHSH game winning probability

Consider Alice and Bob playing in a game, where questions $x, y \in \{0, 1\}$ are sent to them, and they respond with answers $a, b \in \{0, 1\}$ respectively. Alice and Bob win the game if $a + b \pmod 2 = x \cdot y$. The winning probability is given by

$$p_{\text{win}}^{\text{CHSH}} = \frac{1}{4} \sum_{x,y \in \{0,1\}} \sum_{\substack{a,b \\ a+b \mod 2 = x \cdot y}} p(a, b | x, y).$$

For any classical strategy, $p_{\text{win}}^{\text{CHSH}} = \frac{3}{4}$.

If Alice and Bob share an EPR pair, then $p_{\text{win}}^{\text{CHSH}} = \cos^2 \frac{\pi}{8} \approx 0.85$.

5

Quantifying Information

In Chapter 2 we saw an example of a communication task between Alice and Bob in which the goal was to exchange a secret message that remains hidden from an eavesdropper, Eve. For this task we saw the importance of using a large key k that is secretly shared between Alice and Bob but looks completely random from Eve's perspective. In other words, we want that *Eve is ignorant about the key k*.

The ability to generate a shared secret key is a fundamental primitive that enables many other tasks in cryptography beyond secure communication. Moreover, this is a task for which quantum information provides a crucial advantage! Our goal in the next chapters is to develop a method for generating a secret key using quantum communication. In this chapter we accomplish the first step on our road towards this goal: we learn about ways to quantify quantum information, knowledge, and uncertainty. This will allow us to precisely define what it really means for *Eve to be ignorant about the key*!

5.1   When Are Two Quantum States Almost the Same?

We already saw the importance of distinguishing between quantum states for the task of reliable communication: if Bob cannot distinguish between two different states sent by Alice, then he cannot reliably recover the messages that she encoded in those states. The property of two quantum states being distinguishable is not a $0 - 1$ property: two states can be perfectly distinguishable, they can be absolutely indistinguishable (i.e. equal), or they can be "partially distinguishable." In this section we introduce two important measures for quantifying how close two quantum states are to each other.

5.1.1   Trace Distance

Our first measure of closeness is called the *trace distance*. This measure is essential in quantum cryptography as well as in quantum computing in general. To motivate the trace distance, let us suppose that we would like to implement a protocol or algorithm that produces a certain quantum state ρ_{ideal}. Unfortunately, due to imperfections in the design or execution of the protocol we actually end up preparing a different state ρ_{real}. Suppose that this protocol or algorithm is a subroutine that is part of a much larger protocol or computation. How is the larger protocol affected if it is executed on the state ρ_{real} instead of ρ_{ideal}?

Intuitively it is clear that if the states ρ_{real} and ρ_{ideal} are nearly impossible to distinguish with respect to any measurement then it should also not matter much which one is used in the larger protocol. This is because the larger protocol can itself be thought of as a measurement that could be used to distinguish between the two states.

The essential quality of the trace distance is that it directly quantifies how well it is possible to distinguish two states by making the best possible measurement on them. In fact, this sentence can be taken as the definition of the trace distance!

Let's investigate how we could formalize this more precisely. Suppose that we consider a scenario where we are given ρ_{real} or ρ_{ideal} each with a priori probability $1/2$. Suppose that we are challenged to decide which is the case. To answer this challenge we may perform a two-outcome measurement with POVM elements, say, M_{real} and $M_{\text{ideal}} = \mathbb{I} - M_{\text{real}}$. If we perform this measurement the probability of giving the right answer is, on average over the choice of which state we're actually given,

$$
\begin{aligned}
p_{\text{succ}} &= \frac{1}{2}\,\text{tr}\left(M_{\text{real}}\rho_{\text{real}}\right) + \frac{1}{2}\,\text{tr}\left(M_{\text{ideal}}\rho_{\text{ideal}}\right) \\
&= \frac{1}{2} + \frac{1}{2}\,\text{tr}\left(M_{\text{real}}\left(\rho_{\text{real}} - \rho_{\text{ideal}}\right)\right) ,
\end{aligned}
$$

where for the second equality we used that $M_{\text{ideal}} = \mathbb{I} - M_{\text{real}}$. The above holds for any choice of measurement. To find the best choice we can optimize over the choice of M_{real}. Recall from the definition of a POVM that M_{real} is required to be Hermitian and that $0 \leq M_{\text{real}} \leq \mathbb{I}$, i.e. all eigenvalues of M_{real} are real and lie between 0 and 1. This allows us to write the maximum probability of successfully distinguishing between the two states as

$$
p_{\text{succ}}^{\text{max}} = \frac{1}{2} + \frac{1}{2}\max_{0 \leq M \leq \mathbb{I}}\text{tr}\left(M\left(\rho_{\text{real}} - \rho_{\text{ideal}}\right)\right) .
$$

Note that it is always easy to succeed with probability exactly $1/2$ by giving a random answer. So $p_{\text{succ}}^{\text{max}} \geq 1/2$ always, and the second term above is always non-negative. It gives us our definition of the trace distance.

> **Definition 5.1.1 (Trace distance).** *The* trace distance *between two quantum states ρ_0 and ρ_1 of the same dimension is given by*
>
> $$
> D(\rho_0, \rho_1) = \max_{0 \leq M \leq \mathbb{I}}\text{tr}\left(M\left(\rho_0 - \rho_1\right)\right) . \tag{5.1}
> $$

The definition expresses the trace distance as an optimization problem. For calculations, it is much more convenient to have a closed form expression. For this, we use the following.

Theorem 5.1.1 *The trace distance between two quantum states ρ_0 and ρ_1 evaluates to*

$$
D(\rho_0, \rho_1) = \frac{1}{2}\|A\|_1 = \frac{1}{2}\,\text{tr}\left(\sqrt{A^\dagger A}\right) , \tag{5.2}
$$

where $A = \rho_0 - \rho_1$ and $\|A\|_1$ is the Schatten 1-norm *of the matrix A, i.e. the sum of its singular values.*

BOX 5.1   Analytical Expression of the Trace Distance

Given two density matrices ρ_0 and ρ_1, what is the operator M that maximizes the quantity $\mathrm{tr}\,(M\,(\rho_0 - \rho_1))$? To find this out, consider the diagonalization

$$\rho_0 - \rho_1 = \sum_j \lambda_j |u_j\rangle\langle u_j|\,,$$

where $\{\lambda_j\}_j$ are the eigenvalues and $\{|u_j\rangle\}_j$ the eigenvectors. Using cyclicity of the trace we get that for any M,

$$\mathrm{tr}\,(M\,(\rho_0 - \rho_1)) = \sum_j \lambda_j \langle u_j| M |u_j\rangle\,.$$

For any M such that $0 \le M \le \mathbb{I}$, for any unit vector $|u\rangle$ we have that $0 \le \langle u|M|u\rangle \le 1$. Then it is clear that, if at all possible, we should choose M such that $\langle u_j|M|u_j\rangle = 0$ whenever $\lambda_j \le 0$, and $\langle u_j|M|u_j\rangle = 1$ whenever $\lambda_j > 0$. Both conditions are satisfied by choosing M as the projector onto the positive eigenspace of the matrix $\rho_0 - \rho_1$. In other words, if we introduce the set $S_+ = \{j|\lambda_j > 0\}$ then an optimal M is given by $M_{\mathrm{opt}} = \sum_{j \in S_+} |u_j\rangle\langle u_j|$. Finally, observe that for a Hermitian matrix A, $\mathrm{tr}(\sqrt{A^\dagger A})$ is exactly the sum of the singular values of A, i.e. the sum of the absolute values of its eigenvalues. Therefore, (5.2) coincides with (5.1).

If you want to understand how we went from (5.1) in the definition to (5.2) in the theorem, see Box 5.1. Before we continue let's see some examples.

Example 5.1.1 Suppose that ρ_0 and ρ_1 are classical states, so $\rho_0 = \sum_x p_x |x\rangle\langle x|$ and $\rho_1 = \sum_x q_x |x\rangle\langle x|$. What is $D(\rho_0, \rho_1)$? Since $\rho_0 - \rho_1$ is diagonal in the standard basis, its singular values are the absolute values of its diagonal coefficients. Thus

$$D(\rho_0, \rho_1) = \frac{1}{2}\|\rho_0 - \rho_1\|_1 = \frac{1}{2}\sum_x |p_x - q_x|\,.$$

This is precisely the *total variation distance* between the distributions (p_x) and (q_x), which is a natural distance measure on distributions. ∎

Example 5.1.2 Now suppose that ρ_0 and ρ_1 are cq-states, of the form $\rho_0 = \sum_x p_x |x\rangle\langle x| \otimes \rho_{0.x}$ and $\rho_1 = \sum_x q_x |x\rangle\langle x| \otimes \rho_{1.x}$. Again, how do we compute $D(\rho_0, \rho_1)$? Now $\rho_0 - \rho_1 = \sum_x |x\rangle\langle x| \otimes (p_x\rho_{0.x} - q_x\rho_{1.x})$ is not diagonal, but it is *block diagonal*. Since the singular values of a block-diagonal matrix are the singular values of the individual blocks, we get that

$$D(\rho_0, \rho_1) = \sum_x \frac{1}{2}\|p_x\rho_{0.x} - q_x\rho_{1.x}\|_1\,.$$

∎

Finally, let's see a "fully quantum" example.

Example 5.1.3 Consider $\rho_1 = |0\rangle\langle 0|$ and $\rho_2 = |+\rangle\langle +|$. To evaluate $D(\rho_1, \rho_2)$ we first calculate

$$\rho_1 - \rho_2 = \begin{pmatrix} 1 & 0 \\ 0 & 0 \end{pmatrix} - \frac{1}{2}\begin{pmatrix} 1 & 1 \\ 1 & 1 \end{pmatrix} = \frac{1}{2}\begin{pmatrix} 1 & -1 \\ -1 & -1 \end{pmatrix}.$$

Therefore, the trace distance is equal to

$$D(\rho_1, \rho_2) = \frac{1}{2} \cdot \frac{1}{2} \operatorname{tr} \sqrt{\begin{pmatrix} 1 & -1 \\ -1 & -1 \end{pmatrix}^2} = \frac{1}{2} \cdot \frac{1}{2} \operatorname{tr} \sqrt{\begin{pmatrix} 2 & 0 \\ 0 & 2 \end{pmatrix}} = \frac{1}{\sqrt{2}}.$$

Another way to do the calculation is to first consider the diagonalization of $\rho_1 - \rho_2$, which can be done by first calculating its eigenvalues by solving for λ in the equation

$$\det\begin{pmatrix} \frac{1}{2} - \lambda & -\frac{1}{2} \\ -\frac{1}{2} & -\frac{1}{2} - \lambda \end{pmatrix} = 0.$$

The solutions are given by $\lambda = \pm\frac{1}{\sqrt{2}}$. One can also find the eigenvector $|e_+\rangle = (x \ \ y)^T$ corresponding to $\lambda = \frac{1}{\sqrt{2}}$,

$$\frac{1}{2}\begin{pmatrix} 1 & -1 \\ -1 & -1 \end{pmatrix}\begin{pmatrix} x \\ y \end{pmatrix} = \frac{1}{\sqrt{2}}\begin{pmatrix} x \\ y \end{pmatrix} \implies \frac{x}{y} = \frac{-1}{\sqrt{2}-1}.$$

The normalization condition gives $x^2 + y^2 = 1$, and the solution is found to be

$$x = \cos\frac{\pi}{8}, \quad y = \sin\frac{\pi}{8}.$$

The optimal measurement operator that distinguishes ρ_1, ρ_2 is then given by $M_{\mathrm{opt}} = |e_+\rangle\langle e_+|$, and

$$\operatorname{tr}\left(M_{\mathrm{opt}}(\rho_1 - \rho_2)\right) = \frac{1}{\sqrt{2}}. \qquad \blacksquare$$

A few properties of the trace distance are often useful:

Proposition 5.1.2 *The trace distance is a metric, that is, a proper distance measure that corresponds to our intuitive notion of distance. More precisely, we have the following properties for all density matrices ρ, σ, τ:*

1. *Non-negativity: $D(\rho, \sigma) \geq 0$, where equality is achieved if and only if $\rho = \sigma$.*
2. *Symmetry: $D(\rho, \sigma) = D(\sigma, \rho)$.*
3. *Triangle inequality: $D(\rho, \sigma) \leq D(\rho, \tau) + D(\tau, \sigma)$.*
4. *Convexity: for all $\{p_i, \rho_i\}_i$ such that $p_i \geq 0$ and ρ_i is a density matrix for all i, and $\sum_i p_i = 1$, it holds that $D(\sum_i p_i\rho_i, \sigma) \leq \sum_i p_i D(\rho_i, \sigma)$.*

Since states that are ε-close to each other in terms of the trace distance cannot be distinguished well, it will later be convenient to have the notion of a set of states that are all ε-close to a particular state ρ. This is often called the ε-ball of ρ.

> **Definition 5.1.2 (ε-ball around ρ).** *Given any density matrix ρ, the ε-ball around ρ is defined as the set of all states ρ' that are ε-close to ρ in terms of trace distance, i.e.*
>
> $$\mathcal{B}^{\varepsilon}(\rho) := \{\rho' \mid \rho' \geq 0, \mathrm{tr}(\rho') = 1, D(\rho,\rho') \leq \varepsilon\}.$$

Exercise 5.1.1 Show that if ρ_1 and ρ_2 are two density matrices with orthogonal support, i.e. $\rho_1\rho_2 = \rho_2\rho_1 = 0$, then $D(\rho_1,\rho_2) = 1$. This is consistent with Box 1.2: orthogonal states can be perfectly distinguished.

QUIZ 5.1.1 *What is the trace distance between $\rho_1 = |+\rangle\langle+|$ and $\rho_2 = |-\rangle\langle-|$?*

(a) 0

(b) 0.5

(c) 1

QUIZ 5.1.2 *In which basis should one measure to distinguish the states $\rho_1 = |+\rangle\langle+|$ and $\rho_2 = |-\rangle\langle-|$ optimally?*

(a) *In the standard basis.*

(b) *In the Hadamard basis.*

(c) *No measurement is required. For those states random guessing is optimal.*

QUIZ 5.1.3 *In which basis should one measure to optimally distinguish the maximally mixed state of a qubit $\rho_1 = \frac{\mathbb{I}}{2}$ and the state $\rho_2 = |+\rangle\langle+|$?*

(a) *In the standard basis.*

(b) *In the Hadamard basis.*

(c) *No measurement is required. For those states random guessing is optimal.*

5.1.2 Fidelity

A second common measure for closeness of states is known as the fidelity. The fidelity has an intuitive interpretation that applies to a situation where we want to verify whether we have managed to produce a desired target state $|\psi\rangle$. Suppose that we want to build a machine that produces $|\psi\rangle\langle\psi|$, yet we are only able to produce some state ρ. Let us suppose that, having prepared ρ, we perform a measurement on it to check whether we have prepared the correct state $|\psi\rangle$. We can do this (theoretically) by performing the two-outcome measurement

$$M_{\mathrm{succ}} = |\psi\rangle\langle\psi|,$$

$$M_{\text{fail}} = \mathbb{I} - |\psi\rangle\langle\psi| \ .$$

The fidelity between the actual output state ρ and the target state $|\psi\rangle$ is defined as a function of the success probability of this measurement by

$$F(|\psi\rangle, \rho) = \sqrt{\text{tr}(M_{\text{succ}}\rho)} = \sqrt{\langle\psi|\rho|\psi\rangle} \ . \tag{5.3}$$

More generally, we define the fidelity between two density matrices as follows. Note that some authors also use the square of (5.3) as the fidelity and it is advisable to carefully check the definition in any paper where you see the fidelity being used.

Definition 5.1.3 (Fidelity). *Given density matrices ρ_1 and ρ_2, the fidelity between ρ_1 and ρ_2 is*

$$F(\rho_1, \rho_2) = \text{tr}\left(\sqrt{\sqrt{\rho_1}\rho_2\sqrt{\rho_1}}\right) \ .$$

For pure states $\rho_1 = |\psi_1\rangle\langle\psi_1|$ and $\rho_2 = |\psi_2\rangle\langle\psi_2|$ the fidelity takes on the simplified form

$$F(\rho_1, \rho_2) = |\langle\psi_1|\psi_2\rangle| \ .$$

If only one of the states $\rho_1 = |\psi_1\rangle\langle\psi_1|$ is pure, we have

$$F(\rho_1, \rho_2) = \sqrt{\langle\psi_1|\rho_2|\psi_1\rangle} \ ,$$

which corresponds to the formula given in (5.3).

Proposition 5.1.3 *For any two quantum states ρ, σ, the fidelity satisfies the following properties:*

1. *Normalization: $0 \leq F(\rho, \sigma) \leq 1$.*
2. *Symmetry: $F(\rho, \sigma) = F(\sigma, \rho)$.*
3. *Multiplicativity under tensor product: $F(\rho_1 \otimes \rho_2, \sigma_1 \otimes \sigma_2) = F(\rho_1, \sigma_1) \cdot F(\rho_2, \sigma_2)$.*
4. *Invariance under unitary operations: $F(\rho, \sigma) = F(U\rho U^\dagger, U\sigma U^\dagger)$.*
5. *Relation to trace distance: $1 - F(\rho, \sigma) \leq D(\rho, \sigma) \leq \sqrt{1 - F^2(\rho, \sigma)}$. Conversely, we also have that $1 - D(\rho, \sigma) \leq F(\rho, \sigma) \leq \sqrt{1 - D^2(\rho, \sigma)}$.*

The third property, multiplicativity under tensor products, is particularly useful. Since the trace distance does not satisfy this condition, when evaluating the distance between states that have a tensor product form it is common to use the fifth property relating trace distance and fidelity in combination with multiplicativity of the fidelity.

5.2 What It Means to Be Ignorant

Suppose that we have designed a quantum protocol such that Alice's outcome in the protocol is a classical n-bit string k that will ultimately form part of a secret key, such as for use in the one-time pad. (In particular, we may also want another party, Bob, to obtain the same k. In this chapter we concentrate on Alice.) If Alice obtains the string k, we can represent it as a quantum state $|k\rangle\langle k|_K$. More generally, if we know that Alice obtains k with probability p_k, then the state of her key is $\sum_k p_k |k\rangle\langle k|_K$.

However, due to possible eavesdropping during the protocol, Alice's string may be correlated with information that is held by a third party, Eve. In general, we can model Eve's so-called *side information* as a quantum state ρ_k^E that depends on k. Then the joint state of Alice's string k and Eve's side information about it can be expressed as

$$\rho_{KE} = \sum_{k \in \{0.1\}^n} p_k |k\rangle\langle k|_K \otimes \rho_k^E , \qquad (5.4)$$

which is an example of a cq-state. Given such a state, how do we quantify the "security" of the string k in system K, in terms of how safely k can be used as a secret key? Informally, we want the classical string in system K to be uniformly random and uncorrelated with Eve's system. Before we arrive at a formal definition, let us first look at a few examples, where for simplicity we consider just a single bit of key.

Example 5.2.1 Consider the state

$$\rho_{KE} = \frac{1}{2} \sum_{k \in \{0.1\}} |k\rangle\langle k|_K \otimes |k\rangle\langle k|_E . \qquad (5.5)$$

Clearly, we have $\rho_K = \mathrm{tr}_E (\rho_{KE}) = \mathbb{I}_K / 2$. This means that if we look only at the key by itself, then it is uniformly random. But clearly Eve knows everything about the key: whenever K is in the state $|k\rangle\langle k|$, then so is E! In this example the information that Eve has is an exact classical copy of k. States of the form (5.5) are called *classically maximally correlated states*. Both systems are diagonal in the standard basis, and both systems are prepared precisely in the same state $|k\rangle\langle k|$ with some probability. ∎

Example 5.2.2 Consider the state $\rho_{KE} = |0\rangle\langle 0|_K \otimes \rho_E$. In this case Eve is clearly uncorrelated with the key, because the state is a tensor product and $\rho_k^E = \rho_E$ does not depend on k. However, ρ_K is certainly not uniform! In fact, the only possible key here is $k = 0$, which is easy to guess for anyone. Completely insecure! ∎

Example 5.2.3 Consider the EPR pair

$$|\mathrm{EPR}\rangle_{KE} = \frac{1}{\sqrt{2}} (|0\rangle_K |0\rangle_E + |1\rangle_K |1\rangle_E)$$

and let $\rho_{KE} = |\mathrm{EPR}\rangle\langle\mathrm{EPR}|_{KE}$. As you know, here $\rho_K = \mathrm{tr}_E (\rho_{KE}) = \mathbb{I}/2$. That is, the key K is uniform. But is it uncorrelated from Eve? Clearly it is not: we know that no matter what basis we measure K in there always exists a corresponding measurement on E that yields the same outcome. This is because for all unitaries U,

$$\begin{aligned}
U_K \otimes U_E^* \ket{\text{EPR}}_{KE} &= (U_K \otimes \mathbb{I}_E)(\mathbb{I}_K \otimes U_E^*) \ket{\text{EPR}}_{KE} \\
&= (U_K \otimes \mathbb{I}_E)((U_K^*)^T \otimes \mathbb{I}_E) \ket{\text{EPR}}_{KE} \\
&= (U_K \otimes \mathbb{I}_E)(U_K^\dagger \otimes \mathbb{I}_E) \ket{\text{EPR}}_{KE} \\
&= (U_K U_K^\dagger \otimes \mathbb{I}_E) \ket{\text{EPR}}_{KE} \\
&= \ket{\text{EPR}}_{KE} \, ,
\end{aligned}$$

where in the second equality we have used that for the state $\ket{\text{EPR}}_{KE}$ and for any U,

$$(\mathbb{I}_K \otimes U_E) \ket{\text{EPR}}_{KE} = (U_K^T \otimes \mathbb{I}_E) \ket{\text{EPR}}_{KE} \, .$$

Therefore, the corresponding measurement on E is simply to measure in the basis defined by U_E^* (i.e. the basis in which U_E^* is diagonalized). ∎

Based on intuition gained from these examples we give the following definition.

Definition 5.2.1 (Ignorant). *Let ρ_{KE} be a cq-state, where K is an n-bit classical register. We say that Eve (holding system E) is* ignorant *about the key K if and only if*

$$\rho_{KE} = \rho_{KE}^{\text{ideal}} = \frac{1}{2^n} \mathbb{I}_K \otimes \rho_E \, . \tag{5.6}$$

That is, the key is uniform *and* uncorrelated *from Eve.*

In any actual implementation, we can never hope to attain the perfection given by the state in Eq. (5.6). However, we can hope to get close to such an ideal state, motivating the following definition.

Definition 5.2.2 (Almost ignorant). *Let ρ_{KE}^{real} be a cq-state, where K is an n-bit string, and $\varepsilon \geq 0$. We say that Eve (holding system E) is ε-ignorant about the key K if*

$$D\left(\rho_{KE}^{\text{real}}, \rho_{KE}^{\text{ideal}}\right) \leq \varepsilon \, , \tag{5.7}$$

where as in Definition 5.2.1 we define $\rho_{KE}^{\text{ideal}} = \frac{1}{2^n} \mathbb{I}_K \otimes \rho_E$.

Why is this a good definition? Recall from the previous section that the trace distance measures how well it is possible to optimally distinguish between two quantum states. We saw that if two states are ε-close in trace distance then no measurement can tell them apart with an advantage more than $\varepsilon/2$. This has important consequences if we want to later use the key in another protocol; for example, in an an encryption protocol such as the one-time pad. Recall from Chapter 2 that an encryption scheme is secret/secure if and only if for all prior distributions $p(m)$ over messages, and for all messages m, we have $p(m) = p(m|c)$, where c denotes the ciphertext. Such a secrecy can be achieved using the one-time pad, if Eve is completely ignorant about the key. You may think of the one-time pad scheme as a type of measurement to distinguish ρ_{KE}^{ideal} and ρ_{KE}^{real}. If the security of the protocol was very different if we used ρ_{KE}^{real} instead of the ideal ρ_{KE}^{ideal}, then any "attack" by an adversary in the protocol would give a means to distinguish the two states. This is precisely ruled out if the states are close in trace distance.

QUIZ 5.2.1 *Suppose that the state of the key and Eve is given by $\rho_{KE} = |0\rangle\langle 0|_K \otimes \rho_E$. Is Eve ignorant about the key K?*

(a) *Yes*
(b) *No*

QUIZ 5.2.2 *Suppose that the state of the key and Eve is given by:*

$$\rho_{KE} = \frac{1}{2}\left(|0\rangle\langle 0|_K \otimes |0\rangle\langle 0|_E + |1\rangle\langle 1|_K \otimes |0\rangle\langle 0|_E\right).$$

Is Eve ignorant about the key?

(a) *Yes*
(b) *No*

5.3   Measuring Uncertainty: The Min-Entropy

In the previous section we measured the quality of Alice's key k by the distance between the state ρ_{KE}^{real} and an "ideal" state ρ_{KE}^{ideal}. In general, this distance may be very close to maximal, even in cases where intuitively the key k in register K is "largely independent" of the quantum state in register E. For example, suppose that K is 100 bits long, and E contains the first two bits of K only. Then, this situation is almost perfectly distinguishable from the "ideal" situation where E has no information at all. (A distinguishing measurement would look at the first two bits of the state and check if they agree with Eve's side information or not; if so, it is very likely that we are in the "real" situation.) However, in such a situation we'd still like to say that, in some respects, we are "in good shape": sure, Eve knows two bits, but there are still 98 bits on which she has no information at all!

This example motivates us to find a more refined measure of uncertainty about the classical string k than the trace distance with the ideal situation. We will achieve this by introducing an appropriate measure of *conditional entropy* of the classical random variable K, conditioned on the quantum side information E. To clarify that we now consider general distributions over strings, and not necessarily strings that we plan to use right away as key, from now on we use the less suggestive letter X to represent the classical information, instead of K in the previous section.

5.3.1 Entropy

Let us first consider a simple scenario where Eve has no side information, but Alice's string is not necessarily uniform either. That is, we look at the classical state

$$\rho_X = \sum_x p_x |x\rangle\langle x|_X , \tag{5.8}$$

which is our way of representing the probability distribution $\{p_x\}_x$ over strings x. How could we measure Eve's uncertainty about X?

When discussing communication tasks, the most useful measure of uncertainty is the Shannon entropy (also called von Neumann entropy in the context of quantum information) defined as

$$H(X) = -\sum_x p_x \log p_x ,$$

where Shannon chose the binary logarithm, motivated by the fact that information to be transmitted is commonly encoded using two bits "0" and "1". For the special case where $x \in \{0, 1\}$ the Shannon entropy is also often called the *binary entropy*, expressed as $h(p_0) = -p_0 \log p_0 - (1 - p_0) \log(1 - p_0)$. We already encountered this measure in Box 4.1, when discussing different ways of measuring the entanglement present in a pure bipartite state.

Is the Shannon entropy also a useful measure of uncertainty in the context of cryptography? To investigate this question consider the following scenario. Suppose Alice purchased a box (possibly from Eve!) which generates a string $x = x_1, \ldots, x_n$ when she presses the "ON" button. If the string was uniformly random, then for all x, $p_x = 1/2^n$ and the Shannon entropy is $H(X) = n$. Suppose now that, while we are promised that x is uncorrelated from Eve, the distribution p_x is *not* uniform. However, we are guaranteed that the entropy is still $H(X) \approx n/2$. But suppose that we know nothing else about the box, except for this fact about the entropy of the distribution under which it generates a string. Would you still be willing to use x as an encryption key?

At first sight the situation might not look too bad. After all, while the string does not have maximum entropy $H(X) = n$, it still has half as much entropy, which for very large n is still large. Intuitively, this should mean that the eavesdropper, Eve, still has a lot of uncertainty about X, shouldn't it?

Let us consider the following distribution as an example:

$$p_x = \begin{cases} \frac{1}{2} & \text{for } x = 11 \ldots 1 \\ \frac{1}{2} \cdot \frac{1}{2^n - 1} & \text{otherwise} . \end{cases} \tag{5.9}$$

 Exercise 5.3.1 Show that the entropy of X with distribution $\{p_x\}_x$ is $H(X) \approx n/2$.

So this distribution has large entropy. But does it have a lot of uncertainty? Note that the probability that the box generates the string $x = 11 \ldots 1$ is $1/2$, independent of the length of the string. This means that if we attempt to use x as an encryption key, Eve will be able to guess the key, and thus decrypt any message encrypted with it, with probability $1/2$. The problem here is that this particular string has high probability of being returned by the box, which is not secure at all.

The example shows that the Shannon entropy is not a good measure of uncertainty for cryptography. Luckily, there exists an alternate measure of entropy which turns out to be much more useful for our purposes. Inituively, instead of measuring the "average" uncertainty as for the case of the Shannon entropy, this new measure considers the "worst-case" uncertainty that is present in a distribution.

> **Definition 5.3.1 (Min-entropy).** *Given a random variable X with distribution $\{p_x\}_x$ the* min-entropy H_{min} *of X is defined as*
>
> $$H_{min}(X) = -\log\left(\max_x p_x\right).$$

For the uniform distribution on n bits, we get $H_{min}(X) = H(X) = n$. However, for the distribution in example in (5.9) we get that $H_{min}(X) = -\log 1/2 = 1$. That is, in this case the min-entropy is tiny, which reflects our observation on lack of security. Looking at it more closely, note that the min-entropy precisely captures our intuitive idea of what it means for Eve to be uncertain about X. In the example Eve could guess the string output by the box with probability $1/2$; this is a constant (independent of n) and so the min-entropy is constant. In general, the best guessing strategy for Eve is to guess the most likely string, so the maximum success probability that she has in guessing the output of the box is precisely $P_{guess}(X) = \max_x p_x$. This observation shows that the min-entropy has a neat operational interpretation as

$$H_{min}(X) = -\log P_{guess}(X).$$

 Remark 5.3.1 You may wonder why the min-entropy is not the right measure of uncertainty for communication tasks. This is because for communication one usually considers the case where the states that are communicated take the form $\rho^{\otimes n}$, where n tends to infinity. That is, when developing his theory of information Shannon considered what happens when the users can repeatedly use the same communication channel a large number of times. In this setting, Shannon's idea for the right way to measure uncertainty was to consider $i(x) := -\log p_x$ as a measure of "surprisal," that is, a measure of the amount of information gained by observing x. This led him to introduce the Shannon entropy as a measure of the *average* surprisal $H(X) = \sum_x p_x i(x)$. When doing cryptography, however, we are always interested in the worst case, not the average case. The min-entropy $H_{min}(X) = \min_x i(x)$ is precisely this "smallest surprisal." Figure 5.1 shows the difference between these quantities for the case of a binary random variable.

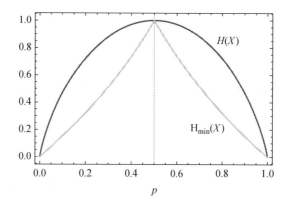

Fig. 5.1   Comparison between Shannon entropy $H(X)$ and min-entropy $H_{min}(X)$ for a binary random variable X.

Exercise 5.3.2   Show that the min-entropy of a discrete random variable X satisfies the following bounds:

$$0 \leq H_{min}(X) \leq H(X) \leq \log |X| \,,$$

where $|X|$ denotes the number of possible values that X can take.

QUIZ 5.3.1 *Consider a device that emits $X = 0$ with probability $p_0 = 3/4$ and $X = 1$ with probability $p_1 = 1/4$. What is the min-entropy of this distribution?*

(a) $H_{min}(X) = 0$
(b) $H_{min}(X) = 2 - \log 3$
(c) $H_{min}(X) = 2 - \frac{3}{4}\log 3$
(d) $H_{min}(X) = 1$

5.3.2   The Conditional Min-Entropy

Let's now consider the general case where Eve has access to a quantum state ρ_x^E that is correlated with x. How can we quantify the uncertainty about X *given the extra quantum state in register E*? Motivated by the interpretation of the min-entropy as a guessing probability, it is natural to introduce the conditional min-entropy $H_{min}(X|E)$ as the maximum probability that Eve manages to guess X, *given access to her quantum register E*. To evaluate the conditional min-entropy we need to answer the following question: Given ρ_x^E with probability p_x, what is Eve's best chance at guessing x by making a measurement on E? This is a generalization of the problem of distinguishing quantum states that we considered earlier, where there can be more than two quantum states to distinguish between.

Definition 5.3.2 (Conditional min-entropy). *Consider a cq-state ρ_{XE}. The conditional min-entropy $H_{min}(X|E)$ is defined as*

$$H_{min}(X|E)_{\rho_{XE}} := -\log P_{guess}(X|E) , \tag{5.10}$$

where $P_{guess}(X|E)$ is the maximum probability that Eve guesses x. Precisely,

$$P_{guess}(X|E) := \max_{\{M_x\}_x} \sum_x p_x \operatorname{tr}\left(M_x \rho_x^E\right) , \tag{5.11}$$

where the maximization is taken over all POVMs $\{M_x \geq 0 \mid \sum_x M_x = \mathbb{I}\}$ on E. When it is clear from context, we omit the subscript ρ_{XE}, i.e. we write $H_{min}(X|E) = H_{min}(X|E)_{\rho_{XE}}$.

Note that the definition of the min-entropy involves a maximization over all possible POVMs. In general, this could be hard to compute! When $x \in \{0,1\}$ takes on only two values then the problem has a solution, which we already figured out in Section 5.1.1: in this case the guessing probability P_{guess} is directly related to the trace distance $D(\rho_0^E, \rho_1^E)$.

Example 5.3.1 Consider the state $\rho_{XE} = \frac{1}{2}|0\rangle\langle 0|_X \otimes |0\rangle\langle 0|_E + \frac{1}{2}|1\rangle\langle 1|_X \otimes |+\rangle\langle +|_E$. Then the conditional min-entropy $H_{min}(X|E) = -\log P_{guess}(X|E)$ where

$$P_{guess}(X|E) = \max_{\substack{M_1, M_2 \geq 0 \\ M_1 + M_2 = \mathbb{I}}} \frac{1}{2} \operatorname{tr}(M_0|0\rangle\langle 0|_E) + \frac{1}{2} \operatorname{tr}(M_1|+\rangle\langle +|_E)$$

$$= \max_{0 \leq M \leq \mathbb{I}} \frac{1}{2} \operatorname{tr}(M|0\rangle\langle 0|_E) + \frac{1}{2} \operatorname{tr}(|+\rangle\langle +|_E) - \frac{1}{2} \operatorname{tr}(M|+\rangle\langle +|_E)$$

$$= \frac{1}{2} + \frac{1}{2} \max_{0 \leq M \leq \mathbb{I}} \operatorname{tr}\left(M(|0\rangle\langle 0|_E - |+\rangle\langle +|_E)\right)$$

$$= \frac{1}{2} + \frac{1}{2} D(|0\rangle\langle 0|_E, |+\rangle\langle +|_E) . \qquad \blacksquare$$

If X can take more than two possible values then it is in general difficult to compute $P_{guess}(X|E)$ by hand. However, the optimal success probability can be expressed as a *semidefinite program* (SDP). An SDP is a convex program that generalizes linear programs (LP).[1] In particular, the optimum of an SDP can in general be approximated efficiently, in time polynomial in the dimension of the states ρ_x^E. Programming languages oriented towards linear algebra, such as Matlab or Julia, generally have packages that allow you to do this. The fact that the conditional min-entropy can be evaluated numerically is important for cryptographic applications, because this computation can be needed to set the parameters of certain components of a cryptographic protocol, as we'll see later when discussing the use of randomness extractors for privacy amplification. In Section 6.4.4 we'll also see a different technique which

---

1 Semidefinite programming is an important optimization technique in quantum information, as well as a useful mathematical tool. While it is outside the scope of this book to introduce this technique, we will encounter it again briefly in Section 10.5.

BOX 5.2   The Chain Rule

The Shannon entropy also has a conditional variant, which is simply defined as $H(A|B) = H(AB) - H(B)$. With this definition it is easy to verify that the conditional Shannon entropy has a very useful property, which is called the chain rule: $H(A|BC) = H(AB|C) - H(B|C)$. This equation can be verified by expanding all terms using the definition, and observing cancellations.

Unfortunately the conditional min-entropy does not satisfy such a nice chain rule. However, it satisfies the following "partial" chain rule:

$$H_{\min}(A|BC) \geq H_{\min}(AB|C) - \log|B| , \tag{5.12}$$

where we used $|B|$ to denote the dimension of the system B (i.e. in the case of qubits, the number of qubits of B is $\log|B|$). This relation is valid whenever A, B, C are arbitrary quantum registers. We will frequently use the following consequence of it: if X is a classical register and E a quantum register, then

$$H_{\min}(X|E) \geq H_{\min}(X) - \log|E| , \tag{5.13}$$

which says that giving a small number of bits, or qubits, to an adversary (the register E) cannot increase too much their knowledge about a certain partial secret X. Equation (5.13) follows from (5.12) because $H_{\min}(XE) \geq H_{\min}(X)$ as a special case of the data-processing inequality (see Box 6.1).

allows us to approximate $P_{\text{guess}}(X|E)$, as well as determine a near-optimal guessing measurement.

Exercise 5.3.3   Show that for any cq-state ρ_{XE} we have

$$0 \leq H_{\min}(X|E) \leq \log|X| .$$

QUIZ 5.3.2 *Consider a cq-state*

$$\rho_{XE} = \frac{1}{4}|0\rangle\langle 0|_X \otimes |+\rangle\langle +|_E + \frac{3}{4}|1\rangle\langle 1|_X \otimes |-\rangle\langle -|_E . \tag{5.14}$$

What is the conditional min-entropy $H_{\min}(X|E)$ *of this state?*

(a) $H_{\min}(X|E) = 0$

(b) $H_{\min}(X|E) = 2 - \log 3$

(c) $H_{\min}(X|E) = 2 - \frac{3}{4}\log 3$

(d) $H_{\min}(X|E) = 1$

BOX 5.3   Smoothed Min-Entropy

Due to imperfections in a protocol or algorithm we often do not exactly produce the state ρ_{XE} that we want: rather, we may only manage to produce a state ρ'_{XE} that is close, and we may not even know the exact form of ρ'_{XE} (other than the fact that it is ε-close to ρ_{XE}). Due to this uncertainty it is usually more physically relevant to look at the smoothed min-entropy, which gives the maximum value of $H_{\min}(X|E)$ over all states $\rho'_{AE} \in \mathcal{B}^\varepsilon(\rho_{AE})$. Formally, the smoothed conditional min-entropy $H^\varepsilon_{\min}(X|E)$ is defined as

$$H^\varepsilon_{\min}(X|E)_\rho = \max_{\rho' \in \mathcal{B}^\varepsilon(\rho)} H_{\min}(X|E)_{\rho'} \, .$$

Here we used the trace distance as the notion of closeness between ρ' and ρ. Sometimes other measures, such as the fidelity, are used. Up to small adjustments in the parameter ε, the exact choice of a distance measure does not change the qualitative properties of the smooth conditional min-entropy.

Optional Section: A General Quantum Conditional Min-Entropy

In full generality we may be interested in quantifying uncertainty about *quantum* information, i.e. in a setting where the system X is also quantum. We use A, instead of X, to label such a quantum system. How should we measure the conditional min-entropy $H_{\min}(A|E)$? To gain intuition on how this could be defined, think of the guessing probability as a way of quantifying how close it is possible for Eve, holding system E, to put herself in a state that is *maximally correlated* with the classical system X, by guessing it correctly and holding on to her guess. A natural quantum extension of this idea is to get as close as possible to the maximally entangled state between A and E while only allowing one to perform operations on E. Here is a more formal definition.

Definition 5.3.3 (Quantum conditional min-entropy). *For any bipartite density matrix* ρ_{AE}, *let* $|A|$ *be the dimension of A and define*

$$\mathrm{Dec}(A|E) = \max_{\Lambda_{E \to A'}} F\big((\mathbb{I}_A \otimes \Lambda_{E \to A'})\rho_{AE}, |\phi\rangle\langle\phi|_{AA'}\big)^2 \, ,$$

where A' is a system that has the same dimension as A and

$$|\phi\rangle_{AA'} = \frac{1}{\sqrt{|A|}} \sum_{a=1}^{|A|} |a\rangle_A \otimes |a\rangle_{A'}$$

is the maximally entangled state between A and A', the maximization is performed over all quantum maps Λ mapping system E to A' (see Definition 3.3.1), and the function F is the fidelity from Definition 5.1.3. Then the conditional min-entropy of A, conditioned on E, is

$$H_{\min}(A|E) = -\log\big(|A| \cdot \mathrm{Dec}(A|E)\big) \, .$$

Remark 5.3.2 An alternative way to express the quantum conditional min-entropy is as

$$H_{\min}(A|E) = \max_{\sigma_B} \sup \left\{ \lambda \in \mathbb{R} \,\middle|\, \rho_{AB} \leq 2^{-\lambda} \mathbb{I}_A \otimes \sigma_B \right\}. \tag{5.15}$$

The equivalence between the two definitions is not obvious, but it can be shown using duality of semidefinite programming.

5.4   Uncertainty Principles: A Bipartite Guessing Game

Now that we have a solid measure of ignorance, or uncertainty, how do we make use of it? Our goal is to find situations where Alice can certify that a certain string x that she generated must be at least partially unknown to an eavesdropper, Eve. In this section and the next we start building some intuition on how to achieve this by studying simple "games" in which Eve's uncertainty can be controlled. For now, we study these games for their own sake, but later, we will see how their analysis can be used to show security of more complex cryptographic protocols for secret key distribution.

Intuitively, the fundamental principle of quantum mechanics that will allow us to achieve our goal is called the *uncertainty principle*. Very informally, this principle allows us to limit how well Eve can predict the outcomes of different incompatible measurements on Alice's state. The games that we study in this section and the next provide different ways to formalize the uncertainty principle.

As a first step, we consider eavesdroppers that only have the ability to store and process classical information. Because Eve has less power, the setting will be easier to analyze. We will see in the next section how to handle the general setting of a quantum eavesdropper. To set things up, consider the following guessing game.

> **Definition 5.4.1 (Guessing game – Alice and Eve).** *Suppose two parties, Alice and Eve, play the following game (Figure 5.2).*
>
> 1. *Eve prepares a qubit in an arbitrary state ρ_A and sends it to Alice.*
> 2. *Alice chooses a bit $\Theta \in \{0,1\}$ uniformly at random.*
> 3. *If $\Theta = 0$ then Alice measures ρ_A in the standard basis. If $\Theta = 1$ then she measures in the Hadamard basis. She obtains a measurement outcome $X \in \{0,1\}$.*
> 4. *Alice announces Θ to Eve.*
> 5. *Eve wins if she can correctly guess the bit X.*

Suppose that you play the role of Eve in this guessing game, and that Alice plays exactly as described in the game. How should you choose the state ρ_A to maximize your chances of success? For example, you could choose $\rho_A = |0\rangle\langle 0|$. If Alice chooses $\Theta = 0$ then you can predict the outcome – since she measures in the standard basis it is simply "0". But if she chooses $\Theta = 1$, then she measures in the Hadamard basis

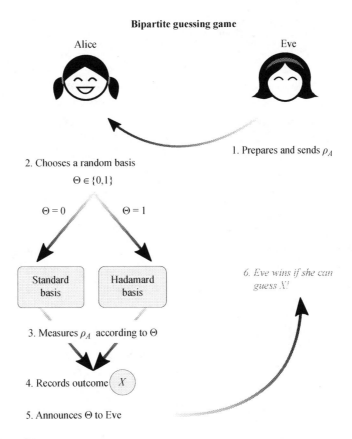

Fig. 5.2   The guessing game between Alice and Eve. Eve prepares a quantum state and sends it to Alice, who chooses randomly to measure in the standard basis or in the Hadamard basis. Eve then tries to guess Alice's measurement outcome, given the basis Alice chose.

and her outcome is uniformly random from your point of view – the best you could do in that case is make a random guess. To avoid this you could choose $\rho_A = |+\rangle\langle+|$, but then the situation is inverted: while in case $\Theta = 1$ you have perfect information, in case $\Theta = 0$ there is nothing better that you can do than make a random guess.

Intuitively the difficulty is that, because the two measurements that Alice can perform are in "incompatible" bases, there does not seem to be a state that Eve can prepare such that she would be able to know a priori what the outcome should be in both bases. Let's push this a bit further and analyze formally what is the best that Eve can do in the game. In general, Eve's maximum success probability is

$$P_{\text{guess}}(X|\Theta) = p(\Theta = 0) \cdot P_{\text{guess}}(X|\Theta = 0) + p(\Theta = 1) \cdot P_{\text{guess}}(X|\Theta = 1)$$

$$= \frac{1}{2} \cdot \left(P_{\text{guess}}(X|\Theta = 0) + P_{\text{guess}}(X|\Theta = 1) \right), \tag{5.16}$$

where the second equality holds since Alice makes her choice of basis Θ uniformly at random.

Note that Eve has to make a guess for each case, $\Theta = 0$ and $\Theta = 1$, when she is asked. Since we assumed that Eve has only classical information, we can record her

---

BOX 5.4   Uncertainty Principle

In general, an uncertainty principle states that, given a certain collection of measurement operators, there does not exist a quantum state such that the outcome of measuring the state using any of the measurements is deterministic. The famous Heisenberg uncertainty principle applies to the position and momentum measurements on a quantum particle such as a photon, and states that no photon can be in a state that has a determined position and momentum. In this section we showed an uncertainty principle for the standard and Hadamard bases: we proved that there does not exist a single-qubit state that provides a deterministic outcome in both bases. This is because otherwise Eve could win in the bipartite guessing game with certainty by preparing that state! Moreover, we even proved a quantitative bound on the minimal amount of uncertainty required in any state: on average over the two bases, the unpredictability of the measurement outcome must be at least the min-entropy of $\frac{1}{2} + \frac{1}{2\sqrt{2}}$, which is approximately 0.2.

---

guess for each of the two possibilities ahead of time. By symmetry, we can assume that her guess in each case is "$X = 0$." Thus her maximum success probability is the maximum over all ρ_A of the chance that "$X = 0$" is actually the correct guess. In other words, continuing from (5.16) we get

$$P_{\text{guess}}(X|\Theta) = \frac{1}{2} \cdot \left(\text{tr}(\rho_A |0\rangle\langle 0|) + \text{tr}(\rho_A |+\rangle\langle +|) \right)$$
$$= \frac{1}{2} \cdot \text{tr}\left(\rho_A (|0\rangle\langle 0| + |+\rangle\langle +|) \right). \tag{5.17}$$

To determine the maximum value that this expression can take, we need to maximize over all possible choices of ρ_A made by Eve. Since the expression is linear in ρ_A, and every density matrix is a convex combination of pure states, the maximum will always be attained at a pure state. If $\rho_A = |\psi\rangle\langle\psi|_A$ is pure, then we get

$$P_{\text{guess}}(X|\Theta) = \frac{1}{2} \langle \psi|_A \left(|0\rangle\langle 0| + |+\rangle\langle +| \right) |\psi\rangle_A. \tag{5.18}$$

The maximum over all pure states $|\psi\rangle_A$ of this expression is precisely the largest eigenvalue of the 2×2 matrix $|0\rangle\langle 0| + |+\rangle\langle +|$. It is not hard to do the computation and obtain that $\lambda_{\max} = 1 + \frac{1}{\sqrt{2}}$. Therefore, we have that $P_{\text{guess}}(X|\Theta) = \frac{1}{2} + \frac{1}{2\sqrt{2}} \approx 0.85 < 1$.

Exercise 5.4.1   Write down explicitly the state ρ_A that Eve should prepare in order to succeed in the guessing game with probability exactly $\frac{1}{2} + \frac{1}{2\sqrt{2}}$.

Let's now consider a more general scenario where Eve may keep classical information about ρ_A. In other words, we allow Eve to prepare an arbitrary cq-state $\rho_{AC} = \sum_c p_c \rho_c^A \otimes |c\rangle\langle c|_C$ according to some distribution $\{p_c\}_c$, and send the qubit in A to Alice while keeping the classical system C. Let us convince ourselves that this scenario does not make any difference: it does not help Eve win with higher probability in the game. Indeed, by linearity the guessing probability conditioned on C is given by the average

$$P_{\text{guess}}(X|\Theta C)_{\rho_{AC}} = \sum_c p_c\, p_{\text{guess}}(X|\Theta)_{\rho_c^A} \,. \tag{5.19}$$

Since we have already computed the maximum possible value of $p_{\text{guess}}(X|\Theta)_{\rho_c^A}$, over all possible ρ_c^A, we get that here also

$$P_{\text{guess}}(X|\Theta C)_{\rho_{AC}} \leq \frac{1}{2}\left(1 + \frac{1}{\sqrt{2}}\right) \approx 0.85 \,.$$

The quantity $P_{\text{guess}}(X|\Theta C)$ allows us to directly compute the conditional min-entropy of Alice's outcome X, since by definition

$$\mathrm{H}_{\min}(X|\Theta C) = -\log P_{\text{guess}}(X|\Theta C) \approx 0.22 \,.$$

Let us make one more step and give Eve yet more power. Suppose that she can now create an arbitrary quantum state ρ_{AE}, possibly entangled, and send only the qubit in A to Alice.

Exercise 5.4.2 Show that if Eve can keep entanglement then there is a strategy that allows her to always win the game with probability 1. *[Hint: what's your favorite two-qubit state?]*

The exercise shows that our assumption that Eve only keeps classical information was not only for convenience: it is also necessary for security! As soon as we allow Eve to maintain entanglement with ρ_A then she may be able to guess Alice's outcome X *perfectly*.

How do we get around this? Here is the key: remember from Chapter 4 that entanglement is monogamous! In order to limit Eve's knowledge about Alice's measurement outcomes we will use *two* aspects of quantum mechanics:

- Uncertainty: If Eve has no (or little) entanglement with Alice, then she cannot predict the outcomes of two incompatible measurements. As we showed in this section, it is difficult for her to guess Alice's measurement outcomes, i.e. $P_{\text{guess}}(X|E\Theta) < 1$.
- Monogamy: If we ensure that there is a large amount of entanglement between Alice and some additional party Bob, then we know that Eve can have only very little entanglement with either Alice or Bob.

QUIZ 5.4.1 *Alice and Eve play the bipartite guessing game. What is Eve's optimal guessing probability if she prepares the state $\rho_A = |0\rangle\langle 0|$ for Alice?*

(a) $\frac{1}{2}$

(b) $\frac{3}{4}$

(c) $\frac{1}{2}\left(1+\frac{1}{\sqrt{2}}\right)$

(d) 1

QUIZ 5.4.2 *What is her optimal guessing probability if $\rho_A = |+\rangle\langle+|$?*

(a) $\frac{1}{2}$

(b) $\frac{3}{4}$

(c) $\frac{1}{2}\left(1+\frac{1}{\sqrt{2}}\right)$

(d) 1

5.5   Extended Uncertainty Relation Principles: A Tripartite Guessing Game

In order to make use of the monogamy property of entanglement, we consider an extension of the guessing game from the previous section. This time we impose no constraint on the presence or absence of entanglement between Alice and Eve. Instead, we introduce a third party, Bob, who Alice trusts. In particular, to show security against Eve, Alice and Bob may join forces to make an estimate of the amount of entanglement that Eve may be sharing with Alice. To do so, they perform an "entanglement test" to ensure that they share entanglement between themselves, in a way that will then guarantee that the entanglement between Alice and Eve is small. Here is the new formulation of the guessing game.

Definition 5.5.1 (Tripartite guessing game – Alice, Bob, and Eve). *Suppose that three parties, Alice, Bob, and Eve, play the following game (Figure 5.3).*

1. *Eve prepares an arbitrary state ρ_{ABE} such that A and B are both qubits. She sends qubit A to Alice and qubit B to Bob.*
2. *Alice chooses a bit $\Theta \in \{0,1\}$ uniformly at random.*
3. *If $\Theta = 0$, then Alice measures ρ_A in the standard basis; if $\Theta = 1$, then she measures in the Hadamard basis. She obtains a measurement outcome $X \in \{0,1\}$ and records it.*
4. *Alice announces Θ to both Bob and Eve.*
5. *Given Θ, Bob measures ρ_B in the basis Θ and obtains an outcome $\tilde{X}$. Eve measures ρ_E in any way that she likes, and makes a guess X_E.*
6. *Eve wins the game if $X_E = X = \tilde{X}$.*

Suppose that you try to design the best strategy for Eve in this game. How well can you do? We can express Eve's success probability as

$$p_{\text{succ}} = p(X = \tilde{X} = X_E) = \sum_{\Theta \in \{0,1\}} p_\Theta p(X = \tilde{X} = X_E | \Theta)$$

$$= \frac{1}{2} \sum_{\theta \in \{0,1\}} \text{tr}\left(\rho_{ABE}\left(\sum_{x \in \{0,1\}} |x\rangle\langle x|_\theta^A \otimes |x\rangle\langle x|_\theta^B \otimes M_{x|\theta}^E\right)\right), \quad (5.20)$$

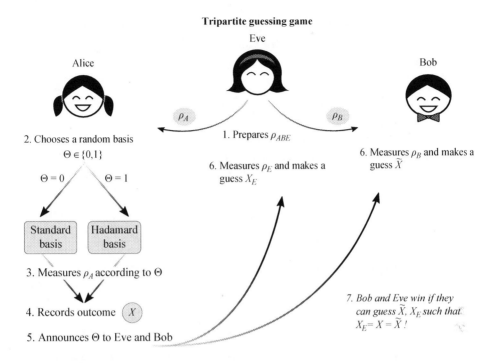

Fig. 5.3   A tripartite guessing game where Eve gets to prepare the global state ρ_{ABE}. She send the qubits A and B to Alice and Bob respectively. Alice measures randomly in either the standard or Hadamard basis and obtains an outcome X. Bob measures in the same basis as Alice and obtains an outcome $\tilde{X}$. Eve provides a guess X_E. We say that Eve wins the game if $X_E = X = \tilde{X}$.

where we used superscripts A, B, and E to denote the systems on which we perform the measurements, and $|x\rangle_\theta$ to denote basis element x in the basis θ. Of course, the difficulty is that we don't know anything a priori about the state ρ_{ABE} or the measurement $\{M^E_{x|\theta}\}_x$ with outcomes x that Eve will perform on E depending on the basis θ. We only know that this must be a quantum state, and Eve can only make measurements that are allowed by the laws of quantum mechanics. Since any POVM can be realized as a projective measurement using a potentially larger ancilla (see Box 3.2), we can without loss of generality assume that Eve's measurements are projective. Indeed, giving her access to a smaller space only makes things more difficult for Eve, and in a security analysis we are always allowed to make the adversary more (but not less!) powerful.

How do we bound the expression in (5.20)? In the previous section, when we considered a purely classical Eve, we were able to express the optimum as a simple eigenvalue problem. Here, if we fix Eve's measurements then again we obtain an eigenvalue problem

$$\max_{\rho_{ABE}} \mathrm{tr}\left(\rho_{ABE} \left(\frac{1}{2} \sum_\Theta \Pi_\Theta \right) \right) , \tag{5.21}$$

where

$$\Pi_\Theta = \sum_{x \in \{0.1\}} |x\rangle\langle x|^A_\Theta \otimes |x\rangle\langle x|^B_\Theta \otimes M^E_{x|\Theta} . \tag{5.22}$$

Note that Π_Θ is a projector because for any Θ, $|x\rangle\langle x|_\Theta$ are orthogonal projectors for $x \in \{0,1\}$, and so are $M^E_{x|\Theta}$ by the assumption that Eve's measurements are projective. The optimum of (5.21) is the largest eigenvalue of Π_Θ. How do we get a handle on this quantity?

Two Tools from Linear Algebra

To make progress we use two tricks from linear algebra. To write them down let us first introduce a shorthand for the maximization problem above. Extending the work that we did to evaluate (5.17) in the previous section, we see that in general the *operator norm* of an operator O is

$$\|O\|_\infty = \max_\rho \mathrm{tr}\left(\rho O\right) ,$$

where the maximization is taken over all ρ such that $\mathrm{tr}(\rho) \leq 1$. When O is Hermitian we can just maximize over all quantum states ρ, that is, ρ satisfying $\rho \geq 0$ and $\mathrm{tr}(\rho) = 1$. Note that this means we can reduce the maximization problem (5.21) to studying

$$\left\|\frac{1}{2}\sum_{\theta\in\{0,1\}}\Pi_\theta\right\|_\infty .$$

For simplicity, in the remainder of the section we omit the subscript ∞ and simply write $\|O\| = \|O\|_\infty$. Here are the two facts we will use:

1. For any two projectors Π_0 and Π_1, we have

$$\|\Pi_0 + \Pi_1\| \leq \max\{\|\Pi_0\|, \|\Pi_1\|\} + \|\Pi_0\Pi_1\| .$$

2. If $\Pi_0 \leq P$ and $\Pi_1 \leq Q$,[2] then $\|\Pi_0\Pi_1\|^2 \leq \|PQP\|$.

Assuming these two facts from linear algebra, let us see how we can bound Eve's probability of winning in the tripartite guessing game. Using the first item,

$$\max_{M^E}\left\|\frac{1}{2}\sum_{\theta\in\{0,1\}}\Pi_\theta\right\|_\infty = \max_{M^E}\frac{1}{2}\left\|\sum_{\theta\in\{0,1\}}\Pi_\theta\right\|_\infty$$
$$\leq \frac{1}{2}\left(1 + \max_{M^E}\|\Pi_0\Pi_1\|\right) , \qquad (5.23)$$

where we have used that $\|\Pi_0\|, \|\Pi_1\| \leq 1$ for any measurements M^E that Eve could make. It remains to analyze $\|\Pi_0\Pi_1\|$. For this we use the second item, for some smart choice of P and Q. Note that since all measurement operators $M^E_{x|\theta} \leq \mathbb{I}$ and $|x\rangle\langle x|_\theta \leq \mathbb{I}$, we have that

$$\Pi_0 \leq \sum_{x\in\{0,1\}}|x\rangle\langle x|^A_0 \otimes |x\rangle\langle x|^B_0 \otimes \mathbb{I}^E ,$$
$$\Pi_1 \leq \sum_{x\in\{0,1\}}|x\rangle\langle x|^A_1 \otimes \mathbb{I}^B \otimes M^E_{x|1} .$$

---

2 Recall that $A \leq B$ means that $B - A \geq 0$, i.e. $B - A$ is a positive semidefinite matrix.

Let P and Q be the operators on the right-hand sides above, respectively. Using the fact that $\langle x|y\rangle = 0$ if $x \neq y$ in the same basis, and that $\sum_y M_{y|1}^E = \mathbb{I}$ for any quantum measurement Eve may make, we get

$$PQP = \sum_{x,y,z} |x\rangle\langle x|_0^A |y\rangle\langle y|_1^A |z\rangle\langle z|_0^A \otimes |x\rangle\langle x|_0^B |z\rangle\langle z|_0^B \otimes M_{y|1}^E$$

$$= \sum_{x,y} \frac{1}{2} |x\rangle\langle x|_0^A \otimes |x\rangle\langle x|_0^B \otimes M_{y|1}^E$$

$$= \frac{1}{2} \sum_x |x\rangle\langle x|_0^A \otimes |x\rangle\langle x|_0^B \otimes \sum_y M_{y|1}^E$$

$$= \frac{1}{2} \sum_x |x\rangle\langle x|_0^A \otimes |x\rangle\langle x|_0^B \otimes \mathbb{I}^E .$$

This gives $\|PQP\| \leq 1/2$. Using the second trick and plugging into Eq. (5.23) we get

$$p_{\text{succ}} \leq \frac{1}{2}\left(1 + \frac{1}{\sqrt{2}}\right) = \frac{1}{2} + \frac{1}{2\sqrt{2}}, \tag{5.24}$$

which is the same number that we obtained for the much simpler game in which Eve was entirely classical!

Exercise 5.5.1 Identify explicitly Eve's optimal strategy in the game: find a state ρ_{ABE} and measurement operators $M_{x|\Theta}$ that give her a success probability of $\frac{1}{2} + \frac{1}{2\sqrt{2}}$.

Remark 5.5.1 Using even more linear algebra it is possible to show that when playing the game n times "in parallel," i.e. allowing Eve to prepare a single ρ_{ABE} where A and B are n qubits, and Alice and Bob both measure all their qubits individually using an independent random basis choice for each qubit, then

$$p_{\text{succ}}^{n \text{ rounds}} \leq \left(\frac{1}{2} + \frac{1}{2\sqrt{2}}\right)^n .$$

Moreover, this bound can be achieved by Eve by preparing a tensor product state $\rho_{ABE}^{(n)} = (\rho_{ABE}^{(1)})^{\otimes n}$, where $\rho_{ABE}^{(1)}$ is an optimal choice of state for the single-qubit version of the game.

QUIZ 5.5.1 *Alice, Bob, and Eve play the tripartite guessing game. They share the state $\rho_{AEB} = |\psi\rangle\langle\psi|_{AE} \otimes |0\rangle\langle 0|_B$, where $|\psi\rangle_{AE} = \frac{1}{\sqrt{2}}(|00\rangle_{AE} + |11\rangle_{AE})$ is a maximally entangled state. What is the optimal probability that Eve guesses Alice's outcome?*

(a) $\frac{1}{2}$

(b) $\frac{3}{4}$

(c) $\frac{1}{2}\left(1 + \frac{1}{\sqrt{2}}\right) \approx 0.85$

(d) 1

QUIZ 5.5.2 *What is the optimal probability that Bob guesses Alice's outcome?*

(a) $\frac{1}{2}$

(b) $\frac{3}{4}$

(c) $\frac{1}{2}\left(1+\frac{1}{\sqrt{2}}\right)\approx 0.85$

(d) 1

QUIZ 5.5.3 *Consider again the same guessing game between Alice, Bob, and Eve. Now they share the GHZ state* $|\mathrm{GHZ}\rangle_{ABE}=\frac{1}{\sqrt{2}}\left(|000\rangle_{ABE}+|111\rangle_{ABE}\right)$. *Can both Eve and Bob guess Alice's measurement outcome in the standard basis with full certainty?*

(a) *Yes*

(b) *No*

CHAPTER NOTES

The quantum conditional min-entropy and its use for showing security in quantum cryptography are first put forward in R. Renner's PhD thesis (Security of quantum key distribution. *International Journal of Quantum Information*, **6**(01):1–127, 2008), where the "partial" chain rule and other simple properties are shown. The formulation of this measure as a semidefinite program and the connection with the guessing probability are made by R. Konig, R. Renner, and C. Schaffner (The operational meaning of min-and max-entropy. *IEEE Transactions on Information Theory*, **55**(9):4337–4347, 2009).

Uncertainty relations in the presence of quantum side information are shown in the paper by M. Berta, et al. (The uncertainty principle in the presence of quantum memory. *Nature Physics*, **6**(9):659–662, 2010) and many follow-up works. The formulation using a tripartite guessing game presented here appears in the work of M. Tomamichel, et al. (A monogamy-of-entanglement game with applications to device-independent quantum cryptography. *New Journal of Physics*, **15**(10):103002, 2013), where some of the applications to cryptography discussed later in the book are introduced.

PROBLEMS

5.1 Trace distance

Imagine that Alice and Bob try to create a shared EPR pair $|\text{EPR}\rangle$. Sadly, they are not very good at this yet and instead they create the shared state

$$\rho_{AB} = (1 - p) \, |\text{EPR}\rangle\langle\text{EPR}| + \frac{p}{4}\mathbb{I} \, ,$$

where $0 \leq p \leq 1$ is some noise parameter. What is the trace distance between this state and the ideal state $|\text{EPR}\rangle\langle\text{EPR}|$, as a function of p?

5.2 Min-entropy

Consider the following density matrices:

   I. $\rho_X = |00\rangle\langle00|$
  II. $\rho_X = \frac{1}{2}|00\rangle\langle00| + \frac{1}{2}|11\rangle\langle11|$
 III. $\rho_X = \frac{3}{4}|0\rangle\langle0| + \frac{1}{4}|1\rangle\langle1|$
 IV. $\rho_X = \frac{3}{4}|+\rangle\langle+| + \frac{1}{4}|-\rangle\langle-|$
  V. $\rho_X = \frac{1}{4}|00\rangle\langle00| + \frac{1}{4}|11\rangle\langle11| + \left(\frac{1}{4} - \varepsilon\right)|01\rangle\langle01| + \left(\frac{1}{4} + \varepsilon\right)|10\rangle\langle10|$

What is the min-entropy of each of them?

5.3 Guessing game

Imagine that Alice and Eve play the following guessing game. In the game they initially share some state ρ_{AE}. Alice produces a uniformly random bit θ. She measures her qubit in the standard basis if $\theta = 0$ and measures in the Hadamard basis if $\theta = 1$, obtaining a bit x as the measurement outcome. She then announces θ to Eve. Eve's goal is to guess the bit x. We know that Eve can guess perfectly if ρ_{AE} is an EPR pair. Imagine that this is the case, so $\rho_{AE} = |\text{EPR}\rangle\langle\text{EPR}|$. Alice wants to foil Eve, so before measuring she first applies some random unitary U to her qubit and then measures. Of course Eve, being really smart, gets wind of this so she will know what unitary Alice has used before measuring. Thus they share the state

$$|\phi_U\rangle = (U_A \otimes \mathbb{I}_E)\frac{1}{\sqrt{2}}(|00\rangle + |11\rangle),$$

and Eve knows both θ and U.

1. What is the guessing probability of Eve?
2. What if Eve does not know U? What is the state ρ_{AE} and what is Eve's winning probability?

5.4 Guessing with three bases

In this problem we investigate a slightly more complex version of the bipartite guessing game between Alice and Eve. In this game we assume that Eve can prepare a qubit state $|\psi\rangle$ which she sends to Alice; Eve is not allowed to keep any quantum side information. Alice will generate a uniform random number $\theta \in \{0,1,2\}$. If $\theta = 0$ she measures in the standard basis, if $\theta = 1$ she measures in the Hadamard basis, and if $\theta = 2$ she measures in the basis

$$|0_Y\rangle = \frac{1}{\sqrt{2}}(|0\rangle + i|1\rangle), \quad |1_Y\rangle = \frac{1}{\sqrt{2}}(|0\rangle - i|1\rangle).$$

In all three of the cases she obtains a bit x. She then announces the value of θ to Eve. Eve's goal is again to guess x knowing what state she prepared and the value of θ. Let's begin by calculating Eve's average winning probability when she gives Alice a few simple states.

1. What is Eve's probability of winning the game for the state $|\psi\rangle = |0\rangle$?
2. What is Eve's probability of winning the game for the state $|\psi\rangle = |+\rangle$?

Of course, these states do not yield an optimal winning probability even for the standard guessing game. Instead Eve now tries to play the game with the state that gave her the optimal winning probability for the standard guessing game, namely

$$|\psi\rangle = \frac{1}{\sqrt{2+\sqrt{2}}}(|0\rangle + |+\rangle)$$

3. What is Eve's probability of winning using this state?

Note that the winning probability for the "standard optimal state" is lower than it is in the standard game. This would suggest that there is a state that can perform even better.

4. Which of the following states has the highest winning probability?

    I. $|\psi\rangle = \frac{1}{\sqrt{2+\sqrt{2}}}(|0\rangle + |0_Y\rangle)$

    II. $|\psi\rangle = \frac{1}{\sqrt{4+2\sqrt{2}}}(|0\rangle + |+\rangle + |0_Y\rangle)$

    III. $|\psi\rangle = \sqrt{\frac{3}{5}}\left(\sqrt{\frac{2}{3}}|0\rangle + \sqrt{\frac{1}{3}}|1_Y\rangle\right)$

It turns out that the state you found in the last question is optimal. Let's compare the winning probability of this "three bases" guessing game to the standard guessing game.

5. Is the maximal winning probability of the "three bases" guessing game *lower* or *higher* than that of the standard guessing game? Can you think of an intuitive reason for why this would be the case?

CHEAT SHEET

Trace distance

$$D(\rho_{\text{real}}, \rho_{\text{ideal}}) := \max_{0 \leq M \leq \mathbb{1}} \text{tr}\left[M\left(\rho_{\text{real}} - \rho_{\text{ideal}}\right)\right]$$

$$= \frac{1}{2}\text{tr}\left[\sqrt{A^\dagger A}\right], \qquad A = \rho_{\text{real}} - \rho_{\text{ideal}}.$$

Properties:

1. $D(\rho, \rho') \geq 0$ with equality iff $\rho = \rho'$.
2. $D(\rho, \rho') = D(\rho', \rho)$.
3. $D(\rho, \rho') + D(\rho', \rho'') \geq D(\rho, \rho'')$.
4. $D(\sum_i p_i \rho_i, \sigma) \leq \sum_i p_i D(\rho_i, \sigma)$.

Fidelity

$$F(\rho, \rho') := \text{tr}\left[\sqrt{\sqrt{\rho}\rho'\sqrt{\rho}}\right].$$

If $\rho = |\psi\rangle\langle\psi|$, then $F(\rho, \rho') = \sqrt{\langle\psi|\rho'|\psi\rangle}$.
Relation to trace distance: $1 - F \leq D \leq \sqrt{1 - F^2}$.

Min-entropy

Unconditional: $\text{H}_{\min}(X) = \text{H}_{\min}(\rho_X) = -\log \max_x p_x$.
Conditional: For a cq-state ρ_{XE}, $\text{H}_{\min}(X|E) := -\log P_{\text{guess}}(X|E)$, where

$$P_{\text{guess}}(X|E) := \max_{\{M_x\}_x} \sum_x p_x \text{tr}\left[M_x \rho_x^E\right], \{M_x \geq 0 \mid \sum_x M_x = \mathbb{1}\}.$$

Properties:

1. $0 \leq \text{H}_{\min}(X|E) \leq \text{H}_{\min}(X) \leq \log|X|$, but only for cq-states! For quantum register X, $\text{H}_{\min}(X|E)$ can be negative.
2. $\text{H}_{\min}(X|E) \geq \text{H}_{\min}(X) - \log|E|$.

A secret key

A key K with d_K possible values is secret from Eve iff it is *uniform and uncorrelated* from Eve, i.e. the joint state ρ_{KE} is of the form

$$\rho_{KE} = \frac{\mathbb{1}_K}{d_K} \otimes \rho_E.$$

6

From Imperfect Information to (Near) Perfect Security

Now that we know how to measure information, knowledge, and ignorance, let's discuss how to *amplify* them! Given a (partial) secret X, about which an eavesdropper, Eve, has some information E, is there a way to amplify the secrecy, or *privacy*, of X? This is the goal of *privacy amplification*. Privacy amplification is an essential component of many cryptographic protocols; in particular, it forms the final step in the quantum key distribution protocols we'll see in later chapters.

In this chapter, we discuss the formal definition of privacy amplification and examine how it can be realized using a beautiful object from theoretical computer science called a *randomness extractor* – itself well worth studying in its own right!

6.1 Privacy Amplification

Imagine (as usual!) that Alice and Bob want to use cryptography to exchange messages securely. For this they have access to a public classical communication channel: they can send each other any classical messages they like, *but* the channel is public: the malicious eavesdropper, Eve, may be listening in on the entire communication. Our only cryptographic assumption on the channel is that it is *authenticated*, meaning that when Alice (or Bob) receives a message she has the guarantee that it came directly from Bob (or Alice). (We discuss this assumption in more detail at the end of the next chapter, see Section 7.5.)

To ensure privacy, Alice and Bob would like to use cryptography: they know (as you do!) that the one-time pad is unconditionally secure, so the only thing they need is to come up with a shared secret key. Moreover, Alice and Bob being old-time friends, they already have a lot of shared secrets, such as the flavor of the first ice-cream cone they shared. By putting all these secrets together and translating them in a string of bits, they're pretty confident that they can come up with some value, call it $x \in \{0,1\}^n$, that's fairly secret ... but only "fairly" so. Unfortunately they're not fully confident about which parts of x can be considered a secret, and which may have leaked. Alice might have told her best friend, Mary, about the ice cream. She definitely wouldn't have told Mary about her (embarrassing) all-time favorite cheeky cartoon, but then her little brother John might know about this. Is there a way for Alice and Bob to somehow "boil down" the secrecy that x contains, throwing away some of the bits and combining the others to create a "perfectly secret" key – all this without knowing a priori which bits of x are secure and which may potentially have been leaked?

The answer is yes! This is precisely what privacy amplification will do for them. Let's describe this task more precisely. Imagine that two mutually trusting parties, Alice and Bob, each hold a copy of the same string of bits x, which we'll call a "weak secret." This secret is taken from a certain distribution p_x, which we can represent through a random variable X; later on we'll call X the "source." The distribution of X itself is not known, but the sample x is available to both parties. An eavesdropper has side information E that may be correlated with X; for example, E could be the first bit of X, the parity of X, or an arbitrary quantum state ρ_x^E. Given this setup, the goal for Alice and Bob is to each produce the same string z, which could be shorter than x but must be such that the distribution of z (represented via a random variable Z) is (close to) uniform, even from the point of view of the eavesdropper.

To summarize using symbols, privacy amplification is implemented by a function PA such that

$$\rho_{XE} = \sum_x p_x |x\rangle\langle x|_X \otimes \rho_x^E \xrightarrow{\ \mathrm{PA}_X \otimes \mathbb{I}_E\ } \rho_{ZE} \approx_\varepsilon \frac{\mathbb{I}_Z}{|Z|} \otimes \rho_E . \qquad (6.1)$$

One possible way to achieve this is to simply "throw away" X and replace it by a uniformly random string $Z \sim U_m$, where we use U_m to denote the uniform distribution on $\{0,1\}^m$. While this works, it is not really satisfying: where do we get that random string anyways? The goal is to use X! Now if we do want to use X, we will need to make assumptions on it. For example, if X is a fixed string, or if X is random but $E = X$, then there is zero randomness or privacy to start with and there is nothing we can do. But in less extreme cases, we can sometimes do interesting things, as the following warm-up exercises show.

Exercise 6.1.1   Suppose that $X \in \{0,1\}^3$ is uniformly distributed, and $E = X_1 \oplus X_2 \in \{0,1\}$. Give a protocol for privacy amplification that outputs two secure bits (without any communication). What if $E = (X_1 \oplus X_2, X_2 \oplus X_3) \in \{0,1\}^2$, can you still do it? If not, give a protocol extracting just one bit.

Exercise 6.1.2   Suppose the eavesdropper is allowed to keep any two of the bits of X as side information. Give a protocol for Alice and Bob to produce a Z which contains a single bit that is always uniformly random, irrespective of which two bits of X are stored by the eavesdropper. How about an E that contains two bits – can they do it?

In the next section we focus on a seemingly simpler task of *extracting randomness*. As we'll see, this task is very closely connected to privacy amplification, but at first it looks simpler; in particular, it only talks about Alice – Bob will wait for a little bit!

6.2   Randomness Extractors

In the task of randomness extraction there is a single party, Alice, who has access to an n-bit string x with distribution X. We call X the *source*. X is unknown, and it may be correlated with an additional system E over which Alice has no control. For example, E could contain some information about the way in which the source was generated, or some information that an adversary has gathered during the course of an earlier protocol involving X. Alice's goal is to produce a new string Z that is close to uniform and uncorrelated with E.

Now, of course Alice could dump X and create her own uniformly random Z, say by measuring a $|0\rangle$ qubit in the Hadamard basis. To make the problem interesting we won't allow any quantum resources to Alice. She also doesn't have that much freely accessible randomness – maybe she can get some, but it will be limited and costly. Alice's goal is to leverage what she has to the best she can: she wants to *extract* randomness from X, not import it from some magical elsewhere!

6.2.1   Randomness Sources

 Let's see some concrete examples of sources X, and how it is possible to extract uniform bits from them.

i.i.d. Sources

The simplest case of a randomness source is an *i.i.d. source*, where the term i.i.d. stands for *independent and identically distributed*. A (classical) i.i.d. source $X \in \{0,1\}^n$ has a distribution $\{p_x\}$ which has a product form: there is a distribution $\{p_0, p_1\}$ on a single bit such that for all $(x_1, \ldots, x_n) \in \{0,1\}^n$,

$$\Pr(X = (x_1, \ldots, x_n)) = \Pr(X_1 = x_1) \cdots \Pr(X_n = x_n) = p_{x_1} \cdots p_{x_n} .$$

Such sources are sometimes called *von Neumann* sources, since they were already considered by von Neumann.[1]

Can we extract uniformly random bits from an i.i.d. source? As a warm-up, let's consider how we could obtain a nearly uniform bit from a source such that each bit X_i is 0 with probability $p_0 = 1/4$ and 1 with probability $p_1 = 3/4$. Suppose we let $Z = X_1 \oplus X_2 \oplus \cdots \oplus X_n \in \{0,1\}$ be the parity of all n bits of X. Our goal is to show that $\Pr(Z = 0) \approx 1/2 \pm \varepsilon$ for reasonably small ε. Intuitively, taking the parity works because it is sensitive to all inputs and so, even though each individual input is biased, overall we should get a bit that looks pretty uniform.

- Let's first consider $n = 2$. How well does our strategy work? We can compute

$$\Pr(Z = 0) = \Pr(X_1 = 0 \wedge X_2 = 0) + \Pr(X_1 = 1 \wedge X_2 = 1)$$
$$= p_0^2 + p_1^2 = 1/16 + 9/16 = 0.625 ,$$

---

1 If you are curious about the history of randomness extraction, look up the von Neumann extractor online!

and using a similar calculation we find $\Pr(Z = 1) = 0.375$. Not quite uniform, but closer than what we started with!

- We can continue with $n = 3$. In this case,

$$\Pr(Z = 0) = \Pr(X_1 = 0 \wedge (X_2 \oplus X_3 = 0)) + \Pr(X_1 = 1 \wedge (X_2 \oplus X_3 = 1))$$
$$= p_0 \cdot 0.625 + p_1 \cdot 0.375 = 0.4375 \, .$$

Even better!

Exercise 6.2.1   Continue the calculation above for increasing values of n. Using a recurrence relation, show that the bias of Z, i.e. the quantity $|\Pr(Z = 0) - \Pr(Z = 1)|$, goes to zero as n grows. At what rate? Do you find our procedure efficient?

Independent Bit Sources

A slightly broader class of sources are *independent bit sources*. As their name suggests, such sources are characterized by the condition that each bit is chosen independently; however, the distribution could be different for different bits. Clearly, any i.i.d. source is also an independent bit source, but the converse does not hold.

Exercise 6.2.2   Show that there exists an independent two-bit source X such that $\Pr(X = (0,0)) = \Pr(X = (1,1)) = 3/16$, but there is no i.i.d. source satisfying the same condition.

It turns out that taking the parity of all the bits in the string generated by an independent bit source still results in a bit that is increasingly close to uniform as $n \rightarrow \infty$, provided each bit from the source is not fully biased to start with.

Exercise 6.2.3   Let X be an independent n-bit source such that $\delta < \Pr(X_j = 0) < 1 - \delta$ for some $\delta > 0$ and all $j \in \{1,\ldots,n\}$. Give an upper bound on the distance from uniform of the parity of the bits of X, as a function of the parameters n and δ.

Bit-Fixing Sources

Bit-fixing sources are a special case of independent sources where each bit of X can be of one of two kinds only: either the bit is completely fixed, or it is uniformly random. For example, the three-bit source X such that $\Pr(X = (1,0,0)) = \Pr(X = (1,1,0)) = 1/2$, with all other probabilities being 0, is a bit-fixing source: the first bit is fixed to 1, the second is uniformly random, and the third is fixed to 0.

You can verify for yourselves that, just as for the previous two types of sources we considered, taking the parity of all bits from a bit-fixing source gives a uniformly random bit. This time, we do even better: as long as at least one of the bits from the source is not fixed, the parity is (exactly) uniformly random.

General Sources

The randomness sources we just discussed all have something in common: they produce a string in which each bit is chosen *independently*. What if we relax this condition?

Consider a tricky example, called an *adversarial* bit-fixing source: this is the same as a bit-fixing source, except the value taken by the fixed bits can depend on the previous bits. For example, the three-bit source X such that $\Pr(X = (1,0,0)) = \Pr(X = (1,1,1)) = 1/2$, with all other probabilities being 0, is an adversarial bit-fixing source: the first bit is fixed to 1, the second is uniformly random, and the third is fixed to either 0 if the second was a 0, or 1 if the second was a 1. To see that this kind of source can be much more tricky, first check that our earlier choice of Z as the parity of all the bits of X no longer works on the example. Indeed, in this case Z is always equal to 1! However, taking the parity of the first two, or the first and last, bits does work.

Exercise 6.2.4   Show that for any fixed choice of a subset of bits, there exists an adversarial bit-fixing source such that only one bit is fixed, but nevertheless the parity of the bits in the chosen subset is a constant – arbitrarily far from uniform!

As you can imagine, there is a whole universe of possible kinds of sources. How do we classify them? For the purposes of extracting randomness, we aim to measure the inherent uncertainty of the source, or in other words its *entropy*, as this is the quality that we aim to extract from it. It turns out that the min-entropy provides just the right measure of extractable randomness. We will see in the next section (see Box 6.1) why this is the case. For now, let's give a first definition.

| **Definition 6.2.1.** *A random variable X is called a k-source if* $\mathrm{H}_{\min}(X) \geq k$. |

Before we move on, we should realize that there is something crucial missing from this definition. Remember that we're going to apply the idea of randomness extraction to a cryptographic task, privacy amplification. But we forgot to account for the eavesdropper! The process of randomness extraction is not going to happen in a void: we ought to take into account the possibility of an additional system E that may be correlated with X. Call E the *side information*. X is a classical string of bits, but E may be quantum. How do we model this? Using the same approach as in the previous chapter, we do this by introducing a cq-state ρ_{XE}, which in general takes the form

$$\rho_{XE} = \sum_x |x\rangle\langle x|_X \otimes \rho_x^E,$$

where each ρ_x^E is positive semidefinite and $\mathrm{tr}(\rho_{XE}) = \sum_x \mathrm{tr}(\rho_x^E) = 1$.

Remark 6.2.1 Using side information gives us a convenient way to model any source X as the result of an initially *uniform* string about which the adversary has gained

partial information. For example, you can represent a bit-fixing source as a uniform source correlated with a system E which contains some of the bits of X.

Exercise 6.2.5 Let X be an independent source, where the i-th bit X_i has distribution $\{p_i, 1 - p_i\}$. Show that there exists a pair of correlated random variables (Y, Z) on $\{0, 1\}^n \times \{0, 1\}^n$ such that Y is uniformly distributed in $\{0, 1\}^n$ but for any $z \in \{0, 1\}^n$ the random variable $V = Y_{|Z=z}$ is such that V_i has the same distribution as X_i if $z_i = 0$, and as $1 - X_i$ if $z_i = 1$.

Let's update our definition.

| **Definition 6.2.2.** *A cq-state ρ_{XE} is called a k-source if* $H_{\min}(X|E) \geq k$. |

Can we construct extraction procedures that produce uniformly random bits from any k-source, without knowing anything else about the source?

6.2.2 Strong Seeded Extractors

In all examples we've seen so far we applied a fixed function, call it Ext, to the source X; for example, we considered $\text{Ext}(X) = X_1 \oplus \cdots \oplus X_n$. Such a function is known as a deterministic extractor, meaning that it is just one fixed function $Z = \text{Ext}(X)$ that does not introduce any randomness beside what is already present in X.

Ideally we'd like to show that it is possible to extract randomness from any k-source using such a deterministic function. Unfortunately this is not possible: there is no fixed deterministic procedure that can be used to extract even a *single* bit of randomness from every possible k-source, even when k is almost maximal, $k = n - 1$! This is a bit disappointing, but let's understand why.

Lemma 6.2.2 *For any function* $\text{Ext} : \{0, 1\}^n \to \{0, 1\}$ *there exists an* $(n - 1)$-*source X such that $\text{Ext}(X)$ is constant.*

Proof Let $b \in \{0, 1\}$ be such that $|S_b| \geq 2^n/2 = 2^{n-1}$ with $S_b = \{x \mid \text{Ext}(x) = b\}$. Note that there must exist such a b. Choose a subset $S' \subseteq S_b$ such that $|S'| = 2^{n-1}$. Define X by the following distribution:

$$p_x = \begin{cases} 1/2^{n-1} & \text{if } x \in S', \\ 0 & \text{otherwise}. \end{cases}$$

Clearly, $H_{\min}(X) = n - 1$, but $\text{Ext}(X) = b$ is a constant! ∎

Have we reached the end of the road – are we stuck to designing special-purpose functions that only work for this or that special kind of source, as we did with independent sources? Luckily there is a way out, but we're going to need an additional resource: a little extra randomness. This extra randomness will be called the *seed* of the extractor. Think of the seed as a second input $Y \in \{0, 1\}^d$ to which Alice has access and is promised to be uniformly random and independent of X and E. This gives us the notion of a *seeded extractor*. A seeded extractor is a function $\text{Ext} : \{0,1\}^n \times \{0,1\}^d \to \{0,1\}^m$ which takes two inputs, the source X and the seed Y, and returns an output Z that, informally, should be close to uniform, as long as the source X has high enough (conditional) min-entropy. To express this condition mathematically, we imagine applying the extractor to the registers X and Y of a ccq-state:

$$\rho_{XYE} = \sum_x |x\rangle\langle x|_X \otimes \left(\frac{1}{2^d}\sum_y |y\rangle\langle y|_Y\right) \otimes \rho_x^E \,,$$

where the register Y is initialized as a uniformly random string. Applying Ext and storing the result in a register Z, we obtain

$$\rho_{XYZE} = \frac{1}{2^d}\sum_{x,y} |x\rangle\langle x|_X \otimes |y\rangle\langle y|_Y \otimes |\text{Ext}(x,y)\rangle\langle \text{Ext}(x,y)|_Z \otimes \rho_x^E \,.$$

Since we usually consider that at this point the source X and the seed Y are "gone," we can trace them out to obtain the cq-state

$$\begin{aligned}
\rho_{ZE} = \rho_{\text{Ext}(X,Y)E} &= \frac{1}{2^d}\sum_{x,y}|\text{Ext}(x,y)\rangle\langle\text{Ext}(x,y)|_Z \otimes \rho_x^E \\
&= \sum_z |z\rangle\langle z|_Z \otimes \left(\frac{1}{2^d}\sum_y\sum_{x:\text{Ext}(x,y)=z}\rho_x^E\right),
\end{aligned} \tag{6.2}$$

where for the second line we made a change of variables. Our goal is to make sure that Z is almost uniformly random and uncorrelated with E, i.e. the trace distance between ρ_{ZE} and the "ideal" state

$$\rho_{ZE}^{ideal} = \frac{1}{2^m}\mathbb{I}_Z \otimes \rho^E \,, \quad \text{where} \quad \rho^E = \sum_x \rho_x^E \,,$$

is as small as possible. We are now ready to give the formal definition.

Definition 6.2.3. *A (k,ε)-weak seeded randomness extractor is a function* $\text{Ext} :$ $\{0,1\}^n \times \{0,1\}^d \to \{0,1\}^m$ *such that for any k-source ρ_{XE},*

$$D\left(\rho_{\text{Ext}(X,Y)E}, \frac{\mathbb{I}}{2^m}\otimes\rho_E\right) \leq \varepsilon \,, \tag{6.3}$$

where $Y \sim U_d$ is uniformly distributed and independent of X and E, and $\rho_{\text{Ext}(X,Y)E}$ is defined in (6.2).

If the seed is perfectly uniform, why don't we just return it as our output: define $\text{Ext}(X,Y) = Y$? Well, this satisfies the definition. So maybe there is something wrong with the definition? Remember that our goal is to *extract* randomness from X, and that additional uniform randomness should not be considered free. So we want to

> ## BOX 6.1   The Data Processing Inequality
>
> Let's see why the min-entropy is an upper bound on the amount of extractable randomness. In the process we will show a useful entropy inequality, the data processing inequality. To show why, recall that $H_{min}(X|E) = -\log P_{guess}(X|E)$. Suppose now that we apply some function f to X. How hard is it to guess $f(X)$ given E, i.e. what's $P_{guess}(f(X)|E)$? Clearly, since one way to guess $f(X)$ is to guess X, and then apply f to our guess, we have $P_{guess}(f(X)|E) \geq P_{guess}(X|E)$. However, this is equivalent to
>
> $$H_{min}(f(X)|E) \leq H_{min}(X|E) \,,$$
>
> which is called the data-processing inequality. This inequality means that also the output of the extractor, which for fixed seed y is obtained as a function $f(X) = \text{Ext}(X,y)$, must have min-entropy at most $H_{min}(X|E)$. This implies that the output $\text{Ext}(X,Y)$, conditioned on $Y = y$, can be uniform on at most $H_{min}(X|E)$ bits!

keep Y as small as possible, even though X, and k, could be very large, in which case we'd like to maintain a long output (large m) with only a little help from the seed (small d).

A second motivation for keeping the seed small comes from remembering our goal of achieving privacy amplification. We'll see in the next section how to use extractors to solve this problem. For now, let us simply point out that our solution of an extractor outputting its seed would be similar to asking Alice and Bob to throw away their initial secret X and share a fresh random string Y – which would of course be besides the point, since coming up with a shared uniformly random string is the problem that they are trying to solve in the first place!

This motivates a stronger definition of extractor, which is the one we'll use from now on.

> **Definition 6.2.4.** A (k,ε)-strong seeded randomness extractor *is a function* $\text{Ext} : \{0,1\}^n \times \{0,1\}^d \to \{0,1\}^m$ *such that for any k-source* ρ_{XE},
>
> $$D\left(\rho_{\text{Ext}(X,Y)YE}, \frac{\mathbb{I}}{2^m} \otimes \rho_{YE}\right) \leq \varepsilon \,, \tag{6.4}$$
>
> *where* $Y \sim U_d$ *is uniformly distributed and independent of X and E.*

The important difference between the stronger requirement (6.4) and (6.3) is that in (6.4) we did not trace out the seed of the extractor. Remembering the operational interpretation of the trace distance, (6.4) means that the largest probability with which an "adversary" can distinguish the two states $\rho_{\text{Ext}(X,Y)YE}$ and $\frac{\mathbb{I}}{2^m} \otimes \rho_{YE}$ is at most $\frac{1}{2} + \frac{1}{2}\varepsilon$. Because here the adversary is also given the seed Y, it can only help it distinguish the two states, and so (6.4) is a stronger condition compared to (6.3).

QUIZ 6.2.1 *Suppose* Ext : $\{0,1\}^n \times \{0,1\}^d \to \{0,1\}^m$ *is a* (k,ε)-*strong extractor. For which of the following random variables X and Y is it true that* $\left\| \rho_{Ext(X,Y)YE} - I \otimes \rho_{YE} \right\|_1 \leq \varepsilon$? *Mark all that apply.*

(a) *X is uniform and independent of Y and E, and Y is such that* $H(Y|E) \geq k$.
(b) *X is such that* $H(X|E) = k-1$, *and Y is uniform and independent of X and E*.
(c) *X is such that* $H(X|E) \geq k$, *and Y is uniform and independent of X and E*.
(d) *X is such that* $H(X|E) \geq k$, *and Y is uniform with* $H(Y|X) < d$.

6.3 Solving Privacy Amplification Using Extractors

So how do we use extractors to solve privacy amplification? You can probably guess how this can be done. Let Ext be a (k,ε)-strong seeded randomness extractor. Here is a simple protocol:

1. Alice and Bob share a weak secret X, which may be correlated with an eavesdropper holding quantum side information E.
2. Alice chooses a random seed Y for the extractor, and computes $K_A = \text{Ext}(X,Y)$. She sends Y to Bob over a public communication channel.
3. Upon receiving Y, Bob sets $K_B = \text{Ext}(X,Y)$.

First, note that this protocol is always correct: Alice and Bob output the same string, $K_A = K_B$. Is it secure? Remember the criterion (6.1) we introduced to define security of privacy amplification. Note also that here, at the end of the protocol, Eve has access to her original side information E, but also to any communication exchanged over the public channel: precisely the seed Y. So the condition becomes

$$X : \mathrm{H_{min}}(X|E)_\rho \geq k \quad \overset{\text{PA}}{\longmapsto} \quad Z = \text{Ext}(X,Y) : \rho_{ZYE} \approx_\varepsilon \frac{\mathbb{I}_Z}{|Z|} \otimes \rho_{YE} \,,$$

which is precisely the requirement of a (k,ε)-strong extractor!

Now all that remains to do is show how to construct a good enough extractor. Let's get to it!

6.4 An Extractor Based on Hashing

Much research has gone into constructing randomness extractors, and they have found many applications throughout computer science and mathematics. The quality of an extractor is measured by the parameters it achieves, and different applications require different trade-offs. The main targets consist in extracting as much randomness as possible (large m) using the smallest possible seed (small d) and with the best possible error (small ε), all from arbitrary sources with min-entropy (at least) k.

Going back to the intuition we developed on the examples, we saw that taking the parity of a random subset of the bits of the source often (but not always) provides a good way to extract a bit of randomness. Now that we know about seeds, to hedge our bets we could use the seed to specify the subset of bits whose parity is taken. This way, for any given source, we can hope that most seeds will give us subsets that are random for that source. In this section we'll see a way to make this intuition work. For this we'll have to make a little detour and learn about certain families of hash functions.

6.4.1 Two Universal Families of Hash Functions

Informally, a hash function is a function that maps long strings to shorter strings, with the property that the output of the hash function tends to be "well-distributed." What this means depends on the application we have in mind for the hash function – indeed, the term "hash function" can be interpreted in many different ways. The only standard requirement, as its name indicates, is that a hash function should not increase the length of its input! An additional reasonable requirement, which formalizes the "well-distributed" aspect, is the following.

> **Definition 6.4.1 (1-universal family).** *A family of hash functions* $\mathcal{F} \subseteq \{f : \{0,1\}^n \to \{0,1\}^m\}$, *where* $m \leq n$, *is called* 1-universal *if for every string* $x \in \{0,1\}^n$ *and* $z \in \{0,1\}^m$ *we have*
>
> $$\Pr_{f \in \mathcal{F}}(f(x) = z) = \frac{1}{2^m} . \tag{6.5}$$

Read the definition carefully: in (6.5) both x and z are fixed, and the probability is taken over a uniformly random function from the family. The condition is equivalent to saying that for any fixed x the random variable $F(x)$, where F is uniformly distributed over all f in $\mathcal{F}$, is uniformly distributed in $\{0,1\}^m$. Let's see an example of a 1-universal family of hash functions.

Exercise 6.4.1 For any $y \in \{0,1\}^n$ let $f_y : \{0,1\}^n \to \{0,1\}^n$ be defined by $f_y(x) = x \oplus y$, where the parity is taken bitwise. Show that the family of functions $\mathcal{F} = \{f_y, y \in \{0,1\}^n\}$ is 1-universal.

You may want to convince yourself that a family of 1-universal hash functions is already sufficient to construct a *weak* seeded extractor. To do this, use the seed to select a random function from the family, and output the value of the function evaluated on the source. More formally, define $\mathrm{Ext}(x,y) = f_y(x)$. The property of 1-universality ensures that the output will be uniformly distributed, even if the input is fixed. However, recall our earlier criticism: in this case it is apparent that we are "cheating," and that all the randomness is coming from the seed. Indeed, it turns out that the property of 1-universality is not sufficient to obtain a *strong* seeded extractor. We'll need the following stronger property.

BOX 6.2   Finite Fields

For q a prime power, a finite field with q elements is a finite set $\mathbb{F}_q$ of size q equipped with addition and multiplication laws that satisfy certain natural requirements. The detailed requirements will not be important for us: the only thing that matters is that a field is a set whose elements can be added and multiplied in the usual way, and such that all elements, except the zero, have a multiplicative inverse.

An example of a field is the real numbers $\mathbb{R}$ with addition and multiplication, but it is not finite. Another example is for $q = 2$, we have $\mathbb{F}_2$ the finite field with two elements. As a set, $\mathbb{F}_2 = \{0,1\}$. The operations are addition, which is taken modulo 2 (so $1+1=0$), and standard multiplication. As it turns out, this is the only field with two elements, and in fact for any prime power q there is a unique finite field $\mathbb{F}_q$ of size q. For example, there is a unique field $\mathbb{F}_3 = \{0,1,2\}$, a unique field $\mathbb{F}_4 = \mathbb{F}_{2^2}$, etc. It is a good exercise to explicitly work out the multiplication table for these two examples.

Definition 6.4.2 (2-universal family). *A family of hash functions $\mathcal{F} = \{f : \{0,1\}^n \to \{0,1\}^m\}$ is called* 2-universal *if for every two strings $x, x' \in \{0,1\}^n$ with $x \neq x'$, and any two $z, z' \in \{0,1\}^m$, we have*

$$\Pr_{f \in \mathcal{F}} (f(x) = z \wedge f(x') = z') = \frac{1}{2^{2m}} . \tag{6.6}$$

Condition (6.6) in the definition would be satisfied if $f(x)$ and $f(x')$ were *jointly* chosen uniformly and independently at random in $\{0,1\}^m$. This is a stronger condition than (6.5): we now require that the pair of random variables $(F(x), F(x'))$, for F uniformly distributed over $f \in \mathcal{F}$, are jointly uniform.

Exercise 6.4.2   Show that any 2-universal family of functions is automatically 1-universal.

Exercise 6.4.3   Show that the family of hash functions from Exercise 6.4.1 is *not* 2-universal.

You can check that for any $m \leq n$ the set of all possible functions $f : \{0,1\}^n \to \{0,1\}^m$ is 2-universal. But it is too big: it has size $|\mathcal{F}| = 2^{m2^n}$, so that selecting a function at random from the set would require a seed length $d = m2^n$!

Let's see a much more efficient construction. For this construction we need to use finite fields, which we recap briefly in Box 6.2. Here we'll use $q = 2^n$, so $\mathbb{F}_q$ is the finite field with 2^n elements. For any $(a,b) \in \mathbb{F}_q^2$ let

$$f_{a,b} : \mathbb{F}_q \to \mathbb{F}_q , \qquad f_{a,b}(x) = ax + b ,$$

where addition and multiplication are done in $\mathbb{F}_q$. Then $\mathcal{F} = \{f_{a,b}, (a,b) \in \mathbb{F}_q^2\}$ is a 2-universal family of only $q^2 = 2^{2n}$ hash functions. To show this we need to verify that equation (6.6) from the definition holds. So let's fix distinct $x \neq x' \in \mathbb{F}_q$ and two $z, z' \in$

$\mathbb{F}_q$. What is the probability, over a uniformly random choice of (a,b), that $f_{a,b}(x) = z$ and $f_{a,b}(x') = z'$? The two conditions are equivalent to $ax + b = z$ and (taking the difference) $a(x' - x) = z' - z$, thus $a = (z' - z)/(x' - x)$, where the condition $x \neq x'$ and the fact that $\mathbb{F}_q$ is a field allow us to perform the division. This equation determines a unique possible value for a. Moreover, once a is fixed there is a unique possible value for b: $b = z - ax$ (this shouldn't be a surprise, since we started with two linear equations and two unknowns). Out of 2^{2n} possibilities, we end up with a single one: $\Pr_{a,b}(f_{a,b}(x) = z \wedge f_{a,b}(x') = z') = 2^{-2n}$, as desired.

One last technicality: recall that our goal was to construct a 2-universal family of functions $f : \{0,1\}^n \to \{0,1\}^m$, for arbitrary n and $m \leq n$, whereas what we have managed to construct so far are functions from $\mathbb{F}_q \to \mathbb{F}_q$. Since $|\mathbb{F}_q| = q = 2^n$, the domain of f can be identified with $\{0,1\}^n$ in an arbitrary way. The range of f may be bigger than $\{0,1\}^m$, but there is a simple solution: throw away the last $(n-m)$ bits of $f(x)$! We'll let you verify that this works, i.e. it preserves the property of 2-universality.

6.4.2 The 2-Universal Extractor

Equipped with an arbitrary 2-universal family of hash functions, we define an extractor as follows.

Definition 6.4.3 (2-universal extractor). *Let* $\mathcal{F} = \{f_y : \{0,1\}^n \to \{0,1\}^m, y \in \{0,1\}^d\}$ *be a 2-universal family of hash functions such that* $|\mathcal{F}| = 2^d$. *The associated 2-universal extractor is*

$$\mathrm{Ext}_{\mathcal{F}} : \{0,1\}^n \times \{0,1\}^d \to \{0,1\}^m, \qquad \mathrm{Ext}_{\mathcal{F}}(x,y) = f_y(x).$$

You can think of $\mathrm{Ext}_{\mathcal{F}}$ as using its seed y to select a function from the family $\mathcal{F}$ uniformly at random, and then returning the output of the function when evaluated on the source X. How good is this extractor? The key result required to analyze it is known as the *leftover hash lemma*.

Theorem 6.4.1 (Leftover hash lemma) *Let n and $k \leq n$ be arbitrary integers, $\varepsilon > 0$, $m = k - 2\log(1/\varepsilon)$, and $\mathcal{F} \subseteq \{f : \{0,1\}^n \to \{0,1\}^m\}$ a 2-universal family of hash functions. Then the 2-universal extractor $\mathrm{Ext}_{\mathcal{F}}$ is a (k, ε)-strong seeded randomness extractor.*

In the previous section we saw how to construct a 2-universal family with 2^{2n} functions, meaning that the seed length of the 2-universal extractor is $2n$. This is relatively long, and in particular it is longer than the source. While this can be a drawback in some applications for which the randomness required to produce the seed is particularly costly, for our application to privacy amplification, and especially later to quantum key distribution, it is not a significant limitation. Much more important for us is the dependence of the output length on the initial min-entropy, which will ultimately govern the length of key that we are able to produce. In this respect the 2-universal construction is essentially optimal, a good reason to use it!

6.4.3 Analysis with No Side Information

We first prove the leftover hash lemma, Theorem 6.4.1, in the case when there is no side information. This will be a good warm-up for the general case, with quantum side information, whose analysis will follow the same structure.

 Proof The proof proceeds in two steps. In the first step we reduce our ultimate goal, bounding the error of the extractor, i.e. the trace distance between the extractor's output and the uniform distribution, to bounding a different quantity called the *collision probability*. In the second step we show that the collision probability is sufficiently small to imply the desired bound on the error of the extractor. ■

(i) From Trace Distance to Collision Probability

Our goal is to bound $D(\rho_{\text{Ext}(X,Y)Y}, 2^{-(m+d)}\mathbb{I})$, where X has min-entropy at least k and Y is uniformly distributed over d-bit strings. The joint distribution of $(Z = \text{Ext}(X,Y), Y)$ is given by

$$p_{zy} = \Pr((\text{Ext}(X,Y),Y) = (z,y)) = 2^{-d} \sum_{x:f_y(x)=z} p_x . \tag{6.7}$$

Since we are measuring the trace distance between two classical distributions, the trace distance reduces to the total variation distance (see Example 5.1.1) and we get

$$D(\rho_{\text{Ext}(X,Y)Y}, 2^{-(d+m)}\mathbb{I}) = \frac{1}{2} \sum_{z,y} \left| 2^{-d} \sum_{x:f_y(x)=z} p_x - 2^{-d-m} \right|$$

$$\leq 2^{\frac{m}{2}-1} \left(2^{-d} \sum_{z,y} \left| \sum_{x:f_y(x)=z} p_x - 2^{-m} \right|^2 \right)^{1/2}$$

$$= 2^{\frac{m}{2}-1} \left(2^d \sum_{z,y} p_{zy}^2 - 2^{-m} \right)^{1/2} ,$$

where for the second line we applied the Cauchy–Schwarz inequality (see Box 6.3). This completes our first step. The quantity $CP(ZY) = \sum_{z,y} p_{zy}^2$ is called the collision probability of (Z,Y), and we turn to bounding it next.

(ii) A Bound on the Collision Probability

Using the definition (6.7) and expanding the square,

$$\sum_{z,y} p_{zy}^2 = 2^{-2d} \sum_{y,z} \sum_{\substack{x,x': \\ f_y(x)=f_y(x')=z}} p_x p_{x'}$$

$$= 2^{-2d} \sum_{y,z} \left(\sum_{\substack{x \neq x': \\ f_y(x)=f_y(x')=z}} p_x p_{x'} + \sum_{x:f_y(x)=z} p_x^2 \right)$$

$$= 2^{-(d+m)} \sum_{x \neq x'} p_x p_{x'} + 2^{-d} \sum_x p_x^2$$

$$\leq 2^{-(d+m)} + 2^{-(d+k)} .$$

Here the crucial step is in bounding the summation over $x \neq x'$ when going from the second to the third line: we are using the property of 2-universality to argue that for

---

BOX 6.3   The Cauchy–Schwarz Inequality

The Cauchy–Schwarz inequality is a simple inequality which is incredibly useful in analysis. There are many different possible formulations for the inequality. The most standard one states that for any two sequences of complex numbers, (a_i) and (b_i), the following holds:

$$\left| \sum_i a_i^* b_i \right| \le \left(\sum_i |a_i|^2 \right)^{1/2} \left(\sum_i |b_i|^2 \right)^{1/2}.$$

Sometimes this is expressed more succinctly using vectors as $|\vec{a}^* \cdot \vec{b}| \le \|\vec{a}\|\|\vec{b}\|$, where $\|\cdot\|$ is the Euclidean norm. We will also use a version of the inequality for matrices. It states that for any two complex matrices A and B of the same dimension $m \times n$,

$$|\operatorname{tr}(A^\dagger B)| \le (\operatorname{tr}(A^\dagger A))^{1/2}(\operatorname{tr}(B^\dagger B))^{1/2}.$$

This matrix version can be shown by applying the "usual version" to the matrix coefficients (A_{ij}) and (B_{ij}), because $\operatorname{tr}(A^\dagger B) = \sum_{i,j} A_{ij}^* B_{ij}$.

---

any $x \ne x'$ there is a fraction exactly 2^{-m} of all f_y that map both x and x' to the same value. To bound the second term in going from the second-last to last lines we used $\sum_x p_x^2 \le \max_x p_x = 2^{-H_{\min}(X)}$ and the assumption $H_{\min}(X) \ge k$.

Plugging this back into the bound on the trace distance from (i) we obtain

$$D(\rho_{\mathrm{Ext}(X,Y)Y}, 2^{-(d+m)}\mathbb{I}) \le 2^{\frac{m-k}{2}-1}.$$

The right-hand side is less than ε if $k \ge m + 2\log(1/\varepsilon)$, proving the theorem.

QUIZ 6.4.1 *What is the collision probability of a source that always outputs the same string? Equivalently, what is $CP(\{1\})$?*

(a) 0
(b) $\frac{1}{2}$
(c) 1

QUIZ 6.4.2 *What is the collision probability of a uniformly random source on n bits? Equivalently, what is $CP(\{p_j\})$, where $p_j = 2^{-n}$ for $j \in \{1, \ldots, 2^n\}$?*

(a) 2^{-2n}
(b) 2^{-n}
(c) $2^{-n/2}$

6.4.4 The Pretty-Good Measurement and Quantum Side Information

We would like to extend the proof in the previous section to the case where the source X is correlated with some quantum side information E, that is, $\rho_{XE} = \sum_x |x\rangle\langle x| \otimes \rho_x^E$ is an arbitrary cq-state such that $H_{\min}(X|E) \geq k$. Before diving into this, let's make a small detour by considering the related problem of optimally distinguishing between a set of quantum states. Note that this section is technically more advanced than most of the book, and it is not required to read it before continuing with the following chapters. We still encourage you to give it a try!

The Pretty-Good Measurement

Let $\rho_{XE} = \sum_x |x\rangle\langle x| \otimes \rho_x^E$ be a cq-state. What is the optimal probability with which Eve, holding the quantum system E, can successfully guess x? We've seen this problem already: the answer is captured by the guessing probability,

$$P_{\text{guess}}(X|E)_\rho = \max_{\{M_x\}} \sum_x \text{tr}(M_x \rho_x^E) , \qquad (6.8)$$

where the maximum is taken over all POVM $\{M_x\}$ on E. But what is the best POVM? If $x \in \{0,1\}$ takes only two values you know the answer: in this case we can write

$$\text{tr}(M_0\rho_0^E) + \text{tr}(M_1\rho_1^E) = \text{tr}\left(\frac{M_0 + M_1}{2} \cdot \left(\rho_0^E + \rho_1^E\right)\right) + \text{tr}\left(\frac{M_0 - M_1}{2} \cdot \left(\rho_0^E - \rho_1^E\right)\right)$$

$$\leq \frac{1}{2} + \frac{1}{2}D(\rho_0^E, \rho_1^E) ,$$

and moreover the last inequality is an equality if M_0 and M_1 are the projectors on the positive and negative eigenspaces of the Hermitian matrix $\rho_0^E - \rho_1^E$ respectively.

When $|X| > 2$, unfortunately the situation is a bit more murky. The problem of finding the optimal measurement can be solved efficiently with a computer by expressing the optimization problem (6.8) as a *semidefinite program*, a generalization of linear programs for which there are efficient algorithms. But what we'd really like is a nice, clean mathematical expression for what the optimal measurement is, so that we can work with it in our proofs! No such simple closed form is known. However, what we can do is find a simple measurement that always achieves *close* to the optimum: the *pretty-good measurement*.

So what is this "pretty-good" measurement? To get some intuition let's first look at the case where the states ρ_x^E are perfectly distinguishable; for example, $\rho_x^E = p_x |x\rangle\langle x|$ is simply a classical copy of X. Then it is clear what we should do: measure in the computational basis, and recover x! Observe that in this case the POVM elements M_x are directly proportional to ρ_x: we can think of the states as "pointing" in some direction correlated with x, and it is natural to make a measurement along that direction.

Can we generalize this idea? Let's try defining $M_x = \rho_x^E$. This is positive semidefinite, so it satisfies the first condition for a POVM. However, $\sum_x M_x = \sum_x \rho_x^E = \rho^E$

is not necessarily the identity, as required by the second condition. The solution? Normalize!

> **Definition 6.4.4.** *Given a collection of positive semidefinite matrices* $\{\rho_x\}$, *the* pretty-good measurement *(PGM) associated with the collection is the POVM with elements*
>
> $$M_x = \rho^{-1/2}\rho_x\rho^{-1/2},$$
>
> *where* $\rho = \sum_x \rho_x$ *and the inverse is the Moore–Penrose pseudo-inverse, i.e. we use the convention* $0^{-1} = 0$.

Remark 6.4.2 Note how we handled division by zero in the definition. Defining division by zero may seem odd, but this convention makes sense in the context of Hermitian matrices. If ρ is orthogonal to some subspace then the pseudo-inverse ρ^{-1} should also be orthogonal to that subspace. A useful property of this convention is that it makes it so that if P is an orthogonal projection on a space that contains the support of ρ, then $(P\rho P)^{-1} = P\rho^{-1}P$.

How well does the PGM compare to the optimal guessing measurement? Let $\{N_x\}$ be an optimal guessing POVM for Eve. Then by definition

$$
\begin{aligned}
P_{\text{guess}}(X|E) &= \sum_x \text{tr}\left(N_x\rho_x^E\right) \\
&= \sum_x \text{tr}\left((\rho^{1/4}N_x\rho^{1/4})(\rho^{-1/4}\rho_x^E\rho^{-1/4})\right) \\
&\leq \left(\sum_x \text{tr}\left(\rho^{1/2}N_x\rho^{1/2}N_x\right)\right)^{1/2}\left(\sum_x \text{tr}\left(\rho^{-1/2}\rho_x^E\rho^{-1/2}\rho_x^E\right)\right)^{1/2} \\
&\leq \left(\text{PGM}(X|E)\right)^{1/2},
\end{aligned}
$$

where

$$\text{PGM}(X|E) = \sum_x \text{tr}(M_x\rho_x^E) = \sum_x \text{tr}\left(\rho^{-1/2}\rho_x\rho^{-1/2}\rho_x\right) \tag{6.9}$$

is the success probability of the PGM in the guessing task. The second and third lines are the most important here. To go from the first to the second line we inserted factors $\rho^{1/4}$ and $\rho^{-1/4}$ that cancel each other out (using cyclicity of the trace), but are important for normalization. To go from the second to the third line we used the Cauchy–Schwarz inequality (see Box 6.3) twice: first, for each x we apply the matrix version of the inequality, and second, we apply the usual version to the coefficients $a_x = \text{tr}(\rho^{1/2}N_x\rho^{1/2}N_x)$ and $b_x = \text{tr}(\rho^{-1/2}\rho_x^E\rho^{-1/2}\rho_x^E)$. Finally, to get to the last line we used $\sum_x N_x = \mathbb{I}$ to bound the first term, and the definition of the PGM for the second.

QUIZ 6.4.3 *Consider the two states* $\rho_0 = |0\rangle\langle0| = \begin{pmatrix} 1 & 0 \\ 0 & 0 \end{pmatrix}$ *and* $\rho_1 = |+\rangle\langle+| =$ $\frac{1}{2}\begin{pmatrix} 1 & 1 \\ 1 & 1 \end{pmatrix}$. *Suppose that a referee flips a fair coin, and gives you* ρ_0 *if she flipped heads and* ρ_1 *if she flipped tails. Your goal is to guess the referee's coin flip. What is your best possible probability of success if you start by measuring the state you received in the standard basis?*

(a) $\frac{1}{2}$

(b) $\frac{1}{2} + \frac{\sqrt{2}}{4}$

(c) $\frac{3}{4}$

(d) $\frac{\sqrt{2}}{2}$

QUIZ 6.4.4 *Continuing the previous question, what is your probability of success if you use the PGM associated with the states* ρ_0 *and* ρ_1?

(a) $\frac{1}{2}$

(b) $\frac{1}{2} + \frac{\sqrt{2}}{4}$

(c) $\frac{3}{4}$

(d) $\frac{\sqrt{2}}{2}$

Proof of the Leftover Hash Lemma with Quantum Side Information

Proof The proof follows the same structure as the proof we saw for the case with no side information, but it is slightly more involved technically. We will use the following inequality: for any positive Hermitian σ and positive semidefinite τ such that $\text{tr}(\tau) = 1$ and the support of τ contains the support of σ,

$$\|\sigma\|_1 \leq \text{tr}\big((\tau^{-1/4}\sigma\tau^{-1/4})^2\big)^{1/2} . \tag{6.10}$$

This inequality can be shown using some standard matrix analysis techniques, and we take it for granted; see the chapter notes for a pointer to a proof.   ∎

(i) From Trace Distance to Collision Probability
Our goal is to bound $D(\rho_{\text{Ext}(X,Y)YE}, 2^{-(m+d)}\mathbb{I} \otimes \rho_E)$, where Y is uniformly distributed and X is such that $H_{\min}(X|E) \geq k$. We can write

$$\rho_{\text{Ext}(X.Y)YE} = \sum_{z,y} |z\rangle\langle z| \otimes |y\rangle\langle y| \otimes \rho_{zy}, \qquad \text{with} \qquad \rho_{zy} = 2^{-d}\sum_{x:\, f_y(x)=z} \rho_x .$$

Note that our normalization makes it so that

$$\sum_{z,y} \text{tr}(\rho_{zy}) = 2^{-d}\sum_{x,y} \text{tr}(\rho_x) = \text{tr}(\rho) = 1 .$$

Since the state $\rho_{\text{Ext}(X,Y)YE}$ is a ccq-state, using the definition of the trace distance (see also Example 5.1.2) we can expand

$$
D(\rho_{\text{Ext}(X,Y)YE}, 2^{-(d+m)}\mathbb{I} \otimes \rho_E) = \frac{1}{2} \sum_{z,y} \left\| \rho_{zy} - 2^{-(d+m)}\rho \right\|_1
$$

$$
\leq 2^{\frac{m+d}{2}-1} \left(2^{-(m+d)} \sum_{z,y} \text{tr}\left((\rho^{-1/4}(\rho_{zy} - 2^{-m}\rho)\rho^{-1/4})^2 \right) \right)^{1/2}
$$

$$
= 2^{\frac{m}{2}-1} \left(2^d \sum_{z,y} \text{tr}\left(\rho_{zy}\rho^{-1/2}\rho_{zy}\rho^{-1/2} \right) - 2^{-m} \right)^{1/2},
$$

where for the second line we first applied (6.10) for each (y,z) with $\sigma = \rho_{zy} - 2^{-(d+m)}\rho$ and $\tau = \rho$, and then the usual Cauchy–Schwarz inequality. Do you recognize the expression in the last line? Using the notation from (6.9), we have

$$
\text{PGM}(Z|YE) = 2^d \sum_z \text{tr}\left(\rho_{zy}\rho^{-1/2}\rho_{zy}\rho^{-1/2} \right),
$$

so the sequence of equations above show that

$$
D(\rho_{\text{Ext}(X,Y)YE}, 2^{-(d+m)}\mathbb{I} \otimes \rho_E) \leq 2^{\frac{m}{2}-1} \left(\text{PGM}(Z|YE) - 2^{-m} \right)^{1/2}.
$$

We have thus managed to relate the distance from uniform to the advantage of the PGM over random guessing (which would succeed with probability 2^{-m}). We can understand this step of the proof as a reduction from arbitrary attacks of an adversary to the extractor, whose optimal success probability is expressed in the first line, to attacks of a very specific form, where the adversary, given a sample (z,y), measures its side information using the PGM associated with the family of states $\{\rho_{zy}\}$. The square root factor on the right-hand side represents the fact that the PGM is quadratically far from optimal. What is the point of losing this square root? The PGM has a crucial advantage, which we are going to use in the second step of the proof: it has a form of "linearity" in the sense that the PGM operators associated with the family of states $\{\rho_{zy}\}$ can be obtained by summing up PGM operators associated with the states $\{\rho_x\}$. Let's see how this works in our favor.

(ii) A Bound on the Collision Probability

Proceeding exactly as in the case with no side information, we can calculate

$$
\text{PGM}(Z|YE) - 2^{-m} = 2^{-d} \sum_{y,z} \sum_{\substack{x,x': \\ f_y(x)=f_y(x')=z}} \text{tr}(\rho_x\rho^{-1/2}\rho_{x'}\rho^{-1/2}) - 2^{-m}
$$

$$
= 2^{-d} \sum_{y,z} \Big(\sum_{\substack{x \neq x': \\ f_y(x)=f_y(x')=z}} \text{tr}(\rho_x\rho^{-1/2}\rho_{x'}\rho^{-1/2})
$$

$$
+ \sum_{x: f_y(x)=z} \text{tr}(\rho_x\rho^{-1/2}\rho_x\rho^{-1/2}) \Big) - 2^{-m}
$$

$$
= 2^{-m} \sum_{x \neq x'} \text{tr}(\rho_x\rho^{-1/2}\rho_{x'}\rho^{-1/2}) + \sum_x \text{tr}(\rho_x\rho^{-1/2}\rho_x\rho^{-1/2}) - 2^{-m}
$$

$$
\leq \text{PGM}(X|E).
$$

Using the 2-universal hashing property, we have managed to relate the advantage over random of the PGM in guessing Z, to the success probability of the PGM to guess X directly. But the last expression is, by assumption, at most $2^{-H_{\min}(X|E)}$, since the guessing probability achieved from using the PGM cannot be more than the optimal one. Together with the bound proven in step (i) we finally obtain

$$D(\rho_{\mathrm{Ext}(X.Y)Y}, 2^{-(d+m)}\mathbb{I}) \leq 2^{\frac{m-k}{2}-1},$$

precisely the same bound as when there was no side information at all.

CHAPTER NOTES

Classical extractors, studied without side information, have a long history in theoretical computer science and pseudorandomness. For many uses of them and their relation to other combinatorial objects we can recommend the book by S. P. Vadhan, *Pseudorandomness* (Foundations and Trends in Theoretical Computer Science, 7(1–3), 2012).

The task of privacy amplification is first introduced by C. H. Bennett, G. Brassard, and J.-M. Robert (Privacy amplification by public discussion. *SIAM Journal on Computing*, 17(2):210–229, 1988), who explain how to solve it in the case of a classical eavesdropper. The relevance of "quantum-proof" extractors, that is, classical extractors that are secure for general classical-quantum sources, to privacy amplification against quantum eavesdroppers is pointed out in R. Renner (Security of quantum key distribution. *International Journal of Quantum Information*, 6(01):1–127, 2008). The two-universal extractor is the first extractor to have been proven secure against quantum adversaries, in R. Konig, U. Maurer, and R. Renner (On the power of quantum memory. *IEEE Transactions on Information Theory*, 51(7):2391–2401, 2005). Subsequently, better extractors have also been proven secure, in particular with a logarithmic key length by A. De, et al. (Trevisan's extractor in the presence of quantum side information. *SIAM Journal on Computing*, 41(4):915–940, 2012).

The pretty-good measurement is defined and analyzed in a paper by P. Hausladen and W. K. Wootters (A 'pretty good' measurement for distinguishing quantum states. *Journal of Modern Optics*, 41(12):2385–2390, 1994). For a proof of the trace inequality (6.10), we refer to Lemma 5.1.2 in R. Renner.

PROBLEMS

6.1 Using the pretty-good measurement

Alice sends Bob one of the three states

$$\rho_0 = |0\rangle\langle 0| = \begin{pmatrix} 1 & 0 \\ 0 & 0 \end{pmatrix}, \ \rho_1 = \frac{1}{2}\mathbb{I} = \frac{1}{2}\begin{pmatrix} 1 & 0 \\ 0 & 1 \end{pmatrix}, \ \rho_2 = |1\rangle\langle 1| = \begin{pmatrix} 0 & 0 \\ 0 & 1 \end{pmatrix}$$

with equal probability. Bob wants to determine which state Alice sent him. In this problem we work out the success probability of different possible strategies for Bob. A strategy is a POVM $\{B_0, B_1, B_2\}$ and its success probability is

$$p_{\text{succ}}(B) = \frac{1}{3}\text{tr}(B_0\rho_0) + \frac{1}{3}\text{tr}(B_1\rho_1) + \frac{1}{3}\text{tr}(B_2\rho_2),$$

i.e. it is the probability that Bob correctly guesses the state sent by Alice, assuming that Alice sends him each of the three states with equal probability 1/3.

The first strategy that we consider consists in Bob measuring in the Hadamard basis.

1. Bob sets $B_0 = |+\rangle\langle+|$, $B_1 = 0$, and $B_2 = |-\rangle\langle-|$. That is, if he measures $|+\rangle$ then he guesses that the state sent was $\rho_0 = |0\rangle\langle 0|$ and if he measures $|-\rangle$ he guesses that the state sent was $\rho_2 = |1\rangle\langle 1|$. What is Bob's success probability?

After trying that procedure, Bob decides to switch to measuring ρ in the standard basis.

2. Bob sets $B_0 = |0\rangle\langle 0|$, $B_1 = 0$, and $B_2 = |1\rangle\langle 1|$. What is his new success probability?

Bob decides that he's done with ad hoc approaches and wants to use a measurement that will be somewhat reliable.

3. What is Bob's success probability if he uses the pretty-good measurement?

Bob wants to know whether he's found the optimal measurement. To help find this out, he will apply the following fact. Suppose that σ is a positive semidefinite matrix (not necessarily of trace 1) such that $p_i \rho_i \leq \sigma$ for each i. Then the optimal success probability of a distinguishing measurement on the ensemble $\rho = \sum_i p_i \rho_i$ is at most $\mathrm{tr}\,\sigma$. (Recall that $A \leq B$ means that $B - A$ is positive semidefinite.)

4. What is the best upper bound Bob can derive from this fact?
5. Can you prove the above-claimed fact?

6.2 Deterministic extractors on bit-fixing sources

We saw that no deterministic function can serve as an extractor for all random sources of a given length. This doesn't rule out the possibility that a deterministic extractor can work for some restricted class of sources! In this problem we'll construct deterministic extractors that work for some specific sources.

Let n be even and fix $t < n/2$. Define the following sources on $\{0,1\}^n$.

I. X_0 is $100\cdots00$ on the first t bits and uniformly random on the last $n - t$ bits.
II. X_1 is uniformly random over the set of strings with an even number of 0's.
III. X_2 is uniformly random over the set of strings where the first $n/2$ bits are the same as the last $n/2$ bits.

1. What is the min-entropy $H_{\min}(X_0)$?
2. What is the min-entropy $H_{\min}(X_1)$?
3. What is the min-entropy $H_{\min}(X_2)$?

Now consider the following functions.

- $f_0(x) =$ the XOR of the first t bits of x.
- $f_1(x) = x_L \cdot x_R$, where $x = (x_L, x_R)$ are the left and right halves of x and $\cdot$ denotes inner product modulo 2.
- $f_2(x) =$ the XOR of all of the bits of x.

We're interested in which functions serve as extractors on which sources. For example, $f_0(X_0)$ is always equal to 1, while $f_2(X_0)$ is equal to a uniform random bit.

4. Which of the following random variables have positive entropy?
   I. $f_1(X_1)$
   II. $f_1(X_2)$

    III. $f_2(X_0)$
    IV. $f_2(X_1)$
    V. $f_2(X_2)$

Now suppose that Alice and Bob share a classical secret $X \in \{0,1\}^n$ which they are using to hide communications from Eve. Alice and Bob make an error and, as a result, Eve learns $t < n$ bits of X. If Alice and Bob knew which bits Eve learned, then they could throw those bits out and keep the rest of the bits to use as their secret. However, if they don't know which are the leaked bits, then things get trickier. If t is much smaller than n, then Alice and Bob still have lots of information that Eve does not; in particular, we have $\mathrm{H}_{\min}(X|E) = n - t$. How can they make use of this without generating new shared randomness?

(You may notice that this is exactly the problem of extracting randomness from a bit-fixing source, as introduced in the chapter.)

5. Suppose Alice and Bob take the XOR of all of their bits (including the ones that Eve has learned!), producing just one output bit. What is the correlation between this bit and the bits that Eve has learned?
6. What is the largest t such that the XOR function manages to extract a bit of randomness?

Alice and Bob now want to extract many bits of randomness instead of just one. Their idea is to take subsets of the bits and treat each subset as its own bit-fixing source. The trouble that they run into is that they don't know which bits Eve will learn.

7. What is the largest number of independent subsources they can make such that it is possible for them to securely extract one bit of randomness from each subsource?

6.3 Conditional min-entropy

It is well known, and not hard to verify by direct calculation, that the conditional Shannon entropy satisfies an exact "chain rule," which is the equality $H(XY) = H(X) + H(Y|X)$. This equality holds for any random variables X, Y, even correlated. In this problem, we'll investigate the chaining properties of conditional min-entropy.

Recall the definition of conditional min-entropy as $\mathrm{H}_{\min}(X|E) = -\log p_{\mathrm{guess}}(X|E)$, where

$$p_{\mathrm{guess}}(X|E) = \sum_e \Pr(E = e) \cdot p_{\mathrm{guess}}(X|E = e) \, .$$

As a warm-up, let's consider a very simple two-bit classical source XY with the following distribution:

$$p(XY = 00) = \frac{1}{4}, \ p(XY = 01) = \frac{1}{4}, \ p(XY = 10) = \frac{1}{4}, \ p(XY = 11) = \frac{1}{4} \, .$$

1. Compute $\mathrm{H}_{\min}(XY)$ for this source.
2. Compute $\mathrm{H}_{\min}(X) + \mathrm{H}_{\min}(Y|X)$ for this source.

Now let's consider a two-bit classical source XY with the following distribution:

$$p(XY = 00) = \frac{1}{2}, \ p(XY = 01) = \frac{1}{8}, \ p(XY = 10) = \frac{1}{4}, \ p(XY = 11) = \frac{1}{8} \, .$$

3. Compute $H_{min}(XY)$ for this source.
4. Compute $H_{min}(X) + H_{min}(Y|X)$ for this source.

Finally, consider the following distribution:

$$p(XY = 00) = \frac{3}{8},\ p(XY = 01) = \frac{1}{4},\ p(XY = 10) = \frac{5}{16},\ p(XY = 11) = \frac{1}{16}.$$

5. Compute $H_{min}(XY)$ for this source.
6. Compute $H_{min}(X) + H_{min}(Y|X)$ for this source.

7

Distributing Keys

Let's start distributing keys! We will approach this objective in a series of steps, ending up with the famous BB'84 quantum key distribution (QKD) protocol in the next chapter. In this chapter, we introduce the task of key distribution, give some example protocols in simplified settings, and discuss the problem of information reconciliation. Before we describe the task of key distribution in detail, let's first discuss the meaning of "honest" and "dishonest" parties in a cryptographic protocol.

7.1 Honest and Dishonest

Imagine that several parties engage in some communication protocol. For example, the parties may want to send each other some information, or one of them may want to search a database that is in the possession of another, or they may want to participate in an auction for some public goods, etc. The goal of cryptography is to protect the "honest" parties from the "dishonest" ones during such an interaction. What does it mean to be honest or dishonest?

> **Definition 7.1.1 (Honest and dishonest (informal)).** *A communication protocol between several parties specifies the actions that each party is intended to take in the protocol. These actions may depend on an* input *that the party is supposed to have, and each party may produce an* output *at the end of the protocol.*
>
> - *A party is called* honest *if they follow the protocol precisely. That is, the party initializes their part of the protocol in the correct way, executes all steps as dictated, and returns an output as expected.*
> - *A party is called* dishonest *or* malicious *if they do not follow the protocol. Instead, they can deviate from the protocol in an arbitrary way. This includes, for example, refusing to answer a question, sending some unsolicited information, etc.*
> - *All malicious parties are simultaneously controlled by an entity that we call the* adversary. *If there are two or more malicious parties, the same adversary can make them coordinate their actions in a dishonest way.*

When designing a cryptographic protocol we have to ask ourselves what kind of malicious parties the protocol should protect against. In other words, whether there are some limits to what a malicious party can do to break the protocol, and how many resources they have at their disposal. To be precise we should say "what are the limits," and not "whether there are limits," because there are always limits – for example, if the adversary is allowed to control all users in the protocol then surely no meaningful security can be obtained. Concretely, a minimal assumption that is often left implicit in cryptographic protocols is that an honest party, let us call her Alice, sits in an impenetrable lab that the adversary does not have control over. In other words, Alice can perform local computations without the adversary's knowledge. Only when Alice sends information out of her own lab along a communication channel with another party does the adversary have an opportunity to intercept or tamper with the protocol execution. Here we make that assumption as well. In Chapter 9 we will see that by making use of quantum information it is possible to weaken this demand.

7.2   Secure Key Distribution

In this book we consider many different cryptographic tasks, but the centerpiece is uncontestably the task of key distribution. In this task there are two honest protagonists, Alice and Bob, and one malicious eavesdropper, Eve. Alice and Bob each have secure labs that they have control over and Eve cannot peek into. Moreover, they have access to a variety of communication channels. However, in general Eve may also have access to these channels. A key distribution protocol is a collection of actions for the honest users Alice and Bob such that at the end of the protocol two conditions should hold. First of all, if all goes well Alice and Bob each possess the same key K, which is simply some string of bits. Second, no one else than them should have any information about K. This "no one else" is embodied by the eavesdropper, Eve (who can behave as she likes), and so we require that from the point of view of Eve the key K appears uniformly random.

Let's make these requirements more precise in order to arrive at a formal definition of security for key distribution that we can work with. We consider the two requirements in sequence. The first requirement is called *correctness*: whatever Alice and Bob return, call it K_A for Alice and K_B for Bob, these should always be the same: $K_A = K_B$. As it turns out, this requirement is already too strong, for two reasons. First of all we should always allow a chance that the protocol *aborts*. For example, Eve could completely block the communication channels. In this case Alice and Bob would be forced to "give up." So we can only require that *whenever Alice and Bob both return a key K_A and K_B, then these keys are the same*. To model the case where they abort we introduce a new symbol $\perp$; so if $K_A = \perp$ or $K_B = \perp$ then this means that Alice or Bob decided to abort, respectively. (They do not necessarily both always abort at the same time, as Alice could suspect something but fail to communicate her suspicion to Bob, for example because Eve is interfering with the communication channel.) Even this condition is too strong unfortunately. This is because even for a very good protocol we need to allow a chance that things go wrong but the honest users still don't detect it. We will use a parameter ε_c to denote the probability that this happens, and strive to design protocols that get the smallest possible ε_c. We can now give the first half of the formal security definition of key distribution.

Definition 7.2.1 (ε_c-correctness). *A key distribution protocol between Alice and Bob is ε_c-correct if the following holds. Let K_A and K_B denote the user's outcomes in the protocol. Then*

$$\mathrm{Prob}(K_A \neq \perp \wedge K_B \neq \perp \wedge K_A \neq K_B) \leq \varepsilon_c \ .$$

Next we consider the *secrecy* requirement. This is the requirement that "Eve has no information about the key." How do we formalize such an assumption? Observe that at the end of the protocol we can always write the joint state of the key K_A, in register K, and Eve's quantum state, in register E, as a cq-state ρ_{KE}. If Eve has no information about K_A, as we saw in Chapter 5 this means that $\rho_{KE} = 2^{-\ell}\mathbb{I}_K \otimes \rho_E$, where ℓ is the length of K_A in bits. We call this state the "ideal" state. Now, as for correctness it would be unrealistic to require that the final state of the protocol is always exactly the ideal state. This is because in an actual implementation of the protocol there will always be some small errors, such as noise on the quantum channel, which introduce imperfections in the final outcome; as long as these imperfections don't seriously compromise security it makes sense to allow them. To accommodate this we only require the final state to be very close to the ideal state: ε_s-close in trace distance. Second, we can only require that this is the case when the eavesdropper is not doing something too crazy, i.e. when the chance that she makes the protocol abort is not too high.[1] Here is the formal definition.

Definition 7.2.2 (ε_s-secrecy). *A key distribution protocol is ε_s-secret if the following holds. Let $\mathrm{Pr}(\text{abort})$ denote the probability that either Alice or Bob returns $\perp$. Then it should be the case that*

$$(1 - \mathrm{Pr}(\text{abort}))\big\|\rho_{KE}^{\mathrm{real}} - \rho_{KE}^{\mathrm{ideal}}\big\|_1 \leq \varepsilon_s \ , \tag{7.1}$$

where $\rho_{KE}^{\mathrm{real}}$ is the joint state of Alice's output K_A and the eavesdropper in an execution of the protocol and $\rho_{KE}^{\mathrm{ideal}} = \frac{\mathbb{I}_K}{2^\ell} \otimes \rho_E$.

Observe that (7.1) is equivalent to saying that either $\mathrm{Pr}(\text{abort})$ is very close to 1, in which case the equation is satisfied, or it is not very close to 1 and in that case we require that $\rho_{KE}^{\mathrm{real}} \approx \rho_{KE}^{\mathrm{ideal}}$.

In Box 7.1 we describe some essential assumptions that limit the power of the eavesdropper in all key distribution protocols that we will consider. In addition, let's now discuss the type of communication channel that Alice and Bob may have access to in the protocol. Here are some channel types that we may consider. For each type of channel, we describe what access the eavesdropper, Eve, has to communication made over that channel.

---

1 This is because we can always consider an eavesdropper that, for example, always forces the protocol to abort unless some very specific conditions are satisfied, which would somehow guarantee that the key is some fixed value such as 0^ℓ, a sequence of ℓ zero bits.

BOX 7.1   Assumptions

Definition 7.2.2 refers to "the eavesdropper." It is quite important to clarify exactly what is the power that this eavesdropper may have. For key distribution we make the following assumptions:

1. All parties are bound by the laws of quantum physics. Even though this may seem obvious, it is worth stating explicitly. At the end of the day our security proof will model all possible actions of the eavesdropper, and the framework we will use for this is quantum mechanics. If quantum mechanics is wrong or incomplete, our security proof may not hold against adversaries that make use of unexpected physical effects. In particular, we don't consider relativistic effects, black holes, and the like. (Luckily — we'd need quite a few more books to set these up!)
2. The users Alice and Bob behave honestly, i.e. as described in the protocol. They have access to private labs that are perfectly shielded from the eavesdropper. All computations performed in their labs, classical or quantum, are done perfectly. This includes the generation of random numbers, preparation of qubits, measurements, etc., whenever required by the protocol. (Later we will discuss a weakening of this requirement where the qubit preparation and measurement devices may make small errors.)
3. The eavesdropper has access to all communication that takes place between the users, and can intercept and modify messages at will (with one important limitation: see the description of the authenticated channel below). In addition, the eavesdropper may make use of an arbitrarily large classical or quantum computer.

1. A classical channel: Alice and Bob can send classical bits in either direction over this channel. Eve has complete access to the channel. In particular, she can read all messages, copy them, modify them, and even impersonate Alice (or Bob).
2. A classical *authenticated* channel (CAC): A classical communication channel with one extra guarantee: For any message sent on that channel, Alice and Bob are promised that the message originated from Bob or Alice respectively, and moreover that it has not been altered in any way. This channel is not secret, because Eve can still read all the messages that travel on it, but Eve cannot impersonate Alice or Bob or alter messages traveling over the channel.
3. A classical *secret* channel: A classical communication channel such that Eve cannot learn any information about the messages traveling over the channel. While she cannot hope to gain any information about any messages sent by the users, Eve can still impersonate them to send fake messages (or delete or replace messages that they send, without reading them).
4. A classical *secret and authenticated* channel: A classical communication channel combining both guarantees above.

5. A *quantum communication* channel: A channel where Alice may send quantum information (in particular, in the form of qubits) to Bob and vice versa, such that Eve has full access to all the quantum communication.

Concretely, the protocols that we consider in this chapter and the following ones will always assume that Alice and Bob have access to (1) a classical authenticated channel and (2) a quantum communication channel. The assumption that the classical channel is authenticated is an important one and we discuss it in more detail in Section 7.5. Before we get to the real protocols, let us start by making our life easier by assuming that Alice and Bob have access to some special kinds of communication channels such that the eavesdropper's access is further restricted.

> **QUIZ 7.2.1** *Alice wants to send one bit of information to Bob and she does not require secrecy. However, Alice wants to make sure Bob knows she has sent the bit herself. She has two options: to send this classical bit over the classical authenticated channel, or to encode it in a qubit and send it to Bob over the quantum communication channel. Only one use of one of those channels is allowed. Which channel should she use?*
>
> **(a)** *Classical authenticated channel*
> **(b)** *Quantum communication channel*

7.3   Distributing Keys Given a Special Classical Channel

In this section we consider how Alice and Bob can generate a key from a very special classical channel, as this includes many of the essential ideas we will need to analyze our real QKD protocols in the next chapters. As always in key distribution, we assume that Alice and Bob have access to a CAC. In addition, here we allow them to use a special channel which has the property that Eve cannot completely learn all messages going across: her ability to eavesdrop is somehow *guaranteed* to be limited. Let $0 \leq q \leq 1$ be a parameter. The channel is called the *binary symmetric channel*, which we denote as $Eve\text{-}BSC(q)$ (Figure 7.1). It has the following properties: whenever Alice sends a bit $b \in \{0,1\}$ across the $Eve\text{-}BSC(q)$,

- Bob correctly receives b and
- Eve obtains the bit b correctly with probability q, otherwise with probability $1-q$ she receives the bit $1-b$ (but she does not know whether the bit is correct or not).

Can we design a correct and secure key distribution protocol using such a channel? At least for some values of q the answer is yes. For example, if $q = 1/2$ then you can see that Eve always learns a uniformly random bit, i.e. nothing, and so a secure key can be exchanged directly. What about other values of q? Let's estimate how much information Eve can gain about a message sent by Alice to Bob on the $Eve\text{-}BSC(q)$. Suppose that Alice chooses a message $x = x_1, \ldots, x_n \in \{0,1\}^n$ uniformly at random and sends it to Bob. By definition of $Eve\text{-}BSC(q)$, Eve will obtain a string $e \in \{0,1\}^n$

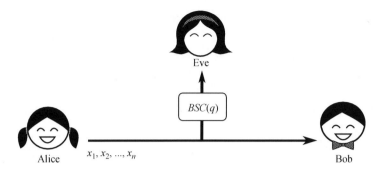

Fig. 7.1 Distributing keys over a special classical channel.

such that on average about qn entries of e are correct and $(1-q)n$ entries have been flipped. Bob, however, receives x exactly. So Alice and Bob have the same string, but Eve has some amount of information about it which can be quantified as a function of the parameter q. In this situation, how can Alice and Bob *extract* a secure key out of their partially secret common information?

If you didn't read Chapter 6 on your way here, now might be a good time to do so! In that chapter we introduced the task of privacy amplification, which is precisely what Alice and Bob have to do here. Moreover, we also gave a method to solve privacy amplification: apply a randomness extractor! This suggests the following protocol.

Protocol 3 (Key distribution using a binary symmetric channel) Let q be the parameter of the *Eve–BSC*. Let integers n, ℓ be chosen such that $\ell \leq n(-\log q) - 2\log(1/\varepsilon_s)$. Let Ext : $\{0,1\}^n \times \{0,1\}^d \to \{0,1\}^\ell$ be a 2-universal randomness extractor (see Section 6.4.2). In the protocol,

1. Alice chooses a string $x = x_1, \ldots, x_n \in \{0,1\}^n$ uniformly at random and sends each bit x_j, $j = 1, \ldots, n$, to Bob over the *Eve–BSC(q)*.
2. Alice picks a uniformly random seed $r \in \{0,1\}^d$ and computes $k_A = \mathrm{Ext}(x,r)$.
3. Alice sends r to Bob over the CAC.
4. Bob computes $k_B = \mathrm{Ext}(x,r)$.
5. Alice returns k_A and Bob returns k_B.

Remember that we need to establish two things for this to be a valid QKD protocol. First, we want that the protocol is ε_c-correct, that is, Alice and Bob output the same key (except for some small probability of failure). Second, we want to show that the protocol is ε_s-secure. To see that the protocol is correct, note that the special channel is such that Bob receives all bits correctly. That is, he obtains $x = x_1, \ldots, x_n$ without error. Because r was sent over the CAC, Bob is guaranteed to receive the correct value. In particular, he knows which function $\mathrm{Ext}(\cdot, r)$ to apply to x, and so his value $k_B = \mathrm{Ext}(x,r)$ is such that $k_A = k_B$ with certainty. So, this protocol is 0-correct.

Why would the protocol be ε_s-secure? Let us first note that by definition Eve's probability of guessing each bit correctly is precisely given by q. Let's also assume that $q > 1/2$, so Eve gets the correct value with probability more than $1/2$. (We can always reduce to this case by deterministically flipping each bit received by Eve.) In this case Eve's best guess for the real bit is clearly the value that she obtained. Thus, if X_i is the random variable associated with Alice's i-th bit, and E_i the value received by Eve, then $P_{\text{guess}}(X_i|E_i) = q$. Since all the bits are chosen and communicated independently,

$$P_{\text{guess}}(X|E) = q^n \,,$$

which can be rewritten as

$$H_{\min}(X|E) = -\log P_{\text{guess}}(X|E) = n(-\log q) \,.$$

If Ext is, for example, the 2-universal extractor from Definition 6.4.3, then by Theorem 6.4.1 we are guaranteed that

$$D\left(\rho_{KRE}, \frac{\mathbb{I}}{2^\ell} \otimes \rho_{RE}\right) \leq \varepsilon_s \tag{7.2}$$

whenever $\ell \leq H_{\min}(X|E) - 2\log(1/\varepsilon_s)$. Here R is the register used to store the seed, or "randomness." Therefore, whenever we choose the parameter ℓ such that

$$\ell \leq n(-\log q) - 2\log(1/\varepsilon_s) \tag{7.3}$$

we obtain a protocol that is ε_s-secure. In cryptography, we typically fix ε_s and ℓ in advance, and then ask how large n has to be in order to achieve the desired key length ℓ with the guarantee ε_s. Here, we get that $n = \ell/\log(1/q) + \Omega(\log(1/\varepsilon_s))$ suffices.

We pause to notice that for the second phase of the protocol, when Alice sends the seed r to Bob, we did not make use of the Eve–$BSC(q)$. Instead, we used the CAC, which means that Eve can entirely learn the seed r. Yet we still obtain security even under that assumption, thanks to the guarantee provided by Eq. (7.2), which includes the key K: this equation means that even if the eavesdropper holding system E learns which function $\text{Ext}(\cdot, r)$ is applied, they nevertheless cannot learn anything about the key (up to ε_s)! The fact that Eve learns r only later, after having obtained the system E, is crucial. If Eve were to know r ahead of time, then in general she could tailor her entire attack to the knowledge of r. In our scenario where we fix exactly what Eve gets then this doesn't really matter, but when we see the real protocols it will be important that r is only determined after the information from which the key will be created has been exchanged.

Exercise 7.3.1   Consider what happens if Eve gets the bit with probability q, but knows whether her intercept attack was successful. (With the remaining probability $1 - q$, she gets a special symbol $\star$ indicating that the bit was lost.) If we fix ε_s and n, can you obtain a longer or shorter key in this case?

The way we used the extractor in this protocol is general: we see that whenever the protocol is such that the min-entropy $H_{\min}(X|E)$ must be high, and moreover such that Bob has the same information as Alice, then Alice and Bob can always extract a key that has length $\approx H_{\min}(X|E) - O(\log(1/\varepsilon_s))$ which is ε_s-secure against Eve.

Example 7.3.1 Consider another special channel where all the information that Alice sends automatically goes to Eve, *except* that Eve has limited memory and can only store a maximum of S bits in total. If Alice sends a completely random n-bit string X across the channel, then $H_{min}(X) = n$, and Eve's knowledge about X is

$$H_{min}(X|E) \geq H_{min}(X) - \log|E| \geq n - S, \qquad (7.4)$$

where the first inequality is by the partial chain rule for the conditional min-entropy (Box 5.2). We thus see that whenever the length of X is greater than Eve's storage, i.e. $n > S$, Alice and Bob can use an extractor to extract a nonzero amount of secure key. ■

QUIZ 7.3.1 *Alice and Bob communicate over a special classical channel such that Bob correctly receives all the bits from Alice. However, Eve receives a bit b_E that is equal to Alice's bit with probability $q = 1/2$ and with probability $1 - q = 1/2$ is equal to Alice's bit flipped. Is it necessary for Alice and Bob to perform randomness extraction on their strings to reduce Eve's knowledge about the key?*

(a) *Yes, Alice and Bob need to perform randomness extraction, since there is a nonzero probability that Eve received Alice's key bit.*

(b) *No, randomness extraction is not required, because Eve holds no information about the key.*

QUIZ 7.3.2 *Now the channel between Alice and Bob has been modified such that Bob still receives all the bits from Alice, but Eve always receives Alice's bit flipped. Is it now necessary for Alice and Bob to perform randomness extraction on their bit strings to reduce Eve's knowledge about the key?*

(a) *Yes, Alice and Bob need to perform randomness extraction, to reduce Eve's knowledge about the key.*

(b) *No, randomness extraction is not required, because Eve never receives the key bit, so she holds no information about the key.*

(c) *Alice and Bob cannot obtain a key in this scenario, because Eve has as much information as Bob has.*

7.4 Information Reconciliation

In the examples from the previous section we assumed that there are never any errors on the channel connecting Alice and Bob. Clearly, this is extremely unrealistic in any real implementation. If we follow Protocol 3 when using a noisy channel the *correctness* of the protocol will be affected: at the end of the first step Alice and Bob will hold two strings $x_A \neq x_B$ which are not always the same. When applying the extractor

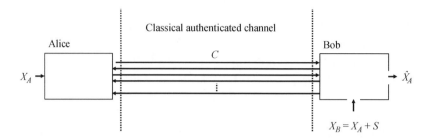

Fig. 7.2   Scheme of a general information reconciliation protocol.

to perform privacy amplification, there is a priori no reason for them to magically obtain the same k_A and k_B.

How can we solve this problem? The key idea is to perform a step of *information reconciliation* on the strings x_A and x_B, prior to privacy amplification. In this step Alice and Bob exchange error-correcting information about x_A, x_B in order to correct errors.

Let's describe the communication scenario more precisely and introduce some notation. After the communication phase, Alice and Bob hold two strings that we model using random variables X_A and X_B respectively. Define S, the *syndrome*, as $S = X_A \oplus X_B$, where we use $\oplus$ to denote the bitwise parity: for example, $010 \oplus 110 = 100$. Note that S is different from zero if and only if there are errors, and moreover S has a 1 in any position where X_A and X_B differ. The goal of information reconciliation is for Bob to correct his string X_B and recover X_A. The information reconciliation protocols we consider are entirely classical and exchange all their messages over the CAC. For any such protocol we let C be the string consisting of all the messages exchanged during the protocol. Let $\hat{X}_A$ be Bob's final output, which he would like to equal X_A (Figure 7.2).

> **Definition 7.4.1 (Information reconciliation).**   *Let X_A and X_B be distributed according to the joint probability distribution $P_{X_A X_B}$.* An information reconciliation *protocol for X_A, X_B is ε-correct and leaks c bits if :*
>
> - $\mathrm{Prob}(\hat{X}_A \neq X_A) \leq \varepsilon.$
> - *The length of the messages exchanged on the CAC is $|C| \leq c$.*

Observe that in the definition we don't require that $\hat{X}_A = X_A$ with certainty. Although we might have liked to, similar to the correctness requirement for key distribution this would have been too strong a condition in general: there's always a chance that things go wrong, and our goal is to design protocols such that this chance is as small as possible. The reason for the second requirement is because any communication exchanged over the CAC leaks to the eavesdropper, Eve. Because in general we don't have a good way to control how this information is related to Alice and Bob's strings X_A and X_B, we take a worst-case approach: worst case, any bit exchanged during the protocol is a bit leaked about X_A. We will then apply the chain rule for the min-entropy (Box 5.2) as

$$\mathrm{H}_{\min}(X|EC) \geq \mathrm{H}_{\min}(X|E) - |C| . \tag{7.5}$$

This equation bounds how much information is lost to Eve during information reconciliation. It is used to estimate how much uncertainty remains in Alice and Bob's strings before they perform privacy amplification as described in the previous section.

If we separate the two goals from Definition 7.4.1 then it is not hard to achieve them. Why is this? Imagine that a reconciliation protocol consists in Alice sending her whole string to Bob over the CAC. This is a great protocol if we only care about correctness, but the leakage $|C| = |X_A|$ is maximal and by (7.5) after reconciliation we would not have any min-entropy left to do privacy amplification. On the other hand, imagine a reconciliation protocol that consists in Alice and Bob doing nothing. Then, for leakage purposes, the protocol is perfect, the leakage is zero, but the strings might never be equal, and so the protocol is only $\varepsilon_c = 1$-correct.

Information reconciliation protocols can be classified depending on their use of the CAC. The most general protocol consists in the exchange of messages in both directions, from Alice to Bob and from Bob to Alice. We call such a protocol a *two-way* or an *interactive* protocol. However, much simpler, and as it turns out often sufficient, protocols would consist of a single message from Alice to Bob. We refer to such protocols as *one-way* reconciliation protocols. Let's describe such a protocol in the next section.

7.4.1 Syndrome Coding

The scheme that we will describe is based on error-correcting codes. Let us first review the definition and main elements of linear error-correcting codes, which are the most widely studied (and used) kind of code. We give the definition for the case of a general finite field (see Box 6.2), although for our purposes it is enough to understand it for the case where $q = 2$. In that case $\mathbb{F}_2^n$ is simply the vector space of all n-dimensional vectors with entries in $\{0,1\}$, with addition of vectors being done coordinate-wise modulo 2, e.g. $(001) + (101) = (100)$. With these preliminaries in place, let's give the definition of a linear error-correcting code.

> **Definition 7.4.2 (Linear code).** *Let $\mathbb{F}_q$ be a finite field of size q and $1 \leq k \leq n$ two integers. An (n,k) q-ary linear code C is a linear subspace of $\mathbb{F}_q^n$ of dimension k; n is called the* length *of the code and k its* dimension.

The individual elements (which are q-ary strings of length n) contained in the subspace defining a linear code are called *codewords*. An (n,k) q-ary code has q^k different codewords. Since we will only be concerned with binary codes, in the following we let $q = 2$.

Since a code is just a subspace, to define it we need a way to characterize a subspace of $\mathbb{F}_2^n$. It is convenient to do this by using an $m \times n$-dimensional *parity-check matrix H*

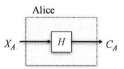

Fig. 7.3   Evaluation of the syndrome map in syndrome coding-based reconciliation.

with entries in $\mathbb{F}_2$, which specifies equations that every codeword should satisfy. The code is then defined as the set of vectors v such that $H \cdot v = 0$. The dimension of the code induced by H is the dimension of the kernel of H, which is $k = n - \text{rank}(H)$. If we take the rows of H to be linearly independent then $k = n - m$ and so $m = n - k$.

The map $s_H : \mathbb{F}_2^n \to \mathbb{F}_2^m$ given by $v \mapsto H \cdot v$ is called the *syndrome map* (see Figure 7.3). The goal of an error-correcting code is to encode information in such a way that it is resistant to errors, and the syndrome map is used to correct errors when they arise. Intuitively this map is useful because it detects when a vector is a valid codeword: if v is in the code then by definition $H \cdot v = 0$. More generally, if $s_H(v) \neq 0$ then the value $s_H(v)$ should give us information about errors, i.e. how to modify v into $w = v + e$ so that the modified vector w is in the code, $s_H(w) = 0$, and e can be interpreted as a small error, i.e. a vector with few 1's.

Example 7.4.1 Let $v = (011)^T$ and $H = \begin{pmatrix} 110 \\ 011 \end{pmatrix}$. Then $s_H(v) = H \cdot v = (10)^T$. This is not the zero vector, and so v is not a codeword. However, $w = v + (100)^T = (111)^T$ satisfies $s_H(v) = 0$ and so it is a valid codeword. In this case, can you write down all vectors in the code? *[Hint: to make sure that you did not forget any codeword, count dimensions.]*   ∎

Let's now show how to use any linear error-correcting code to construct a one-way information reconciliation protocol. The idea is very simple. First, Alice computes the syndrome $C_A = s_H(X_A)$ and sends it to Bob. Let us now consider Bob's actions. We think of Bob's string X_B as a "noisy" version of X_A. His goal is to "decode," i.e. he wants to estimate the error string S such that $X_B = X_A \oplus S$ and then recover $X_A = X_B \oplus S$. Decoding is a little bit more complicated. The first step is that Bob computes the syndrome of X_B, which we call $C_B = s_H(X_B)$. Then Bob computes $C_S = C_B \oplus C_A$. This is the syndrome of the error string, i.e.

$$C_S = s_H(X_A) \oplus s_H(X_B) = s_H(X_A \oplus X_B) = s_H(S)$$

because matrix-vector multiplication is linear. Then, C_S is used by a sub-procedure, which depends on the choice of error-correcting code, that estimates the error string S from C_S and outputs the estimate. Call the estimate returned $\hat{S}$. In general, it will always be the case that $\hat{S}$ is such that $s_H(\hat{S}) = C_S$. However, unless $m = n$ there will be many such strings (precisely, 2^{n-m}) and so we won't always have $\hat{S} = S$. For a good error-correcting code, this will be the case as long as S is "small enough," i.e. it has a small Hamming weight. How small is small enough is a property of the code called its

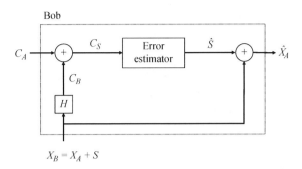

Fig. 7.4   The decoder in syndrome coding-based reconciliation.

"distance." Having recovered the estimate $\hat{S}$, Bob adds it to X_B to obtain $\hat{X}_A = X_B \oplus \hat{S}$ and this is his final output (see Figure 7.4).

Example 7.4.2 Let us go back to Example 7.4.1. We now describe the decoding (Figure 7.4). There are four different syndromes. We design our estimator function as follows:

Syndrome	Error estimate
00	000
01	001
10	100
11	010

Can you guess why we chose this particular map? The idea is that if there are zero or one errors, the estimator will output the correct error estimate. In other words, among all possible error vectors with a given syndrome we always chose the one that has the smallest number of actual errors, i.e. the smallest possible number of 1's. ∎

7.4.2 Some Parameters

Now that we have described the main idea for information reconciliation based on error-correcting codes, let's get a sense of the parameters that can be achieved, so that we can use such schemes in our QKD protocols in the next chapter.

In order to choose an information reconciliation protocol we need to have an idea of the error model, that is how the strings X_A and X_B relate to each other. A natural model is that each symbol of the n-bit strings X_A and X_B is drawn independently from the same joint distribution P_{AB}, where A, B are binary random variables. In this case we can measure how much X_A and X_B typically differ using the conditional entropy

$$H(A|B) = H(A,B) - H(B) \, ,$$

where H is the Shannon entropy. If we further assume that A is uniformly distributed, and B is such that $B = A$ with probability $p = 1 - \delta$ and $B = 1 - A$ otherwise, then by direct calculation we find that $H(A|B) = h(\delta)$, where h is the binary entropy function. In this situation the theory of error-correcting codes tells us that as long as

$$m \geq n \cdot h(\delta) \tag{7.6}$$

then there exists an error-correcting code with parity-check matrix H and syndromes of size m such that for (X_A, X_B) distributed according to P_{AB}^n, with very high probability the string $\hat{S}$ of smallest Hamming weight such that $s_H(\hat{S}) = s_H(X_A \oplus X_B)$ will be precisely $S = X_A \oplus X_B$.[2] This means that in the protocol described in the previous section, if Bob computes $\hat{S}$ to be the smallest weight string whose syndrome is C_S, then he will recover $\hat{X}_A = X_A$ with high probability. In fact, for this to be the case it is sufficient to have the guarantee that $|S| \leq \delta n$ with high probability, i.e. we only need to know that the strings X_A and X_B do not differ too much (it is not necessary to know that each entry of the string has been generated independently according to the same distribution). Finally, it is also known that (7.6) is optimal in the sense that if we take m to be even slightly smaller than this, then in general there will be many possible error strings $\hat{S}$ of small weight that are compatible with a given syndrome C_S, and so it will not be possible to determine which $\hat{S}$ is the correct $S = X_A \oplus X_B$.

In practice, it is not enough that there "exists" an error-correcting code with the right properties. First of all we need to know what that code is, i.e. what is the matrix H, and second we also need that there is an efficient way to perform "syndrome decoding," i.e. recover $\hat{S}$ from C_S. For our purposes we will simply say that such codes do exist, and so whenever we want to perform information reconciliation we will be able to claim that it is possible to do so using leakage that scales as in (7.6), where the parameter δ will be estimated in the protocol. Constructing such codes is by no means easy and it is a major achievement of the theory of error-correcting codes. We refer you to the chapter notes for bibliographical references.

7.5   Everlasting Security

 To end the chapter let us return to an important assumption that is always made when considering QKD protocols: that the communication channel between Alice and Bob, which we've been calling the "CAC," is what its name implies – an authenticated channel. This assumption is used to guarantee that, although the eavesdropper may intercept any communication, she cannot "impersonate" Alice or Bob by sending fake messages on the channel. If we allowed this, it could have dramatic consequences on the correctness of the protocol: for example, Eve could modify the syndrome information sent from Alice to Bob in information reconciliation and thereby make Bob think that he recovered $\hat{X}_A = X_A$ when this is absolutely not the case. Once we see concrete protocols in the next chapter you will see that if Eve could break the authentication she could make much more devastating attacks.

So we need a CAC. How reasonable is it to assume that we have it? This question is not only relevant for key distribution; in general, access to an authenticated channel is a prerequisite for a large variety of cryptographic tasks. If we keep in mind that our goal in quantum cryptography is to implement protocols that have the fewest possible assumptions, we owe it to ourselves to examine this one critically. Indeed, there is no method to implement a CAC "out of the blue," unless one already has a shared

---

2 The probability that this is the case can be made very close to 1 by choosing m just a little bit larger than the lower bound in (7.6), e.g. $m \approx n \cdot h(\delta) + \Omega(\log(1/\delta))$ to get a probability $1 - \delta$.

secret; since this is the goal of QKD in the first place we certainly don't want to do that. Without assuming a prior secret, it is possible to show that constructing a CAC *requires* the use of computational assumptions on the power of the eavesdropper. It is beyond the scope of this book to explain how exactly authentication can be implemented; we will simply reveal that the main primitive used to achieve this is called a "digital signature," and it can be implemented based on any trapdoor one-way function (such as the famous RSA function, whose security rests on the hardness of factoring – but of course with quantum adversaries you wouldn't want to rely on such an assumption!).

Is it reasonable to make computational assumptions, when one of the main goals of QKD is to provide information-theoretic security? This requirement is sometimes raised as a criticism against QKD, whose purpose is precisely to enable the secure exchange of a secret key *without* making computational assumptions. A good argument to counter this criticism and justify the use of computational assumptions for the CAC is the property of "everlasting security." As we will see in the next chapter, the key generated in a QKD protocol is secure as long as the CAC remains authenticated *for the duration of the protocol*. During this time it is indeed crucial that Eve is not able to send fake messages. However, once the protocol has ended and Alice and Bob have generated their private key, it is no longer relevant whether the channel remains authenticated or not. So the computational assumption guaranteeing security of the authenticated channel only needs to hold for a few seconds, and the key generated in the protocol will remain forever secure: Eve has no information about it, and will not be able to gain any additional information by breaking a channel that is no longer in use. This property distinguishes QKD from a much more naive protocol that would use standard cryptographic techniques to directly exchange the secret key; for any such method, if the authentication or any other cryptographic assumption used in the protocol is broken even a year later, then the generated key is immediately made vulnerable.

CHAPTER NOTES

The formal security definition for quantum key distribution appears in the PhD thesis of Renato Renner (*Security of Quantum Key Distribution*. ETH Zurich, 2005). For a thorough discussion of the definition, including why it is "universally composable," we refer to a paper by C. Portmann and R. Renner (Cryptographic security of quantum key distribution. arXiv:1409.3525, 2014).

Prior to the idea of using linear error-correcting codes and one-way reconciliation, there existed ad hoc two-way protocols proposed specifically for the task of information reconciliation. The most well known such protocol is Cascade by G. Brassard and L. Salvail (Secret-key reconciliation by public discussion. In *Workshop on the Theory and Application of Cryptographic Techniques*, pp. 410–423. Springer, 1993), which has a reasonably simple description and has the advantage of being easy to implement.

The inequality (7.6) that quantifies the optimal leakage for any one-way information reconciliation protocol is due to D. Slepian and J. Wolf (Noiseless coding of correlated information sources. *IEEE Transactions on Information Theory*, **19**(4):471–480, 1973). While the bound stated in the equality can be achieved in the limit $n \to \infty$, in practice one has to deal with finite values of n, and moreover efficiency considerations in terms of the decoding operation play an important role in the practicality of a protocol. Therefore it is common to accept a leakage that is marginally larger, of the form $\xi \cdot nH(X_A|X_B)$, where $\xi > 1$ is called the reconciliation efficiency. The constant ξ is often chosen $\xi \approx 1.2$ and this allows very efficient implementations; see the paper by M. Tomamichel, et al. (Fundamental finite key limits for information reconciliation in quantum key distribution. In *2014 IEEE International Symposium on Information Theory*, pp. 1469–1473. IEEE, 2014) for more discussion.

In general, it is an important open problem if there are two-way information reconciliation protocols that can achieve an asymptotically smaller leakage than one-way protocols. As we will see in the next chapter, in the context of QKD protocols any bit leaked for information reconciliation is a bit of key lost, and so minimizing this leakage is crucial to get optimized protocols.

For an introduction to the problem of authentication and how to solve it using private- or public-key cryptography, see Chapter 5 in the lecture notes by R. Pass and A. Shelat (*A Course in Cryptography*, 2010. Lecture notes available at www.cs.cornell.edu/courses/cs4830/2010fa/lecnotes).

PROBLEMS

7.1 Generating key using an anonymous message board

Imagine that Alice and Bob have discovered an anonymous message board in the hallway. It allows both Alice and Bob to post messages in such a way that no one can ever find out who the message came from. In particular, any eavesdropper, Eve, cannot learn whether the message came from Alice or from Bob. The message board simply creates a list of messages posted to it, without indicating a sender. Alice and Bob come up with three candidate protocols.

- **Protocol I**

Step 1. Alice and Bob write a random bit on the board.
Step 2. If the bit of Alice is the same as the bit of Bob then they erase and start from step 1.
Step 3. If the bit of Alice is different from Bob's bit then the next bit of their key is Alice's bit.
Step 4. Alice or Bob erases the bits and repeats from step 1 until they have n bits of key.

- **Protocol II**

Step 1. Alice starts by writing two bits on the board.
Step 2. If the second bit is 0 they take the first bit as a key bit and they repeat step 1.
Step 3. If the second bit is 1 they take the XOR of the two bits as a key bit and start from step 1, but now Bob writes instead of Alice.
Step 4. Alice or Bob executes this alternating protocol until they have n bits of key.

- **Protocol III**

Step 1. Alice and Bob each write $k < n$ random strings of n bits on the board in a *random* order.
Step 2. If Alice sees one of her strings followed by a Bob string she XOR's the two strings.
Step 3. If Bob sees one of his strings preceded by an Alice string he XOR's the two strings.
Step 4. Alice and Bob toss all strings that were never XOR'ed.
Step 5. Alice and Bob XOR all remaining strings together, thus obtaining n bits of key.

1. At the end of the day, we want that Alice and Bob both share an n-bit key, but Eve is ignorant about the key. Which of the above protocols generates such a key? (There is only one correct one!)
2. Can you argue why your chosen protocol is secure?
3. Can you come up with a different protocol that generates key?

7.2 Key rate with special channels

In the chapter you saw how Alice and Bob could establish key in the presence of a limited Eve. In particular, you saw a situation where Alice and Bob possessed a channel that allows them to send classical bits such that Eve would obtain the bit with probability q (which is known to Eve!) and would obtain the flipped bit with probability $1 - q$.

1. As a refresher, calculate the amount of min-entropy that Eve would have about a bit that Alice sent to Eve for the following values of q:
   I. $q = 0$
   II. $q = 1/4$
   III. $q = 1/2$

IV. $q = 3/5$
V. $q = 1$

2. For which values of q would we be able to use this channel to create keys?

Now imagine we are in the situation where Eve has a limited classical memory of size k bits. Imagine that Alice sends Bob n bits through a public channel (of which Eve can copy and store k). Let's take, for example, $k = 1000$.

3. What would Eve's min-entropy be (about the string of n bits) in the following situations?

  I. $n < k$
  II. $n = k$
  III. $n = 10k$

7.3 Information reconciliation

In the chapter we describe an information reconciliation protocol based on the parity-check matrix

$$H = \begin{pmatrix} 1 & 1 & 0 \\ 0 & 1 & 1 \end{pmatrix} .$$

This protocol can reliably correct a single bit-flip error on blocks of three bits. If we assume that every key bit distributed is flipped with probability p and remains unchanged with probability $1 - p$, we could derive that the probability of correctly distributing a three-bit string without error correction is

$$p_{\text{succ}} = (1 - p)^3 ,$$

while using the error correction scheme based on H we have

$$p_{\text{succ}} = 1 - 3p^2 + 2p^3 ,$$

which is of course quite a bit better for small p. Now the question is, can we do even better? Here we will look at a simple expansion of the three-bit linear code from the chapter and look at the seven-bit code generated by the parity-check matrix

$$H = \begin{pmatrix} 0 & 0 & 0 & 1 & 1 & 1 & 1 \\ 0 & 1 & 1 & 0 & 0 & 1 & 1 \\ 1 & 0 & 1 & 0 & 1 & 0 & 1 \end{pmatrix} .$$

Here we will investigate the robustness of this code to errors. Let's set a baseline by looking at the probability of successfully distributing a seven-bit string using no error correction when all bits in the string are affected by a binary symmetric channel which flips bits with probability p.

1. What is the probability of successfully distributing an error-free string?

Of course this code is not magical, i.e. we will never be able to reliably correct all errors. To see why this is the case let us look at the error strings $S = 1000000$ and $S' = 0110000$.

2. Which of the following statements is true?

    I. We can never correct both S and S' since their syndromes are the same. Hence the decoder will not be able to reliably distinguish these two errors from their syndromes.

   II. The error S is never correctible since its syndrome is the zero string, which means the decoder can't detect if this error has happened.

  III. The error S' is never correctible since its syndrome is the zero string, which means the decoder can't detect if this error has happened.

3. Now, assume we use the information reconciliation scheme from Section 7.4.1 with the matrix H and a string of seven bits. Assuming the probability of flipping a bit is again given by p, and we can reliably correct single-bit errors, what is the probability that we can successfully distribute an error-free key (up to third order in p)?

4. In the parameter regime $p \in [0, 1/2]$ is this protocol more resistant to noise (giving a higher p_{succ} for a given value of p) than the three-bit protocol given in the chapter?

8

Quantum Key Distribution Protocols

In the previous chapter we saw the definition of a correct and secure key distribution protocol, and we studied a simple example of such a protocol that works in a restricted setting. In this chapter we tackle the real thing: we construct a *quantum* key distribution (QKD) protocol that obtains security using only a public quantum channel (and, as always, a CAC)! Informally, a QKD protocol allows two honest users, Alice and Bob, to harness the advantages of quantum information processing to generate a shared secret key. The most well known, and indeed the first QKD protocol that was discovered, is called BB'84, after its inventors, Bennett and Brassard, and the year in which their paper describing the protocol was published. In this chapter we describe the BB'84 protocol and we introduce the main ideas for showing that the protocol is secure.

8.1   BB'84 Quantum Key Distribution

In the previous chapter we presented a key distribution protocol, Protocol 3, that could be used when Alice and Bob have access to a very special classical channel: a channel such that Eve receives a noisy copy of each bit communicated over the channel, with some guaranteed minimal amount of noise $q > 0$. This assumption facilitated the analysis, but it is not realistic, as in general there is no way to tell that noise has to be applied for Eve (and not for Bob). Our main idea in this chapter is to use a quantum channel to "simulate" the guarantee provided by this classical channel. As in the previous chapter, in addition to a public quantum channel we assume that Alice and Bob have access to a classical authenticated channel (CAC).

8.1.1   The BB'84 Encoding

Suppose first that we use the quantum channel exactly as a classical channel. At the first step, Alice sends the string $x = x_1, \ldots, x_n$ as a quantum state by encoding it in the standard basis. This can be written as $|x\rangle\langle x| = |x_1\rangle\langle x_1| \otimes \cdots \otimes |x_n\rangle\langle x_n|$. Since the basis is fixed it is public knowledge: Eve knows it as well and she can measure the transmitted quantum state in the standard basis to immediately recover x without error. In other words, she can easily copy x on the fly and later use her copy to correctly guess Alice and Bob's entire key. Clearly this is no better than sending x directly over the CAC, and it doesn't work: the message has no privacy at all. Obviously it won't work either if we use any other fixed basis to encode the information.

However, recall that by the no-cloning theorem from Chapter 1 it is impossible to copy *arbitrary* qubits, i.e. qubits that are not deterministically prepared in a fixed basis. This suggests an idea: in addition to her random string x, for each bit x_j of x Alice will randomly choose a basis $\theta_j \in \{0,1\}$ to encode the bit, where as usual 0 is used to designate the standard basis and 1 the Hadamard basis. She will then send the bit x_j encoded in the basis θ_j, which we denote $|x_j\rangle_{\theta_j} = H^{\theta_j}|x_j\rangle$ with H the Hadamard matrix. Since it is so important, let's record this encoding using a definition.

Definition 8.1.1 (BB'84 encoding). *The BB'84 states are $\{|0\rangle, |1\rangle, |+\rangle, |-\rangle\}$. We can write each BB'84 state in the form $|x\rangle_\theta$ with $x \in \{0,1\}$ denoting the encoded bit and $\theta \in \{0,1\}$ the encoding basis, where $\theta = 0$ labels the standard basis and $\theta = 1$ the Hadamard basis.*

The intuition for using this encoding is that, since the eavesdropper does not know the basis in which the bits are encoded, she cannot measure them directly. Moreover, by the no-cloning theorem she cannot copy them perfectly as quantum states either. Indeed, we already used this intuition in Chapter 3 when we considered the use of BB'84 states for the problem of quantum money. But now the setting is different: for example, what if Eve simply keeps the state, and replaces it by some kind of dummy state that she forwards to Bob? In that case, can Bob detect that he did not receive the correct information? For this there should be some kind of check that Bob can perform in the protocol. Finding such a check raises some other issues. For example, does Bob know the correct basis? If so, how did he learn it? If not, how does he recover x? We will describe the BB'84 protocol soon, but you may wish to pause for a moment and think for yourself how you would build on the idea of using BB'84 states to design a complete QKD protocol.

While you're thinking, let us observe that the standard basis and the Hadamard basis are the eigenbases of the Pauli Z and Pauli X matrices respectively. A more complicated set of states than the BB'84 states that one could use consists of the eigenbases of the Pauli matrices X, Y, and Z. This is known as the six-state encoding.

Definition 8.1.2 (Six-state encoding). *The six states are $\{|0\rangle, |1\rangle, |+\rangle, |-\rangle, |+y\rangle, |-y\rangle\}$, where*

$$|\pm y\rangle = \frac{1}{\sqrt{2}}\left(|0\rangle \pm i|1\rangle\right). \tag{8.1}$$

We can write each such state as $|x\rangle_\theta$ with $x \in \{0,1\}$ the encoded bit and $\theta \in \{0,1,2\}$ the encoding basis, where $\theta = 0$ labels the standard basis, $\theta = 1$ the Hadamard basis, and $\theta = 2$ the eigenbasis of the Pauli Y matrix.

Both the four BB'84 states and the six states are used frequently in quantum cryptographic protocols. Here, for simplicity, we focus on the BB'84 states.

8.1.2   The BB'84 Protocol

We are ready to describe the BB'84 protocol! We first give a simplified version of the protocol that can be used when the users expect that their quantum channel is noiseless, meaning that, in the absence of any eavesdropping, Bob can expect to perfectly receive any qubit sent by Alice. In practice there will always be errors on the communication channel, because no transmission can ever be perfect. We will later modify the protocol to allow this.

Protocol 4 (BB'84 QKD (no noise)) The protocol depends on a large integer n publicly chosen by the users (intuitively n is the number of bits of key that they expect to generate at the end). Let $N = 4n$. Alice and Bob execute the following:

1. Alice chooses a string $x = x_1, \ldots, x_N \in \{0,1\}^N$ uniformly at random and a basis string $\theta = \theta_1, \ldots, \theta_N \in \{0,1\}^N$ uniformly at random. Alice sends to Bob each bit x_j by encoding it in a quantum state according to the basis θ_j as $|x_j\rangle_{\theta_j} = H^{\theta_j} |x_j\rangle$.
2. Bob chooses a basis string $\tilde{\theta} = \tilde{\theta}_1, \ldots, \tilde{\theta}_N \in \{0,1\}^N$ uniformly at random. For each $j = 1, \ldots, N$ he measures the j-th qubit received from Alice in the basis $\tilde{\theta}_j$ to obtain an outcome $\tilde{x}_j$. This gives him a string $\tilde{x} = \tilde{x}_1, \ldots, \tilde{x}_N$.
3. Bob tells Alice over the CAC that he has received and measured all the qubits.
4. Alice and Bob tell each other over the CAC their basis strings θ and $\tilde{\theta}$ respectively.
5. Let $S = \{j | \theta_j = \tilde{\theta}_j\}$ denote the indices of the rounds in which Alice and Bob measured in the same basis. Alice and Bob discard the information for all rounds not in S.
6. Alice picks a random subset $T \subseteq S$ by flipping a fair coin for each $i \in S$ to decide if it is selected in T. Alice tells Bob what T is over the CAC.
7. "Matching outcomes" test: Alice and Bob announce x_T and $\tilde{x}_T$ to each other over the CAC, where we denote by x_T the substring of x corresponding to the indices in the test set T, and similarly for $\tilde{x}_T$. They compute the error rate $\delta = \frac{1}{|T|} |\{j \in T \mid x_j \neq \tilde{x}_j\}|$.
8. If $\delta \neq 0$ then Alice and Bob abort the protocol. Otherwise, they proceed to denote $x_{\text{remain}} = x_{S \setminus T}$ and $\tilde{x}_{\text{remain}} = \tilde{x}_{S \setminus T}$ as the remaining bits, i.e. the bits where Alice and Bob measured in the same basis but which they did not use for testing in the previous step.
9. Alice and Bob return x_{remain} and $\tilde{x}_{\text{remain}}$ as their outputs, respectively.

Remark 8.1.1 The first two steps of the protocol are described as taking place one after the other. However, in an actual execution of the protocol Alice can prepare the qubits one by one and Bob can also measure them one by one. This is very appealing since Alice and Bob only need very simple quantum devices – preparing and measuring single qubits is enough, and no quantum storage is required.

There's a lot going on in this protocol! The most important step is the simplest, step 3: this step guarantees that Bob has measured all his qubits *before* the basis choices θ and $\tilde{\theta}$ are announced publicly. As we will see, this is crucial for security.

Exercise 8.1.1 Show that if Alice announces θ at any step prior to step 3 then the protocol is completely insecure. Namely, there is a way for an eavesdropper, Eve, having only access to the CAC and the communication on the quantum channel, to learn both users' entire output in the protocol.

Let's do a quick "back of the envelope" calculation to estimate the number of output bits that are produced in this protocol. At step 5, since Alice and Bob chose $\theta, \tilde{\theta}$ at random, we expect that on average they will discard roughly $|S| \approx N/2 = 2n$ bits. The size of T will be $|T| \approx |S|/2 \approx n$ bits, and so the length of x_{remain} and $\tilde{x}_{\text{remain}}$ is also approximately n bits.[1]

Now, is this protocol correct and secure? Informally, based on the fact that Alice and Bob obtained exactly the same outcomes $x_T = \tilde{x}_T$, we expect that it should also be the case that $x_{\text{remain}} = \tilde{x}_{\text{remain}}$, and so the protocol should be correct. Furthermore, the same condition should intuitively guarantee that Eve has learned very little information about x. This is because, due to the no-cloning principle, any "copying" that she might have attempted while the qubits were flying from Alice to Bob in step 1 would have been detected, because it couldn't have depended on the secret choice of θ. (Here we use the assumption that Alice has a "secure lab," as described in Box 7.1! Otherwise Eve could peek into it and see x and θ right away.)

Of course, this is just intuition and we have to make it precise. But first let's give a more realistic protocol that accounts for the fact that, even without any eavesdropper, Alice and Bob can't expect to receive exactly the same strings – there will always be some kind of error on their quantum communication channel. To keep a correct protocol we allow an error rate $\delta > 0$ at step 8 and introduce an additional step of *information reconciliation*. Moreover, at the last step Alice and Bob also perform *privacy amplification*. This leads to the following protocol.

Protocol 5 (BB'84 QKD (with noise)) The protocol depends on a small constant $\delta_{\max} > 0$ and a large integer n publicly chosen by the users. Let $N = 4(1 + C\delta_{\max})n$, where C is a large constant that can be determined from the security analysis. Let Ext be a two-universal extractor and H a parity-check matrix for a good classical error-correcting code. Alice and Bob execute the following:

1–7. Same as Protocol 4.

8. If the error rate is $\delta > \delta_{\max}$ then Alice and Bob abort the protocol. Otherwise they set $x_{\text{remain}} = x_{S \setminus T}$ and $\tilde{x}_{\text{remain}} = \tilde{x}_{S \setminus T}$ respectively.

9. Alice and Bob perform information reconciliation: Alice sends some error-correcting information $c = Hx_{\text{remain}}$ across the classical authenticated channel to Bob and Bob corrects the errors in his string $\tilde{x}_{\text{remain}}$ to obtain a corrected string $\hat{x}_{\text{remain}}$.

---

1 The *key rate* of this protocol, defined as the ratio of the expected number of key bits produced divided by the total number of qubits exchanged, is approximately $1/4$. In practice, one can perform various optimizations to improve this, such as using fewer rounds for the matching outcomes test in step 7 and biasing Alice and Bob's choice of basis to increase the likelihood that they make the same choice. For clarity we give the simplest possible formulation of the protocol, without such optimizations.

10. Alice and Bob perform privacy amplification: Alice picks a random seed r and computes $k_A = \text{Ext}(x_{\text{remain}}, r)$. She sends r to Bob, who computes $k_B = \text{Ext}(\hat{x}_{\text{remain}}, r)$.

The string x_{remain} obtained by Alice at step 8 is called the *raw key*. It is named like this because, after that point, only classical post-processing operations are performed: first, information reconciliation and then, privacy amplification. Note that we have not made precise the parameters of the information reconciliation subprotocol (the choice of H) or the privacy amplification subprotocol (the choice of Ext). We will discuss these later.

8.1.3   Correctness and Security

In this section we sketch arguments for the correctness and security of the final BB'84 protocol, Protocol 5, focusing on the intuition. Making these arguments precise will occupy the remainder of the chapter.

Let us first argue correctness. Based on observing a certain error rate δ in step 7 the users can conclude that the strings x_{remain} and also $\tilde{x}_{\text{remain}}$ are likely to match in about $(1 - \delta)$ fraction of positions. Intuitively this is because the set T used for testing is chosen uniformly at random and so the fraction of errors inside T and outside it should be approximately the same; we will explain how to prove this formally in Section 8.4. Based on this estimate the users can decide on the exact parameters for the information reconciliation protocol to guarantee that, after information reconciliation, it holds that $\Pr(x_{\text{remain}} \neq \hat{x}_{\text{remain}}) \leq \varepsilon_c$, where ε_c is the target correctness error.

We now consider the secrecy condition. Let $h(\delta) = -\delta \log \delta - (1 - \delta) \log(1 - \delta)$ be the binary entropy function. The crux of the argument is to prove that at the end of step 8 of the protocol, conditioned on not having aborted in that step, the following bound holds

$$\text{H}_{\min}(X_{\text{remain}}|E) \gtrsim n \left(1 - h(\delta_{\max})\right), \tag{8.2}$$

where E designates all information available to the eavesdropper. In this equation the symbol $\gtrsim$ designates that we are ignoring lower-order terms (growing less fast than any linear function of n) and a logarithmic dependence on the probability of not aborting, which we will discuss later.

We will sketch a proof of (8.2) in Section 8.4, and generalize it in the next chapter. Let us see why it is sufficient to show that the protocol is secret. Taking into account the additional information leaked to Eve when performing information reconciliation, we get by the chain rule (Box 5.2) that

$$\text{H}_{\min}(X_{\text{remain}}|EC) \gtrsim n \left(1 - h(\delta_{\max})\right) - |C|. \tag{8.3}$$

Based on the discussion of information reconciliation protocols from the previous chapter (Section 7.4), we know that it is possible to ensure that $|C| \approx h(\delta_{\max})n$. Using what we know about privacy amplification (Section 6.3), it is then possible to choose parameters for the extractor used to perform privacy amplification in the last step of the protocol so as to obtain an output key k_A that is ℓ bits long and ε_s-secret, where

$$\ell \approx (1 - 2h(\delta_{\max}))n - 2\log(1/\varepsilon_s).$$

QUIZ 8.1.1 *In the analysis of our cryptographic protocols we generally imagine that the adversary, Eve, is "all powerful." Consider a scenario where Eve has placed a transmitter in the random number generators of Alice and Bob, such that she can find out what are the random bits that Alice and Bob generate. Can Alice and Bob be guaranteed security against Eve in this case?*

(a) *Yes, it is possible for Alice and Bob to be secure against Eve even in this case.*
(b) *Quantum mechanics allows Alice and Bob to check whether such a transmitter has been placed in the random number generators.*
(c) *No, one of the crucial assumptions for the security of quantum cryptographic protocols is that Eve has no access to the labs of Alice and Bob. That is, it is not possible for Alice and Bob to be secure against Eve in this case.*

QUIZ 8.1.2 *Consider the following scenario. First, Alice prepares an eigenstate of the Pauli matrix X. Second, Eve measures this state uniformly at random in one of the three bases (i.e. in each of them with probability $p_i = 1/3$): standard, Hadamard, and the Y-basis. Third, Eve sends the post-measurement state to Bob. Bob then measures again in the X-basis. What is the probability that Bob's post-measurement state is the same state as the one that Alice prepared?*

(a) $\frac{1}{3}$
(b) $\frac{1}{2}$
(c) $\frac{2}{3}$
(d) $\frac{3}{4}$

QUIZ 8.1.3 *Alice and Bob run the BB'84 protocol but without the step where Bob announces the receipt of the states. Later in the testing stage they find out that their error rate is zero. They conclude that the quantum channel must be noise-free and that there is no eavesdropper. Hence, omitting the step of confirmation of receipt of the states by Bob did not lead to any compromise of security in this case. Is the reasoning of Alice and Bob correct?*

(a) *Yes*
(b) *No*

Before we look in more detail into showing both requirements, ε_c-correctness and ε_s-security, we pause to make sure that we understand what we mean when we say that Eq. (8.2) should hold "conditioned on not having aborted in step 8." In general, we can always represent the state of the entire system of interest, which for our purposes consists of Alice and her random choices, Bob and his random choices, and Eve's quantum state, as a giant quantum state ρ_{ABE}, where the A part also contains x and θ, the B part contains $\hat{x}$ and $\hat{\theta}$, etc. At step 8 we can imagine that each of the users initializes a special "abort" register, and depending on their classical information they write either "0" (for "not abort") or "1" (for "abort") in that register. Because the classical communication channel is authenticated we know that at this step of the protocol both users make exactly the same decision. "Conditioned on not aborting" means that we measure the "abort" register for both users and post-select on the result being "0" for both of them. Here "post-select" means that we renormalize the state, exactly as if the outcome "0" had been "forced." The resulting state is the one on which (8.2) is evaluated. The following example will make this operation of post-selection clear.

Example 8.1.1 Suppose that Alice, Bob, and Eve share the pure state

$$|\psi\rangle_{ABE} = \frac{1}{\sqrt{3}}\left(|00\rangle_A|00\rangle_B|0\rangle_E + |01\rangle_A|01\rangle_B|0\rangle_E + |10\rangle_A|10\rangle_B|1\rangle_E\right).$$

You can see that this state is in a superposition of three states, such that A and B always have the same information, and E has a bit that is equal to their first bit. Now suppose that Alice and Bob, for some reason, each decide to abort when the parity of their two bits is equal to 0. To determine the state "conditioned on not aborting" we first evaluate the abort condition in a new register A' for Alice and B' for Bob to get

$$|\psi'\rangle_{ABE} = \frac{1}{\sqrt{3}}\left(|00\rangle_A|1\rangle_{A'}|00\rangle_B|1\rangle_{B'}|0\rangle_E + |01\rangle_A|0\rangle_{A'}|01\rangle_B|0\rangle_{B'}|0\rangle_E \right.$$
$$\left. + |10\rangle_A|0\rangle_{A'}|10\rangle_B|0\rangle_{B'}|1\rangle_E\right).$$

Finally, we imagine measuring both A' and B' and forcing the outcome to a 0. After renormalization, the state is

$$|\psi_{\text{not abort}}\rangle_{ABE} = \frac{1}{\sqrt{2}}\left(|01\rangle_A|0\rangle_{A'}|01\rangle_B|0\rangle_{B'}|0\rangle_E + |10\rangle_A|0\rangle_{A'}|10\rangle_B|0\rangle_{B'}|1\rangle_E\right).$$

This is the state "conditioned on not aborting." ∎

In general, assuming that Alice and Bob follow the correct actions of the protocol and that Eve has some given strategy, there is a well-defined probability of the protocol aborting in step 8. This is not a parameter that is known by the users (unless they repeat the protocol many times, but even then they wouldn't know if Eve does the same thing each time or not), but it is a well-defined number. This number will appear in the security proofs. Intuitively, this is because if the probability of aborting is very close to 1 then it means that Eve is doing something pretty crazy, and Alice and Bob will detect this craziness with probability close to 1. However, if by lack of luck they do not detect anything then we really can't guarantee any secrecy. This is a common feature of most cryptographic protocols: there is always a chance that things go wrong, and our goal as protocol designers is to minimize this chance. In other words, we want to obtain good security guarantees for probabilities of aborting that are as close to 1 as we can manage.

On a more technical level, the probability of not aborting will arise in the analysis precisely because the entropy on the left-hand side of the inequality (8.2) is evaluated on the state of the users and Eve at step 8, conditioned on not aborting. Due to a very large renormalization in the case that the probability of not aborting is very small, the inequality that we are able to prove in our security analysis will get worse and worse as the probability of not aborting gets smaller.

8.2   A Modified Protocol

To facilitate the task of showing security for the BB'84 protocol, which we tackle in the next section, we make two small modifications to the protocol. Although it will at first appear like these modifications give more power to the eavesdropper, they will make the analysis simpler.

8.2.1   The Purified Protocol

The first modification is straightforward. Consider the following two experiments. In the first experiment Alice chooses $x, \theta \in \{0,1\}$ uniformly at random and returns $|x\rangle_\theta = H^\theta |x\rangle$, an encoding of the bit x in the basis specified by θ. In the second experiment Alice first prepares an EPR pair $|\text{EPR}\rangle = \frac{1}{\sqrt{2}} |00\rangle + \frac{1}{\sqrt{2}} |11\rangle$. She then chooses a $\theta \in \{0,1\}$ uniformly at random and measures the first qubit in the basis $\{|0\rangle_\theta, |1\rangle_\theta\}$, obtaining an outcome $x \in \{0,1\}$. She returns the second qubit.

We claim that the two experiments are absolutely equivalent. To show this there are two things to verify. First, while in the first experiment Alice makes a choice of x uniformly at random, in the second experiment x is determined as the outcome of a measurement on the EPR pair. But we know that, since the reduced density matrix of the EPR pair on the first qubit is the totally mixed state, any basis measurement on that qubit will return each of the two possible outcomes with probability $1/2$. So the distribution of x is identical in the two experiments.

Second, we should check that when Alice obtains outcome x by measuring the first qubit of the EPR pair in the basis θ, the qubit she returns, i.e. the second qubit of the EPR pair, is indeed projected onto the state $|x\rangle_\theta$. In Example 1.4.4 we showed that this is the case for both the standard and Hadamard bases, and in fact it is a property of the EPR state that is valid for any choice of basis measurement on the first qubit.[2] So it is true – the two experiments are indeed equivalent.

Let us then consider an equivalent formulation of the BB'84 protocol in which, instead of directly preparing BB'84 states, Alice first prepares EPR pairs, keeps the first qubit of each pair to herself, and sends the second qubit to Bob. At a later stage

---

2 Up to a transpose, which in the case of complex coefficients, such as for the Y eigenbasis, amounts to exchanging basis elements.

she measures her qubit in a basis $\theta_j \in \{0,1\}$ chosen uniformly at random and records the outcome x_j. This new formulation of the protocol is completely equivalent to the standard one. Even though it may look more complicated, an important advantage of the new formulation is that it allows us to delay the moment in the protocol when Alice needs to make her choice of basis. We can think of this delay as giving less power to Eve: we will now be able to argue more easily that certain actions of the eavesdropper, taken early on in the protocol, could not have depended on Alice's basis choice, since the choice had not yet been made at the time.

Here is the modified protocol in detail. It is called the "purified" BB'84 protocol.

Protocol 6 (Purified BB'84) Choose parameters as in Protocol 5. Perform the following:

1. Alice prepares N EPR pairs $|\text{EPR}\rangle_{AB}$ and sends the second qubit of each pair to Bob.
2. Bob chooses a uniformly random basis string $\tilde{\theta} = (\tilde{\theta}_1, \dots, \tilde{\theta}_N) \in \{0,1\}^N$. He measures the j-th qubit he received from Alice in the basis $\tilde{\theta}_j$ to obtain an outcome $\tilde{x}_j$.
3. Bob tells Alice over the CAC that he received and measured all the qubits.
4. Alice chooses a uniformly random basis string $\theta = (\theta_1, \dots, \theta_N) \in \{0,1\}^N$ and measures each of her qubits in the bases θ to obtain a string $x = x_1, \dots, x_N$. Alice and Bob exchange their basis strings θ and $\tilde{\theta}$ over the CAC.
5–10. Same as Protocol 5.

Notice how we "pushed" Alice's choice of string x and measurement bases θ all the way from step 1 to step 4 of the protocol, without in fact changing anything about the actual outcomes of the protocol or the eavedropper's power.

The idea of considering a purified variant of the BB'84 protocol can be traced back to a different proposal for QKD put forward by Ekert in 1991. Ekert's main insight was that if Alice and Bob were able to test for the presence of entanglement between their qubits, then (intuitively) by the monogamy of entanglement they would be able to certify that their systems are uncorrelated with Eve's. We will explore Ekert's protocol (and prove the intuition correct!) in the next chapter, when we analyze QKD in the so-called "device-independent" setting.

Remark 8.2.1 Even though the purified protocol requires Alice to prepare EPR pairs, this formulation will only be used for the purposes of analysis. From the point of view of any eavesdropper, which protocol Alice and Bob actually implement makes no difference at all, so it is perfectly fine to prove security of the purified protocol but use the original BB'84 protocol in practice. This is convenient because it is much easier to prepare single-qubit BB'84 states than to distribute EPR pairs across long distances.

8.2.2 More Power to the Eavesdropper

The second modification we make to the BB'84 protocol is less benign, and will appear to give much more power to the eavesdropper. But we will see that it is also very convenient for the analysis! Moreover, if we can prove security against stronger eavesdroppers without too much extra effort, why not do it?

The motivation for this second modification is that it is very hard to model the kinds of attacks that Eve might apply to the quantum communication channel between Alice and Bob. For example, she might partially entangle herself with the qubits sent by Alice, creating a joint state ρ_{ABE} on which we, the mathematicians in charge of showing security, have little control.

Exercise 8.2.1 Consider the case of a single EPR pair ($n = 1$). Suppose that Eve initializes an extra qubit in the state $|0\rangle_E$ and applies a CNOT on it controlled on the qubit B that Alice sends to Bob in step 1 of the protocol (Eve then forwards the qubit B to Bob). Compute the joint state ρ_{ABE} that is created by this operation. Compute the probability that Alice and Bob choose the same basis $\theta = \tilde{\theta}$ and obtain $x = \tilde{x}$. Is this a good attack?

Because it is hard to model general intercepting attacks of the form described in the exercise, we will modify the protocol by allowing Eve to prepare an arbitrary state ρ_{ABE}, where the A and B systems are each made of N qubits, and then give A to Alice, B to Bob, and keep E to herself. The protocol from step 2 onwards is unchanged: Alice and Bob will each measure their respective qubits using random choices of bases and proceed from there on. By giving more power to Eve (she prepares the states, instead of Alice) we're preventing ourselves from thinking too hard about having a model for the attacks: in the new setup, Eve can prepare any state she likes!

This may sound crazy: if we let the eavesdropper prepare any state, then why doesn't she choose, say, $\rho_{ABE} = |000\rangle_{ABE}^{\times N}$? Observe that such a state would pass the "matching outcomes" test from step 7 when $\theta_j = \tilde{\theta}_j = 0$ (standard basis), but it would completely fail whenever $\theta_j = \tilde{\theta}_j = 1$ (Hadamard basis). So even though we're allowing Eve to prepare any state she likes, not all states will be accepted by Alice and Bob in the protocol. For example, you can calculate that the state ρ_{ABE} defined above succeeds with probability about $(3/4)^{N/4}$: this is because roughly $N/4$ rounds are used for testing, and for each such round there is a probability $1/2$ that the basis is the Hadamard basis, in which case the probability of a matching outcome is $1/2$.

This simple example shows that the "matching outcomes" test must play an essential role in the security analysis. How powerful is this test? Can it be used to certify that the state handed over by Eve indeed has the correct form, of being (close to) a tensor product of N EPR pairs? If we manage to show this then we'll be in good shape, because in the modified protocol the only step in which Eve can really have a chance to influence the quantum information exchanged by the users is step 1. It may sound surprising that we would be able to achieve this, as the test only involves local measurements: Can local measurements detect entanglement? The answer is yes, and we'll soon see how it works.

8.3   Security of BB'84 Key Distribution

Let's show security! Based on our knowledge of privacy amplification, to show ε_s-security it suffices to prove a bound on the min-entropy of the form given in Eq. (8.2). Because it is the crux of the security proof, let's restate the inequality here:

$$H_{min}(X_{remain}|E) \gtrsim \kappa n . \qquad (8.4)$$

Recall that in this equation X_{remain} denotes the classical string in Alice's possession at step 8 of the protocol, and is called the raw key. E denotes all the information available to the eavesdropper, Eve: her quantum state, which she created at step 1, as well as all the information exchanged by the users over the CAC. Finally, n is the expected length of X_{remain} and κ is a coefficient which we hope to show is as close to 1 as possible. After information reconciliation and privacy amplification, the users will be left with approximately $\kappa n - h(\delta_{max})n - 2\log(1/\varepsilon_s)$ bits of key, where $h(\delta_{max})n$ is the maximum number of bits used for information reconciliation and $2\log(1/\varepsilon_s)$ the number of bits lost due to privacy amplification.

We will give three different methods to show (8.4). Each of the methods has its advantages and disadvantages, and each gives a different insight on *why* the protocol is secure. The most intuitive, but quantitatively weakest, method is the one from the next section, which gives an interpretation of the matching outcomes test as a test for EPR pairs. In Section 8.3.2 we give a method based on the tripartite guessing game from Chapter 5, whose major advantage is that it shows security under "general attacks," which we will define later. Finally, in Section 8.3.3 we give a method based on entropic uncertainty relations that, when fully worked out, gives the best guarantees on the secret key rate; in particular, it will let us get $\kappa = 1 - h(\delta)$ in (8.4).

8.3.1   Locally Implementing a Bell Basis Measurement

We start with a relatively informal argument for security, which will help us build intuition on the role played by the "matching outcomes" test. We will see that this test corresponds to a "virtual" projection of the state shared by Alice and Bob in an EPR pair. First, let's convince ourselves that this is all we need. For this, suppose that we modified the purified BB'84 protocol by adding an initial step as follows:

0. *Upon receiving their N respective qubits from Eve, Alice and Bob jointly measure each pair of qubits using the two-outcome POVM $\{|EPR\rangle\langle EPR|_{AB}, \mathbb{I}_{AB} - |EPR\rangle\langle EPR|_{AB}\}$, where $|EPR\rangle_{AB}$ denotes an EPR pair on Alice and Bob's joint system. If the number of pairs of qubits that were not found to equal $|EPR\rangle_{AB}$ is larger than $\delta_{max}N$ then they abort. Otherwise, they proceed as usual.*

With this modification the protocol is immediately secure. Indeed, after the completion of step 0 Alice and Bob have the guarantee that at least $(1 - \delta_{max})N$ of their shared pairs of qubits are perfect EPR pairs (since they are projected in the post-measurement state $|EPR\rangle$). Clearly, any bit of the raw key obtained from measurements on these states is perfectly uniform and uncorrelated with Eve. In this situation, getting a bound on the min-entropy such as (8.4) does not pose any difficulty.

The problem with step 0 is that it requires Alice and Bob to perform a joint entangled measurement, which they cannot implement locally. Or can they?

Exercise 8.3.1 Suppose we are given a tripartite state ρ_{ABE}, where A and B are each systems of a single qubit. Show that the probability that a measurement of systems A and B in the standard basis results in matching outcomes is exactly $\text{tr}(\Pi_1 \rho_{AB})$, where

$$\Pi_1 = |\text{EPR}\rangle\langle\text{EPR}| + |\psi_{01}\rangle\langle\psi_{01}|, \quad \text{and} \quad |\psi_{01}\rangle = \frac{1}{\sqrt{2}}|00\rangle - \frac{1}{\sqrt{2}}|11\rangle. \tag{8.5}$$

Similarly, show that if the measurement is performed in the Hadamard basis then the probability of obtaining matching outcomes is $\text{tr}(\Pi_2 \rho_{AB})$, with

$$\Pi_2 = |\text{EPR}\rangle\langle\text{EPR}| + |\psi_{10}\rangle\langle\psi_{10}|, \quad \text{and} \quad |\psi_{10}\rangle = \frac{1}{\sqrt{2}}|01\rangle + \frac{1}{\sqrt{2}}|10\rangle. \tag{8.6}$$

Now suppose that ρ_{AB} is any state such that

$$\frac{1}{2}\text{tr}\left(\Pi_1 \rho_{AB}\right) + \frac{1}{2}\text{tr}\left(\Pi_2 \rho_{AB}\right) \geq 1 - \delta,$$

for some $\delta \geq 0$. Using the above, show that the fidelity

$$F\left(\rho_{AB}, |\text{EPR}\rangle\langle\text{EPR}|\right) = \sqrt{\langle\text{EPR}|\rho_{AB}|\text{EPR}\rangle} \geq \sqrt{1 - 2\delta}.$$

[Hint: to show this, imagine measuring ρ_{AB} in the Bell basis. What can you say about the probability of each of the four possible outcomes?]

The exercise suggests that the "matching outcomes" test that Alice and Bob implement in step 7 of Protocol 6 can play a similar role to the imaginary step 0 introduced above, because high success in the test implies high fidelity with an EPR pair. Therefore, the security of Protocol 6 with step 0 implemented should imply the security of the protocol without step 0, but with step 7 instead.

This sketch of a security proof provides the right intuition for security, and it can be worked out precisely. Rather than pursuing this route, we give two other arguments, each with its own advantages and disadvantages.

QUIZ 8.3.1 *Recall our notation for the Bell states:*

$$|\psi_{00}\rangle = \frac{1}{\sqrt{2}}(|00\rangle + |11\rangle), \qquad |\psi_{01}\rangle = \frac{1}{\sqrt{2}}(|00\rangle - |11\rangle),$$
$$|\psi_{10}\rangle = \frac{1}{\sqrt{2}}(|01\rangle + |10\rangle), \qquad |\psi_{11}\rangle = \frac{1}{\sqrt{2}}(|01\rangle - |10\rangle).$$

Suppose that Alice and Bob both measure their qubits in the standard basis, but want to select for opposite outcomes ($|0\rangle$ and $|1\rangle$ respectively or vice versa). Which of the following projectors Π corresponds to this scenario? (Recall that the matching outcomes test is equivalent to a projection onto a subspace spanned by Bell states.)

(a) $|\psi_{01}\rangle\langle\psi_{01}| + |\psi_{10}\rangle\langle\psi_{10}|$

(b) $|\psi_{01}\rangle\langle\psi_{01}| + |\psi_{11}\rangle\langle\psi_{11}|$

(c) $|\psi_{10}\rangle\langle\psi_{10}| + |\psi_{11}\rangle\langle\psi_{11}|$

(d) $|\psi_{10}\rangle\langle\psi_{10}| - |\psi_{11}\rangle\langle\psi_{11}|$

(e) *None of the above, but some other linear combination of* $|\psi_a\rangle\langle\psi_a|$

(f) *No linear combination of* $|\psi_a\rangle\langle\psi_a|$

QUIZ 8.3.2 *Suppose now that Alice measures her qubit in the standard basis while Bob measures his in the Hadamard basis. They want to select for the "same" outcome, i.e.* $(|0\rangle, |+\rangle)$ *or* $(|1\rangle, |-\rangle)$. *Which of the following projectors* Π *corresponds to this scenario?*

(a) $|\psi_{01}\rangle\langle\psi_{01}| + |\psi_{10}\rangle\langle\psi_{10}|$

(b) $|\psi_{01}\rangle\langle\psi_{01}| + |\psi_{11}\rangle\langle\psi_{11}|$

(c) $|\psi_{10}\rangle\langle\psi_{10}| + |\psi_{11}\rangle\langle\psi_{11}|$

(d) $|\psi_{10}\rangle\langle\psi_{10}| - |\psi_{11}\rangle\langle\psi_{11}|$

(e) *None of the above, but some other linear combination of* $|\psi_a\rangle\langle\psi_a|$

(f) *No linear combination of* $|\psi_a\rangle\langle\psi_a|$

QUIZ 8.3.3 *Let Alice and Bob share n qubit pairs in the state* $\rho_{AB}^{\otimes n}$, *and suppose that the matching outcomes test succeeds with probability exactly* $p_j = 0.95$ *on each of the n pairs. What is the largest value of n for which the overlap* $\langle\psi_{00}|^{\otimes n}\rho_{AB}^{\otimes n}|\psi_{00}\rangle^{\otimes n}$ *is guaranteed to exceed* $1/2$?

(a) 2

(b) 4

(c) 6

(d) 8

8.3.2 Security from the Tripartite Guessing Game

Our second proof of security leverages the tripartite guessing game from Chapter 5. Let's remember roughly how that game proceeds; for details see Section 5.5. In the game, Eve prepares an arbitrary state ρ_{ABE} such that A, B are one qubit each, and gives A to Alice and B to Bob. Alice and Bob choose a random basis $\Theta \in \{0, 1\}$, measure their qubit in that basis, and give Θ to Eve. They win if Alice and Bob's outcomes are equal, and furthermore Eve is also able to guess the same outcome by performing a measurement, which may depend on Θ, on E.

Sounds familiar? Well sure it does! This is exactly one round of the matching outcomes test, with the addition that we now also ask Eve to predict Alice and Bob's matching outcomes. Recall that in Section 5.5 we showed that the maximum success probability in this game is $p_{\text{succ}} \leq \frac{1}{2} + \frac{1}{2\sqrt{2}}$. Let's investigate what this means for our

QKD protocol. First, let's reformulate the maximum success probability in the tripartite guessing game as a guessing probability. Let X_A be Alice's outcome, X_E Eve's guess, and Ω the event that Alice and Bob's outcomes match. Then we have

$$
\begin{aligned}
p_{\text{succ}} &= p\big(\Omega \wedge (X_A = X_E)\big) \\
&= p(X_A = X_E | \Omega) p(\Omega) \\
&= p_{\text{guess}}(X_A | E\Theta\Omega) p(\Omega) .
\end{aligned}
$$

Here for the second line we used Bayes' rule, and for the third line we used that X_E is Eve's best guess for X_A, given the information available to her: her quantum state in register E, and the choice of basis Θ. Shuffling terms around, we get the bound

$$
p_{\text{guess}}(X_A | E\Theta\Omega) = \frac{p_{\text{succ}}}{p(\Omega)} \leq \left(\frac{1}{2} + \frac{1}{2\sqrt{2}} \right) \frac{1}{p(\Omega)} .
$$

Using the relation between guessing probability and min-entropy,

$$
H_{\min}(X_A | E\Theta\Omega) \geq -\log\left(\frac{1}{2} + \frac{1}{2\sqrt{2}} \right) - \log \frac{1}{p(\Omega)} .
$$

This is good progress! As we saw earlier, obtaining lower bounds on the conditional min-entropy of Alice's raw key, given the information available to the eavesdropper, is the most important step in showing security. The derivation above was done considering a single round, but by using the n-round version of the guessing game we can similarly get the bound

$$
H_{\min}(X_A | E\Theta\Omega) \geq -n\log\left(\frac{1}{2} + \frac{1}{2\sqrt{2}} \right) - \log \frac{1}{p(\Omega)} , \tag{8.7}
$$

where now X_A is Alice's n-bit outcome string, and Ω is the event that *all* Alice and Bob's outcomes match.

How does (8.7) compare to our target bound (8.4)? First of all, a minor difference is that we wrote the basis choice Θ explicitly, whereas in (8.4) it is included in E; this is just a question of notation.[3] Another minor difference is that here the bound is on the entire X_A, not only X_{remain}. This is easy to deal with because X_{remain} is a substring of X_A, and so by the data-processing inequality a lower bound on the min-entropy on the latter implies a lower bound on the min-entropy of the former.

A more important difference is that here we are also conditioning on Ω, which is the probability that $X_A = X_B$. What do we know about this probability? If we considered the version of the protocol with $\delta_{\max} = 0$, meaning that Alice and Bob abort as soon as they see a difference, then the Ω would roughly be the same as the probability of not aborting. As expected, our bound on the entropy depends on how likely this event is. The difficulty is that the nonabort condition in the real protocol requires that X_A and X_B are close, but not identical. To deal with this we would have to define a new guessing game in which the winning condition is that Alice and Bob's outcomes match in $(1 - \delta)$ fraction of positions, and Eve's outcome matches Alice's outcome (always).

---

3 Recall that in (8.4) E denotes all the information available to the eavesdropper. This includes the basis information Θ, which was exchanged over the CAC by the users.

This game is a bit harder to analyze, but it can be done. The result is that the bound on the success probability becomes

$$p_{\text{win}} \leq \left(2^{h(\delta_{\max})} \left(\frac{1}{2} + \frac{1}{2\sqrt{2}} \right) \right)^n ,$$

which will make an additional $h(\delta_{\max})$ appear on the right-hand side of (8.7).[4]

Finally, we comment on the coefficient in front of n in (8.7). This coefficient is $\log(1/2 + 1/2\sqrt{2}) \approx 0.16$, not 1. This is a bit disappointing – it means that our analysis only guarantees that, at best, we will be able to obtain one bit of secure key for every (approximately) five rounds of communication in the protocol. Unfortunately this is a limitation of our method – as we saw, it *is* possible to succeed with probability $1/2 + 1/2\sqrt{2}$ in the guessing game, not only $1/2$ as would be needed to obtain a coefficient of 1. In principle we could do better by requiring that Alice and Bob have matching outcomes with high probability; this is because in the optimal strategy for the game they only agree with probability $1/2 + 1/2\sqrt{2}$, but in the QKD protocol we expect them to agree with a higher probability. As it turns out, once we do that the repeated version of the game, with n qubits, becomes much harder to analyze. In the next section we explore a different approach which uses this observation and gives a better constant (but has its own drawbacks!).

8.3.3 Security from Uncertainty Relations

 To get the best quantitative bound on the security of the protocol, i.e. the largest possible κ in (8.4), we use *entropic uncertainty relations*. These relations will help us measure very precisely the trade-off between the success probability in the matching outcomes test, which is a measure of correlation between A and B, and the conditional min-entropy, which measures correlation between A and E. The main drawback of our approach is that it will require us to make an assumption about the eavesdropper's behavior, which is referred to as the "i.i.d.," for "independently and identically distributed," assumption. This assumption is explained in detail in Box 8.1.[5]

Let's start by focusing on a single round of the purified BB'84 protocol. Imagine that Eve is trying to defeat the protocol: her task is to prepare a state ρ_{ABE} such that each of A and B is a single qubit, and E can be arbitrary. Moreover, she would like to achieve two properties. First of all, if Alice and Bob decide to measure their qubit in the same basis and compare their outcomes then they should be identical as often as possible (as otherwise the users will detect this and abort the protocol). Let's introduce random variables (Z_A, Z_B) to represent the result of measuring both qubits A and B in the standard basis, and (X_A, X_B) to represent the result of measuring both qubits in the Hadamard basis. These random variables are not all simultaneously well-defined because we can't measure the qubits in both bases, but each pair by itself is well-defined. Let's define

---

4 As expected, the bound is bigger than the one we had before, because the game is now easier.

5 In the literature on QKD, "making the i.i.d. assumption" is synonymous with "showing security under collective attacks."

BOX 8.1   The i.i.d. Assumption

When analyzing a multi-round cryptographic protocol it is common to make an assumption known as the i.i.d. assumption, where i.i.d. stands for "identically and independently distributed."

This assumption is composed of two parts, the "identically" and the "independently." We start with "independently." In the case of the purified BB'84 protocol this is the assumption that Eve prepares the state ρ_{ABE} as a tensor product $\rho_{ABE} = \rho_{A_1 B_1 E_1} \otimes \cdots \otimes \rho_{A_N B_N E}$. This assumption can be used to derive a bound on the general min-entropy, for all rounds together, from a bound on the min-entropy on each round, which would be obtained from the qualitative considerations in Section 8.3.1 or the quantitative arguments in Section 8.3.3. For a general quantum state the min-entropy does not simply add up across rounds. Informally this is because the min-entropy is related to the guessing probability, and in general Eve can make any global measurement on all her quantum information to predict all bits x at once.

The second component is "identically." This means that we also implicitly assume that each $\rho_{A_i B_i E_i}$ is equal. This assumption can be used to obtain information about some of the rounds (the rounds used to get the key) based on data gathered in other rounds (the test rounds).

While the i.i.d. assumption is commonly made and can greatly simplify the analysis, it is important to realize that it is only this, an assumption, and that in general the eavesdropper may not respect it!

$$\delta = \frac{1}{2}\Pr(X_A \neq X_B) + \frac{1}{2}\Pr(Z_A \neq Z_B) \,, \tag{8.8}$$

so that δ is the probability to obtain different outcomes when the choice of the basis is made uniformly at random. Eve's first goal is to minimize this quantity. Second, Eve would like to be able to predict Alice's outcome, again in a random choice of basis (which she gets to learn). So, she would like to minimize the quantity

$$\frac{1}{2}\mathrm{H}_{\min}(X_A|E) + \frac{1}{2}\mathrm{H}_{\min}(Z_A|E) \,. \tag{8.9}$$

Furthermore, recall the interpretation of the conditional min-entropy as a guessing probability: this is precisely (minus the logarithm of) Eve's maximum chance of guessing Z_A, when the chosen basis is the standard basis, or X_A, when it is the Hadamard basis.

We need to show that Eve's goal is impossible: she can make one quantity or the other small, but not both at the same time. This is the same goal as in the previous section, except that in the previous section we combined both objectives in a single one to obtain a simple game that we could analyze.

As we will see later, due to the i.i.d. assumption it turns out to be sufficient to consider analogues of (8.8) and (8.9) where uncertainty is measured using the von

Neumann entropy, as opposed to the min-entropy. Concretely, we replace the success measure (8.8) by

$$\frac{1}{2}H(X_A|X_B) + \frac{1}{2}H(Z_A|Z_B) \,. \tag{8.10}$$

This measure is directly related to (8.8). Using a simple calculation based on the definition of the von Neumann entropy we can check that

$$\frac{1}{2}H(X_A|X_B) + \frac{1}{2}H(Z_A|Z_B) \leq h(\delta_{\max}) \,, \tag{8.11}$$

where h is the binary entropy function and $1 - \delta_{\max}$ is the probability of succeeding in the matching outcomes test using the state ρ_{AB}. This means that the users, who compute δ in the protocol and verify that $\delta \leq \delta_{\max}$, can thereby verify that (8.11) holds. Similarly, we replace (8.9) by

$$\frac{1}{2}H(X_A|E) + \frac{1}{2}H(Z_A|E) \,. \tag{8.12}$$

To bound the average of the two quantities (8.10) and (8.12) we use the following inequality, which is an example of an *entropic uncertainty relation*. This relation states that for *any* state ρ_{ABE} such that each of A and B is a system of a single qubit, we always have that

$$H(X_A|X_B) + H(Z_A|E) \geq 1 \qquad \text{and} \qquad H(Z_A|Z_B) + H(X_A|E) \geq 1 \,. \tag{8.13}$$

In other words, it is impossible to have both $H(X_A|X_B)$ small *and* $H(Z_A|E)$ small! This is a manifestation of the *monogamy of entanglement*: if the state ρ_{ABE} is strongly correlated in the Hadamard basis across A and B, then it must be unpredictable in the standard basis across A and E – and vice versa. Averaging these two inequalities we obtain

$$\left(\frac{1}{2}H(X_A|X_B) + \frac{1}{2}H(Z_A|Z_B)\right) + \left(\frac{1}{2}H(X_A|E) + \frac{1}{2}H(Z_A|E)\right) \geq 1 \,. \tag{8.14}$$

Using (8.11) in (8.14) we get that

$$\frac{1}{2}H(X_A|E) + \frac{1}{2}H(Z_A|E) \geq 1 - h(\delta_{\max}) \,. \tag{8.15}$$

This relation looks very similar to the equation (8.4) that we want to prove, but there are some differences. First of all, the measure of entropy is not exactly the same. Here, we have the conditional von Neumann entropy, and in (8.4) we have the conditional min-entropy. Second, (8.2) applies simultaneously to all outcomes that were not measured, but (8.15) applies to a single round. Finally, in (8.4) the min-entropy is evaluated on the state conditioned on not aborting, whereas here we have not yet taken this conditioning into account.

The first difference is handled by making the i.i.d. assumption. The key leverage that we get from this assumption is that it allows us to use the *quantum asymptotic equipartition property*. This states that, when considering a large number of samples of a random variable X, the min-entropy converges to the von-Neumann entropy:

$$\frac{1}{n}H^\varepsilon_{\min}(X_1 \cdots X_n) \approx_{n \to \infty} H(X)$$

for i.i.d. X, provided the smoothing parameter ε is chosen sufficiently large. (Informally, the smoothing parameter ε on the left-hand side means that when calculating the min-entropy we take the largest possible value among all distributions that are ε-close to $(X_1, \ldots, X_n)$ in total variation distance; see Box 5.3.) Here, we can use this property to argue that $H_{\min}(X_A|E\Theta) \approx nH(X_{A.1}|E\Theta)$, where $X_{A.1}$ is the first bit of the n-bit string X_A.

The i.i.d. assumption lets us easily handle the second difference as well. Finally, the conditioning requires a little care, and we omit the technical details – let's just say that the result is similar to the dependence on $p(\Omega)$ which we already observed in the previous section.

To conclude we note that in general of course the i.i.d. assumption cannot be experimentally justified: in practice, the eavesdropper can do what they want. Hence security "under the i.i.d. assumption" is, arguably, not security at all, and this is the main limitation of our work in this section. In fact, in the most general case, without making the i.i.d. assumption, it is still possible to show (8.4) using a more involved uncertainty relation, which is shown directly for the min-entropy and therefore bypasses the need for the asymptotic equipartition property. If you are interested, we give a pointer in the chapter notes. The i.i.d. assumption notwithstanding, as a result of all our hard work we managed to prove (8.4), with a coefficient almost 1 in front of the n! This is the best that we could hope for.

8.4   Correctness of BB'84 Key Distribution

We end by formally arguing correctness of the protocol. When we make the i.i.d. assumption then correctness is very easy to show, as we already did informally: in that case the probability of succeeding in the matching outcomes test gives an estimate for the number of positions in which the strings x_{remain} and $\tilde{x}_{\text{remain}}$ are expected to differ, and to get correctness it suffices to select an appropriately good information reconciliation protocol. Formally, we would use a Chernoff bound – since we're about to give a more general argument we skip the details.

Showing correctness without making the i.i.d. assumption requires more work. The main issue is that in Protocol 6 the matching outcomes test in step 7 is performed on the rounds T selected for testing, but for the raw key we use the rounds in $R = S\backslash T$: How can we guarantee that information gathered through the tests performed on rounds in T has some implication for the rounds *not* in T? Intuitively this is because the tested rounds are chosen at random, and moreover they are chosen *after* the adversary has created the state ρ_{ABE}. So, even if ρ_{ABE} can be arbitrarily correlated there should be no way to arrange things so that tests in a randomly chosen subset of rounds pass and yet the untested rounds wouldn't have passed.

Let's explain how we can make this intuition precise. Suppose for simplicity that the number $|S|$ of rounds in which Alice and Bob make the same basis choice is exactly $|S| = 2n$, and that T has size $|T| = |S|/2 = n$. For each $j \in S$, introduce an indicator random variable $Z_j \in \{0, 1\}$ such that $Z_j = 0$ indicates success in the matching outcomes test: $Z_j = 0$ if and only if $x_j = \tilde{x}_j$. With this notation the condition verified by Alice and Bob at step 7 of Protocol 6 can be written as $\sum_{j \in T} Z_j \leq \delta|T|$. To select parameters for the information reconciliation protocol, however, they would like to have a bound on $\sum_{j \in S\setminus T} Z_j$ that they can be confident about. How can we do this?

The key idea is to use the fact that T is chosen as a random subset. Intuitively the average number of failures in T should be about the same as the average in the whole of S: indeed, which rounds are included in T or not is chosen at random by Alice, independently from whether the outcomes in those rounds happened to match or not.

The main tool required to make this intuition precise is called a concentration bound. There are many such bounds available. The most widely used are usually referred to as the "Chernoff bound" or "Hoeffding's inequality," which is a generalized version of the Chernoff bound. If you have never heard of them, go look them up! The following is a variant of the Chernoff bound that turns out to be perfectly tuned for our scenario.

Theorem 8.4.1 *Let $m = n + k$ and consider binary random variables $X_1, \ldots, X_m$. (The X_i may be arbitrarily correlated.) Let T be a uniformly random subset of $\{1, \ldots, m\}$ of size k. Then for any $\delta, v > 0$,*

$$\Pr\left(\sum_{j \in T} X_j \le \delta k \ \wedge \sum_{j \in \{1, \ldots, m\}\setminus T} X_j \ge (\delta + v)n \right) \le e^{-2v^2 \frac{nk^2}{(n+k)(k+1)}} . \tag{8.16}$$

To see what the theorem says in our setting, set $m = |S| = 2n$ and $k = n$.[6] Let's also choose $v = \delta$ for convenience. Plugging in these parameters we get the bound

$$\Pr\left(\sum_{j \in T} Z_j \le \delta n \ \wedge \sum_{j \in S\setminus T} Z_j \ge 2\delta n \right) \le e^{-\delta^2 \frac{n^2}{n+1}} , \tag{8.17}$$

which is valid for any choice of $\delta > 0$. Equation (8.17) implies that the probability that the test performed in step 8 passes, but the outcomes obtained in the nontested rounds $R = S\setminus T$ do not match in a fraction larger than 2δ of these rounds, is tiny – exponentially small in n! Writing abort to denote the event that Alice and Bob abort in step 8 of Protocol 6, we can use Bayes' rule to rewrite the bound above as

$$\Pr\left(\sum_{j \in S\setminus T} Z_j \ge 2\delta n \ \middle| \ \neg\text{abort} \right) \le \frac{e^{-\delta^2 \frac{n^2}{n+1}}}{\Pr(\neg\text{abort})} . \tag{8.18}$$

Writing the bound in this way allows us to clarify our earlier discussion around the role of the probability of aborting. As you can see, the bound (8.18) is only good if $\Pr(\neg \text{abort})$ is not too small; if this probability was extremely tiny, then the right-hand side of Eq. (8.18) would suffer a corresponding blow-up. The probability that the protocol does not abort is not something that we can control or test, and it is natural that this probability has to be taken into account when defining security: we should always allow the protocol to have a very small probability of not aborting, in which case no claim can be made on the security.

---

6 There is a subtlety here, which is that only the *expected* size of T is n, but the size of T may vary from one execution of the protocol to another. We gloss over this issue here, and return to it in Section 9.2.2 in the next chapter.

To conclude, assuming that we choose the parameters of information reconciliation such that strings at a relative distance at most 2δ are corrected with probability at least ε_{IR} (where ε_{IR} is a correctness parameter for information reconciliation), it follows that the QKD protocol is $\varepsilon_c = \varepsilon_{IR} + \varepsilon_a$-secure, where $\varepsilon_a = e^{-\delta^2 n}/\mathrm{Pr}(\neg\mathrm{abort})$ is the right-hand side of (8.18). Note that here 2δ can be made arbitrarily close to δ by choosing as small a v as we like (and paying a corresponding increase in the error term ε_a). Moreover, to be precise we should note that small additional error terms should be included to account for our assumption that the test set has size precisely $|T| = n$; we will see how to deal with this in the next chapter (spoiler: it is easier than what we just did, and leads to smaller errors, which justifies us neglecting this minor point so far).

CHAPTER NOTES

The origins of QKD can be traced back to ideas that Stephen Wiesner had in the 1970s (Conjugate coding. *SIGACT News*, **15**:78–88, 1983). The first concrete proposal for a QKD protocol is due to C. H. Bennett and G. Brassard (Quantum cryptography: Public key distribution and coin tossing. In *Proceedings of IEEE International Conference on Computers, Systems and Signal Processing*, pp. 175–179, 1984). Shortly after, A. K. Ekert discovered a different approach (Quantum cryptography based on Bell's theorem. *Physical Review Letters*, **67**(6):661, 1991), which we review in the next chapter. Up to small variations these are the two main QKD protocols studied, and implemented, to date.

The uncertainty relation in Section 8.3.3 is from the paper by M. Berta, et al. (The uncertainty principle in the presence of quantum memory. *Nature Physics*, **6**(9):659–662, 2010). The quantum asymptotic equipartition property is shown in the work of M. Tomamichel, R. Colbeck, and R. Renner (A fully quantum asymptotic equipartition property. *IEEE Transactions on Information Theory*, **55**(12):5840–5847, 2009). For a complete security proof based on the tripartite guessing game, see the paper by M. Tomamichel, et al. (A monogamy-of-entanglement game with applications to device-independent quantum cryptography. *New Journal of Physics*, **15**(10):103002, 2013). For a complete proof of security based on entropic uncertainty relations, see M. Tomamichel and A. Leverrier (A rigorous and complete proof of finite key security of quantum key distribution. arXiv:1506.08458, 2015). For the general non-i.i.d. case, one can use the uncertainty relations presented by M. Tomamichel and R. Renner (Uncertainty relation for smooth entropies. *Physical Review Letters*, **106**(11):110506, 2011), extended as in Corollary 7.4 of M. Tomamichel (A framework for non-asymptotic quantum information theory. arXiv:1203.2142, 2012). Another method, less strong quantitatively but conceptually elegant, to reduce the analysis of a multi-round protocol to the i.i.d. case is to use "de Finetti reductions"; see, for example, the paper by M. Christandl, R. König, and R. Renner (Postselection technique for quantum channels with applications to quantum cryptography. *Physical Review Letters*, **102**(2):020504, 2009).

In this chapter the analysis remains high-level and focuses on the asymptotic setting, where we can assume that the number of rounds N of the protocol goes to infinity; in practice it is crucial to understand the error terms even for moderately small values of N. For this, see, for example, the paper by C. Pfister, et al. (Sifting attacks in finite-size quantum key distribution. *New Journal of Physics*, **18**(5):053001, 2016).

PROBLEMS

8.1 Thinking adversarially

Let's imagine that we are Eve and we observe someone trying to implement a QKD protocol. Because QKD is hard they might try to cut corners in their implementations. In this problem we present three "candidate" protocols for key distribution. It is your job to try to break them! For each protocol, choose the step (labeled by numbers) in which there is a mistake that allows you to break security.

- **Protocol I**

  Step 1. Alice generates bit strings x, θ.

  Step 2. Alice prepares the bits x encoded in the basis θ, and sends the resulting qubits to Bob.

  Step 3. Alice announces the basis string θ.

  Step 4. Bob measures in the bases corresponding to θ and obtains x.

- **Protocol II**

  Step 1. Alice generates bit strings x, θ.

  Step 2. Alice generates two-qubit states $|x_i\rangle |\theta_i\rangle$ with the first qubit in the standard basis and the second in the Hadamard basis.

  Step 3. Alice sends the two-qubit states to Bob.

  Step 4. Bob announces receipt of the states.

  Step 5. Bob generates a string $\hat{\theta}$ and measures the second qubit in either the standard basis or the Hadamard basis depending on $\hat{\theta}$, getting an output string χ.

  Step 6. Alice and Bob announce θ and χ over an authenticated channel.

  Step 7. If $\chi_i = \theta_i$ then Bob measures the corresponding first qubit in the standard basis, obtaining a bit $\hat{x}_i$.

  Step 8. Alice and Bob discard all data where $\chi \neq \theta_i$, and now share the string $\hat{x}$.

- **Protocol III**

  Step 1. Alice creates a string of EPR pairs and sends one half of each to Bob.

  Step 2. Bob generates a string θ and measures his half of each pair according to the value of θ.

  Step 3. Alice generates a string $\hat{\theta}$ and similarly measures her half of the EPR pairs.

  Step 4. Bob announces over an authenticated channel that he received and measured his qubits.

  Step 5. Alice and Bob compare θ and $\hat{\theta}$ over an authenticated channel.

  Step 6. Alice and Bob use the measurement results obtained for each $\theta_i = \hat{\theta}_i$ as their key.

8.2 Min-entropy from the matching outcomes bound

This problem investigates a direct method to lower bound Alice and Bob's key extraction rate based on the probability that the matching outcomes test succeeds. If we assume that the adversary, Eve, prepares n identical and uncorrelated copies of the tripartite state $|\psi_{ABE}\rangle$ and sends the qubits A to Alice and B to Bob, then as shown in Chapter 8 the key extraction rate can be asymptotically lower-bounded by the min-entropy $H_{min}(X|E)$ per round, where X is the outcome of Alice's measurement on her qubit. The goal of this problem is to prove a lower bound on this quantity.

Recall that if Alice measures her qubit in the standard basis, and the resulting post-measurement state on her qubit and Eve's system E is a classical-quantum (cq) state

$$\rho_{XE} = \frac{1}{2}|0\rangle\langle 0| \otimes \rho_E^{Z,0} + \frac{1}{2}|1\rangle\langle 1| \otimes \rho_E^{Z,1},$$

then the optimal guessing probability $P_{\text{guess}}(X|E)$ such that

$$H_{\min}(X|E) = -\log P_{\text{guess}}(X|E)$$

is given by the Helström measurement, for which $P_{\text{guess}}(X|E) = \frac{1}{2} + \frac{1}{4}\|\rho_E^{Z,0} - \rho_E^{Z,1}\|_1$.

The same reasoning holds for any other choice of Alice's basis, notably the Hadamard basis $\{|+\rangle, |-\rangle\}$. In the BB'84 protocol Alice chooses with probability 1/2 one of the two bases in which to measure her qubit. If we denote by $P_{\text{guess}}(X|E, \Theta = X)$ and $P_{\text{guess}}(X|E, \Theta = 1)$ the optimal guessing probabilities for Alice measuring in the standard ($\Theta = 0$) and Hadamard ($\Theta = 1$) bases respectively, the desired lower bound is given by

$$H_{\min}(X|E\Theta) = -\log\left[\frac{1}{2}P_{\text{guess}}(X|E, \Theta = 0) + \frac{1}{2}P_{\text{guess}}(X|E, \Theta = 1)\right].$$

1. Suppose Alice and Bob share a pure EPR pair $|\text{EPR}\rangle$, uncorrelated with Eve's system: $\rho_{ABE} = |\text{EPR}\rangle\langle\text{EPR}|_{AB} \otimes \rho_E$. What is $H_{\min}(X|E)$?

2. Now consider the general case, where $|\psi_{ABE}\rangle$ is an arbitrary state prepared by Eve. Let p be the probability that this state succeeds in the matching outcomes test, when Alice and Bob both measure in the same basis Θ chosen at random. Give coefficients a, b, c such that

$$p = a\langle\psi_{ABE}|X_A \otimes X_B \otimes \mathbb{I}_E|\psi_{ABE}\rangle + b\langle\psi_{ABE}|Z_A \otimes Z_B \otimes \mathbb{I}_E|\psi_{ABE}\rangle + c,$$

where X, Z are the Pauli observables $X = |0\rangle\langle1| + |1\rangle\langle0|$ and $Z = |+\rangle\langle-| + |-\rangle\langle+|$.

3. Let p_X (resp. p_Z) be the probability that the state $|\psi_{ABE}\rangle$ passes the matching outcomes test in the Hadamard (resp. computational) basis, so that $p = \frac{1}{2}(p_X + p_Z)$. By expanding the qubit A in the computational basis, the state $|\psi_{ABE}\rangle$ can be expressed as $|\psi_{ABE}\rangle = |0\rangle \otimes |u_0\rangle_{BE} + |1\rangle \otimes |u_1\rangle_{BE}$, with $\||u_0\rangle_{BE}\|^2 + \||u_1\rangle_{BE}\|^2 = 1$. Give coefficients a', b' such that $\langle\psi_{ABE}|X_A \otimes X_B \otimes \mathbb{I}_E|\psi_{ABE}\rangle = a'\Re(\langle u_0|X_B \otimes \mathbb{I}_E|u_1\rangle) + b'$.

A similar bound can be obtained for p_Z.

Suppose Alice measures her qubit in the computational basis; the post-measurement state on A and E (tracing out B) can be written as $\rho_{AE}^Z = |0\rangle\langle0|_A \otimes \sigma_E^{Z,0} + |1\rangle\langle1|_A \otimes \sigma_E^{Z,1}$. Similarly, if Alice measures in the Hadamard basis we may write the post-measurement state as $\rho_{AE}^X = |+\rangle\langle+|_A \otimes \sigma_E^{X,+} + |-\rangle\langle-|_A \otimes \sigma_E^{X,-}$.

4. Use the previous two questions to determine coefficients α, β such that

$$2p - 1 \leq \alpha F\left(\sigma_E^{X,0}, \sigma_E^{X,1}\right) + \beta F\left(\sigma_E^{Z,+}, \sigma_E^{Z,-}\right)$$

where F denotes the fidelity. *[Hint: observe that $|u_0\rangle_{BE}$ and $|u_1\rangle_{BE}$ considered in the previous question are purifications of $\sigma_E^{Z,0}$ and $\sigma_E^{Z,1}$ respectively, and use Uhlmann's theorem.]*

5. Recall the inequality $D(\rho, \sigma) \leq \sqrt{1 - F(\rho, \sigma)^2}$. Using also the definition of $H_{min}(X|E)$, what is the best lower bound on $H_{min}(X|E)$ as a function of p that you can get?

8.3 Trusted nodes

In this problem, we explore the idea of "trusted nodes" or "trusted repeaters." Let us imagine that Alice and Bob wish to generate a key between them, but are not able to send qubits to each other. However, Alice is capable of using QKD to generate a key with her friend Charlie, and similarly Charlie and Bob are able to produce a key between them. Such a situation could, for example, arise in a situation in which Alice and Bob are themselves too far apart to perform quantum communication according to the current state of the art in quantum technologies; however, Charlie is located in-between them and close enough to use QKD to make a key with both of them individually. In this context, Charlie is known as a trusted node.

1. Explain how Alice and Bob can generate a secure shared key k_{AB} with the help of Charlie.
2. Discuss whether your solution guarantees that Alice and Bob end up with the same key.
3. Explain how Eve can intercept the communication between Alice and Bob, when Charlie collaborates with Eve.

9

Quantum Cryptography Using Untrusted Devices

In this chapter we introduce a variant of the BB'84 quantum key distribution (QKD) protocol from the previous chapter. This variant is due to Ekert and is often referred to as the E'91 protocol for QKD. Since our protocol won't exactly follow Ekert's original proposal we will call it the "DIQKD protocol." The letters DI stand for "device independent." What this means, informally, is that the new protocol's security doesn't rely on Alice and Bob performing trusted measurements on their qubit in each round – in fact, it doesn't even rely on the assumption that the system they measure in each round is a qubit! In other words, we partially drop assumption number 2 in Box 7.1, thereby obtaining a higher level of security than the BB'84 protocol. As a counterpart the Ekert protocol is more difficult to implement, as it requires the users to have the ability to distribute EPR pairs quickly and make rather accurate measurements on them.

The key difference between the DIQKD protocol and the BB'84 protocol is that in the DIQKD protocol we replace the "matching outcomes" test used in BB'84 by a different test. The new test is based on the CHSH game, which we introduced in Chapter 4. Using the monogamous properties of entanglement the test will let us show security in the more general DI setting. Let's see how this works.

9.1 The DIQKD Protocol

Before we describe the protocol, let's see precisely what the notion of device-independent security covers – and does not cover.

9.1.1 Device-Independent Security

The notion of device independence is motivated by the practical difficulty of characterizing the quantum mechanical devices, such as photon emitters or receptors, used in protocols such as BB'84. You remember that this protocol asks Alice to do things such as "prepare a qubit in the Hadamard basis," and Bob to "measure his qubit in the standard basis." When Alice prepares her qubit, and when Bob measures it, can they really trust their equipment to implement the task correctly? What if, for example, Alice's preparation device sometimes in fact creates *two* qubits, instead of a single one, without her noticing; could the additional qubit be intercepted by Eve and provide her with secret information, without Alice or Bob noticing? The following example shows that such misbehavior of Alice and Bob's equipment can indeed pose a serious security risk.

Example 9.1.1 Consider the purified variant of the BB'84 protocol. Suppose that Eve prepares a state ρ_{ABE} of the following form:

$$\rho_{ABE} = \sum_{x,z=0}^{1} |x,z\rangle\langle x,z|_A \otimes |x,z\rangle\langle x,z|_B \otimes |x,z\rangle\langle x,z|_E . \qquad (9.1)$$

Here A and B are each made of two qubits, instead of just one as required in the protocol. Nevertheless, suppose that Alice and Bob don't notice this: after all, a single photon isn't that easy to spot! Suppose further that their measurement devices, instead of measuring in the standard or Hadamard bases, as they think, in fact perform the following:

- When the device is told to measure in the standard basis, it measures the first qubit of the two-qubit system associated with the device, A or B, in (9.1) in the standard basis.
- When the device is told to measure in the Hadamard basis, it measures the second qubit of the two-qubit system associated with the device in (9.1) in the standard basis.

Such devices will perfectly pass all tests performed in the protocol: indeed, you can verify that for the state in (9.1) when the basis choice is the same the outcome is the same, whereas when the bases are different the outcomes are perfectly uncorrelated. But any key extracted from ρ_{ABE} in (9.1) is completely insecure! (Exercise: give an explicit attack for Eve.) ∎

Although the example may look like a bit of a stretch, similar attacks have been implemented in practice. In fact, one of the first real "attacks" on the BB'84 protocol was that the photon receptor used in an early experiment made a different clicking noise when it measured in one of Bob's bases, thereby "leaking" Bob's choice of measurement basis to any eavesdropper within earshot! (This is an example of a failure of the assumption "Bob's laboratory is safe" from Box 7.1.) Many such attacks, often called *side-channel attacks*, have been demonstrated. Some of the most effective are called "detector blinding" attacks, in which the eavesdropper can take complete control of Bob's measurement device by shining a very bright laser right into it (without Bob noticing!). The problem is that while quantum information can in principle bring us great security, it is also very fragile and hence suceptible to unexpected attacks. Is there a way that we can better protect ourselves?

This is the goal of device-independent security. This notion aims to guarantee security even when there may be dramatic failures of Alice and Bob's equipment, and moreover when such failures could be exploited by an adversary. Now, we have to be careful about what we promise exactly. For example, as an extreme case we could imagine that Bob's device contains radio equipment that automatically transmits all its measurement results to Eve: in this case security is compromised, but there is no way for Bob to detect the radio transmitter unless he opens the device. Similarly, if the random number generator used by Alice to make her basis choices is biased, or controlled by Eve, then security cannot hold. The specific kinds of failures that are allowed by a device-independent proof of security have to be specified on a case-by-case basis. For QKD we will make the following assumptions, which refine item 2 from Box 7.1:

2.a Alice and Bob's labs are perfectly isolated: once the protocol starts no information enters or exits their respective labs unless specified by the protocol.

2.b Alice and Bob's random number generators are perfect.

2.c The measurement devices used by Alice and Bob to perform measurements are arbitrary. These devices are initialized in a state ρ_{ABE} which may be chosen by the adversary. At each step of the protocol, each of Alice and Bob's devices makes a measurement when instructed, and always produces an outcome $x \in \{0,1\}$. The measurement that is performed is arbitrary. In particular, the device may have memory and behave differently in each round.

2.d At the end of the protocol the devices are discarded and will never be re-used. They will never fall into Eve's hands.

As you can see, the main novelty in device independence is assumption 2.c, which allows the devices to perform any kind of measurement, on any state; both may have been decided by Eve as part of her "attack." In the analysis of the BB'84 protocol in the previous chapter we allowed Eve to prepare any state for the devices, *but* Alice and Bob still had to receive a single qubit, and they could trust the way that measurements were made on that qubit (indeed, this was instrumental to the use of uncertainty relations). Here we remove that assumption.

We mention that the last assumption, while not crucial in our context, is important when we think about the problem of *composition*, which arises when trying to combine different cryptographic protocols, in sequence or even simultaneously and involving overlapping sets of users; this is because in the DIQKD protocol the devices themselves know Alice and Bob's raw key,[1] and could potentially store it in memory. So it is important that the devices are not re-used in another protocol where Alice and Bob might want to use the key produced with those devices.

QUIZ 9.1.1 *In the device-independent setting, attacks by Eve can be modeled by specifying what kinds of devices she gives to Alice and Bob. Which of the following attacks do we hope our device-independent protocol will protect against?*

(a) *Alice's devices communicate with Bob's devices during the protocol.*

(b) *Eve gets to examine Alice and Bob's devices at the end of the protocol.*

(c) *Alice's devices send information to Eve during the protocol.*

(d) *Eve's laboratory is arbitrarily entangled with Alice and Bob's laboratories at the beginning of the protocol.*

---

1 Recall that the *raw key* is the string of bits obtained by each user as a result of their measurements in the protocol, and before the classical post-processing steps of information reconciliation and privacy amplification.

9.1.2 The Protocol

We are almost ready to describe the DIQKD protocol. As already mentioned, the protocol is based on the CHSH game. However, we need to make a small modification to the game in order to make it useful for QKD. (If you don't remember the CHSH game, now is a good time to check the rules again.) Indeed, in the honest optimal strategy for the CHSH game (Box 4.2) Alice and Bob never use the same basis, and thus they never obtain perfectly correlated outcomes. However, in order to produce a key it will be convenient for them to obtain (almost) perfectly correlated outcomes for at least one choice of a pair of bases. To make this possible in the protocol we think of Bob's device as having three, instead of two, possible measurement settings. Since in the device-independent setting we don't assume that we know what measurements are made, we will label them using a value $\theta \in \{0,1\}$ for Alice, where $\theta = 0$ means that Alice is asking for a standard basis measurement and $\theta = 1$ means a Hadamard basis measurement, and $\tilde{\theta} \in \{0,1,2\}$ for Bob, where $\tilde{\theta} \in \{0,1\}$ correspond to the usual CHSH inputs (for which the ideal device would measure in the basis described in Box 4.2), and the additional value $\tilde{\theta} = 2$ instructs the device to measure in the standard basis. We refer to the values of $\theta, \tilde{\theta}$ as *inputs* that the user introduces into their device, asking for a measurement outcome to be returned. Since Alice's device also measures in the standard basis on input 0, and since the honest devices share an EPR pair, this means that on inputs $(\theta, \tilde{\theta}) = (0,2)$ the devices are expected to produce matching outcomes.

Protocol 7 (DIQKD protocol) The protocol depends on a large integer n and a small parameter $\delta_{\max} > 0$ publicly chosen by the users (intuitively n is the number of bits of key that they expect to generate at the end and $\delta_{\max}$ is an error tolerance). Let $N = 12(1 + C\delta_{\max})n$, where C is a large constant that can be determined from the security analysis. Alice and Bob execute the following:

1. Alice chooses a uniformly random basis string $\theta = \theta_1, \ldots, \theta_N \in \{0,1\}^N$ and sequentially instructs her measurement device to measure in the bases θ. The device returns a string of outcomes $x = x_1, \ldots, x_N \in \{0,1\}^N$.

2. Bob chooses a uniformly random basis string $\tilde{\theta} = \tilde{\theta}_1, \ldots, \tilde{\theta}_N \in \{0,1,2\}^N$ and sequentially instructs his measurement device to measure in the bases $\tilde{\theta}$. The device returns a string of outcomes $\tilde{x} = \tilde{x}_1, \ldots, \tilde{x}_N \in \{0,1\}^N$.

3. Alice and Bob tell each other their basis strings θ and $\tilde{\theta}$ respectively over the CAC.

4. Alice picks a random subset $T \subseteq \{1, \ldots, N\}$ by flipping a fair coin for each $i \in \{1, \ldots, N\}$ to decide if it is selected in T. Alice tells Bob what T is over the CAC. They each set $T' = \{j \in T, \tilde{\theta}_j \in \{0,1\}\}$, $T'' = \{j \in T, \theta_j = 0 \wedge \tilde{\theta}_j = 2\}$, and $R = \{j \notin T, \theta_j = 0 \wedge \tilde{\theta}_j = 2\}$.

5. Alice and Bob announce x_T and $\tilde{x}_T$ to each other over the CAC. They compute the success probabilities $p_{\text{win}} = |\{j \in T', x_j \oplus \tilde{x}_j = \theta_j \wedge \tilde{\theta}_j\}|/|T'|$ and $p_{\text{match}} = |\{j \in T'', x_j = \tilde{x}_j\}|/|T''|$. If $p_{\text{win}} < \cos^2 \pi/8 - \delta_{\max}$ or $p_{\text{match}} < 1 - \delta_{\max}$ they abort.

6. Let x_{remain} and $\tilde{x}_{\text{remain}}$ be Alice and Bob's outcomes restricted to indices in R. Alice and Bob perform information reconciliation and privacy amplification on x_{remain} and $\tilde{x}_{\text{remain}}$.

In the protocol description we have not fleshed out the last step in full detail, because it is identical to the last steps of the BB'84 protocol with noise presented in the previous chapter. The important difference here is step 5, which plays the role of step 7 from the purified BB'84 protocol.

Before we proceed let's check the expected length of key produced by this protocol. Because the values $\theta_i, \tilde{\theta}_i$ are chosen uniformly at random we expect that

$$|R| \approx \frac{1}{6}|\{1,\ldots,N\}\backslash T| \approx \frac{1}{6}\frac{N}{2} = (1+C\delta_{\max})n \ .$$

As we will see, the steps of information reconciliation and privacy amplification lead to a moderate loss in the length of the raw keys x_{remain} and $\tilde{x}_{\mathrm{remain}}$, so that if the constant C is chosen large enough we can count on obtaining roughly n bits of final key.

How can we show security of this protocol? Based on the work done in the previous chapters we already know that it is sufficient to show two things. First of all, we need to show that the entropy $H_{\min}(X_{\mathrm{remain}}|E)$, evaluated on the state of the users at step 6 in the protocol conditioned on not aborting in step 5, is large. Second, we also need to make sure that $X_{\mathrm{remain}} \approx \tilde{X}_{\mathrm{remain}}$, as this will allow us to bound how much information is leaked to Eve in the step of information reconciliation.

If we make the i.i.d. assumption (Box 8.1) then, using the condition $p_{\mathrm{match}} < 1 - \delta_{\max}$ from step 5 and a similar analysis as in Section 8.4 in the previous chapter, it is possible to show that leakage from information reconciliation is of order $h(\delta_{\max})|R|$, which can be made arbitrarily small by taking $\delta_{\max}$ small enough. Therefore we focus on the first condition, guaranteeing uncertainty from Eve. In the previous chapter we saw how this condition, which was summarized in Eq. (8.4), can be achieved using three different methods: a direct method based on interpreting the matching outcomes game as an "entanglement projection test," a method based on guessing games, and a method based on uncertainty relations. Here we focus on the method that generalizes best to the device-independent setting, the use of guessing games, and introduce a new guessing game adapted to the CHSH test used in the protocol. (We will see that we also use ideas from the method based on uncertainty relations.) The first method, which characterizes the entanglement shared by the users, can also be extended, and we will give the main ideas for taking this route in Section 9.3.

9.2   Security of Device-Independent Quantum Key Distribution

We start our security argument by focusing on a single round of the protocol, and analyze the adversary's power to gain information about Alice's output using a new tripartite guessing game.

9.2.1   A CHSH-Based Guessing Game

Let's consider the following guessing game. By studying this game we will be able to show a bound on Eve's uncertainty in the DIQKD protocol. In the game there are three players, Alice, Bob, and Eve. Alice receives an input $\theta \in \{0,1\}$, Bob receives $\tilde{\theta} \in \{0,1,2\}$, and Eve receives no input (if you prefer you can think that her input is always the same). The players have to produce answers $x, \tilde{x}, z \in \{0,1\}$ respectively. They win the game if and only if both of the following conditions hold:

- If $\tilde{\theta} \in \{0,1\}$ then $x \oplus \tilde{x} = \theta \wedge \tilde{\theta}$.
- If $\theta = 0$ and $\tilde{\theta} = 2$ then $x = z$.

Note that the two conditions never apply simultaneously. However, Alice in general doesn't know which condition is going to be checked (because if her input is $\theta = 0$ then both could in principle apply), and this is what makes the game hard: on the one hand Alice wants to play the CHSH game the best she can with Bob, but on the other hand she wants to make sure that Eve has a way of knowing what outcome she'll get. If this sounds impossible, indeed it is! The following exercise asks you to show a bound on the maximum winning probability in the game.

Exercise 9.2.1 Suppose that Alice and Bob play the game according to the optimal CHSH strategy, and Eve always returns a uniformly random $z \in \{0,1\}$. Show that this strategy succeeds with probability

$$p_{\min} = \frac{2}{3} \cos^2 \frac{\pi}{8} + \frac{1}{6} + \frac{1}{6} \cdot \frac{1}{2}$$

in the game. *[Hint: consider all six possible cases for the questions.]*

Exercise 9.2.2 Show that it is impossible to win in this game with probability larger than

$$p_{\max} = \frac{2}{3} \cos^2 \frac{\pi}{8} + \frac{1}{6} + \frac{1}{6} .$$

[Hint: what is the maximum probability for winning in the CHSH game?]

There is a gap between $p_{\min}$ and $p_{\max}$ obtained in the exercises. What is the right answer? As it turns out, the correct maximum is exactly $p_{\min}$. Intuitively this is because, in order to win with probability close to $\cos^2 \pi/8$ in the CHSH part of the game, Alice *has* to measure an EPR pair and hence return random outcomes that Eve couldn't possibly predict with probability more than $1/2$. This is a version of the phenomenon of *monogamy* which we already encountered in Section 4.5 in Chapter 4. Concretely, it is possible to show the following trade-off.

Lemma 9.2.1 (CHSH guessing lemma) *Consider an arbitrary strategy for the players in the CHSH guessing game. Let ω be the probability that the first test passes (conditioned on $\tilde{\theta} \in \{0,1\}$) and γ the probability that the second test passes (conditioned on $\theta = 0$ and $\tilde{\theta} = 2$). Suppose that $\omega \geq \cos^2 \pi/8 - \delta$ for some $0 \leq \delta \leq 1/2$. Then $\gamma \leq 1/2 + 2\sqrt{\delta}$.*

We leave the proof of the lemma as an exercise. There are different possible ways to approach it. We indicate one possible proof strategy, which formalizes the intuition described earlier. The first step is to characterize the state shared by Alice and Bob as being close to an EPR pair using the condition $\omega \geq \cos^2 \pi/8 - \delta$. This can be done by building on the contents of Section 9.3 below. The second step, which is easier, uses that if Alice and Bob share a perfect EPR pair, then Eve has no information about Bob's outcomes.

From the lemma we get that the maximum winning probability in the game is the maximum over all possible $0 \leq \delta \leq 1/2$ of the expression

$$p_{\text{win}} \leq \frac{2}{3}\left(\cos^2\frac{\pi}{8} - \delta\right) + \frac{1}{6} + \frac{1}{6}\left(\frac{1}{2} + 2\sqrt{\delta}\right).$$

You can easily verify that the right-hand side is always less than $p_{\min}$, as claimed. This bound is the analogue of the bound $p_{\text{win}} \leq \frac{1}{2} + \frac{1}{2\sqrt{2}}$ shown on the tripartite guessing game from Chapter 5.

9.2.2 Security under the i.i.d. Assumption

Let's move on to show security of the DIQKD protocol in the i.i.d. setting (Box 8.1). As we saw in the previous chapter, under the i.i.d. assumption, in order to get a bound on the final uncertainty $H_{\min}(X_{\text{remain}}|E)$ it suffices to show a bound on the uncertainty in each round of the protocol as measured by the conditional von Neumann entropy $H(X_i|E\Theta_i = 0)$, where Θ_i is Alice's choice of basis and X_i her measurement outcome in the i-th round of the protocol. In this expression we condition on $\Theta_i = 0$, because this is the only case that is used to create the raw key. Let's see how we can get such a bound using Lemma 9.2.1. First of all, we always have

$$H(X_i|E\Theta_i = 0) \geq H_{\min}(X_i|E\Theta_i = 0), \tag{9.2}$$

because the min-entropy is the "smallest" entropy measure.[2] We also know that $H_{\min}(X_i|E\Theta_i = 0)$ has an interpretation as the maximum probability with which Eve, given access to the quantum system E, can guess the outcome X_i (when $\Theta_i = 0$). This is exactly the quantity γ that is estimated in Lemma 9.2.1. Precisely, from the lemma we get that

$$P_{\text{guess}}(X_i|E\Theta_i = 0) \leq \frac{1}{2} + 2\sqrt{\cos^2\frac{\pi}{8} - \omega_i},$$

where ω_i is the probability that Alice and Bob's outputs in the i-th round satisfy the CHSH conditions, conditioned on their inputs being chosen in $\{0,1\}$. Taking the logarithm and using (9.2) we get

$$H(X_i|E\Theta_i = 0) \geq -\log\left(\frac{1}{2} + 2\sqrt{\cos^2\frac{\pi}{8} - \omega_i}\right) = 1 - O(\sqrt{\delta_i}), \tag{9.3}$$

where $\delta_i = \cos^2\pi/8 - \omega_i$. Equation (9.3) shows that, as expected, the closer the winning probability is to the CHSH optimum, the more uncertainty there is in Alice's outcomes. Equation (9.3) gives the right order asymptotically (for very small δ), and if we don't care too much about parameters, for example the number of rounds of the protocol that are "wasted" for testing, then it is good enough for us. If we want the optimal trade-off, by using more refined optimization techniques it is possible to obtain a more precise bound:

---

2 Among all Rényi entropies; we ask that the reader take the inequality on faith.

$$H(X_i|E\Theta_i) \geq 1 - h\left(\frac{1}{2} + \frac{1}{2}\sqrt{16\omega_i(\omega_i - 1) + 3}\right). \tag{9.4}$$

This bound is better in general because in contrast to (9.3) any value of ω_i larger than $3/4$ gives a positive lower bound on the conditional entropy. This means that as soon as Alice and Bob are able to observe outcomes that surpass the classical optimum winning probability in the CHSH game, they are able to certify that their raw key contains some uncertainty!

To conclude our analysis there are a couple more steps to make. First of all, how can we infer a bound on the quantity ω_i based on data that is collected in the protocol? Second, to measure the amount of key that will eventually be produced we need to be able to estimate the size of R, the set of indices from which the raw key is taken.

Since we are making the i.i.d. assumption, in principle the user's device has a well-defined success probability $\omega = \omega_i$ in the CHSH game. Moreover, this is precisely the quantity that is estimated at step 5 of the protocol. If we assume that the number of rounds selected for testing, $|T|$, is a constant fraction of n, then the quality of the user's estimate for ω can be estimated using the same technique as in Section 8.4. Let's explain how to do this in detail, without even making the untrue assumption that the number of test rounds is constant. To remove that assumption we need to estimate the chance that the number of test rounds deviates by too much from its expectation value. To model the situation we introduce binary random variables $Z_1, \ldots, Z_k$, where $k = |T'|$ (remember that T' is the subset of rounds tested for the CHSH condition), such that Z_j equals 1 if the CHSH condition in round j is satisfied. Then at step 5 of the protocol the users set $p_{\text{win}} = |T'|^{-1} \sum_{j \in T'} Z_j$. Note that this is an "observed" quantity, i.e. it may vary each time we run the protocol, even with the same devices. We would like to know the "true value" ω, i.e. the probability of success of the device in the CHSH game (instead of its average success in any particular run). How different can ω and p_{CHSH} be?

Let's first start by estimating the size of T'. We can think of the inputs for the rounds T' as being selected after the set of rounds T' itself is chosen by Alice: for instance, we could imagine Bob choosing rounds in which $\tilde{\theta}_j = 2$ at random, and Alice choosing a random set T; this defines the set T' but the players still have the freedom to choose specific inputs for those rounds. Since the probability of any given round lying in T is $1/2$, and independently the probability that Bob chooses $\tilde{\theta}_j = 2$ is $1/3$, the expected size of $|T'|$ is $n/6$. To show that the chance that the actual size differs from the expected size by too much is small, we need a simple concentration inequality.

Theorem 9.2.2 (Chernoff bound[3]) *Let $X_1, \ldots, X_n$ be i.i.d. random variables taking values in $\{0, 1\}$, and $\mu = E[X_i]$. Then for all $0 < \alpha < 1$,*

$$\Pr\left(\left|\frac{1}{n}\sum_{i=1}^{n} X_i - \mu\right| > \alpha\mu\right) \leq 2e^{-\frac{\alpha^2 \mu n}{3}}.$$

---

3 Herman Chernoff. A note on an inequality involving the normal distribution. *The Annals of Probability*, **9**(3):533–535, 1981.

If we apply the theorem with $\mu = 1/6$ and $\alpha = 1/4$ we obtain that the probability that $|T'| < n/8$ is at most $e^{-n/(3 \cdot 6 \cdot 16)}$. Let's assume that this is not the case. Then we can apply the same bound once more, with some different α, to obtain

$$\Pr\left(\sum_{j \in T'} Z_j > (1+\alpha)|T'| \Big| \omega \right) \leq 2 e^{-\frac{\alpha^2 \hat{p}_{\mathrm{win}} |T'|}{3}}.$$

Hence, using our lower bound on the size of $|T'|$ as well as $\omega \geq 1/2 - \sqrt{2}/4$ (Exercise: why?),

$$\Pr\left(\omega < \frac{1}{1+\alpha} p_{\mathrm{win}} \right) \leq 2 e^{-\frac{\alpha^2 n}{C}}$$

for some large constant C.

So far we have managed to show that, except with probability exponentially small in n, provided the protocol does not abort in step 5 it must be the case that $\omega \geq p_{\mathrm{win}}/(1+\alpha) \geq \cos^2 \pi/8 - 2\delta$ (if we choose $\alpha = \delta$). Now it is time to apply (9.4). Churning through the numbers we arrive at

$$H(X_j|E\Theta_j) \geq 1 - h(c\sqrt{\delta}), \tag{9.5}$$

for some constant c. To conclude, the last thing that we need to do is to estimate the size of R. Using Theorem 9.2.2 one last time it is easy to show that with very high probability R is almost $N/12$. Using a similar reasoning as in Section 8.3.3 in the previous chapter we arrive at the bound

$$\mathrm{H}_{\min}(X_{\mathrm{remain}}|E) \geq \left(1 - h(c\sqrt{\delta})\right)|R| \gtrsim \left(1 - h(c\sqrt{\delta})\right)\frac{N}{12}.$$

This is not an optimal bound. What is important for us is that it depends linearly on the total number of rounds N, and so we have the guarantee that the protocol generates a linear amount of key. (Remember that to get the final key length we'd also have to subtract the information exchanged for information reconciliation, which as mentioned earlier scales like $h(\delta)|R|$.) In practice various optimizations are possible to improve this rate. In particular, in a real implementation the users would bias their choice of measurement basis so that the pair $(0, 2)$ happens most of the time, and only a comparatively small subset of the rounds are used for testing. This can help make $|R|$ very close to N, instead of $N/12$ here.

We note a final subtlety in the analysis that we have glossed over. Earlier we wrote things like "assuming this holds" when computing bounds on the size of T or the CHSH winning probability. What if these conditions do not hold? What we did show is that conditioned on not aborting both conditions hold, except with probability ε that is exponentially small. What this means is that in fact we have not quite obtained a lower bound on the conditional min-entropy, but what is known as the "smooth" conditional min-entropy $\mathrm{H}_{\min}^{\varepsilon}(X_{\mathrm{remain}}|E)$ (see Box 5.3). What the ε means is that we are not bounding the entropy directly on the state ρ_{XE} from the protocol, but on a state that is very close – the state where all the conditions that "almost surely hold" *actually* hold. While this is a hypothetical state that never arises in practice, because it is so close to the real state it is sufficient to prove security on it: no adversary will ever be able to tell a real execution from an ideal one, except with advantage ε that is exponentially small.

The previous arguments only handle the i.i.d. setting. Using a more technically involved argument it is possible to give bounds that apply in general. Such techniques lie beyond the scope of this book, but we give pointers in the chapter notes. For concreteness, and not insisting on actual parameters, we give a typical formulation for a complete security statement that can be shown about the DIQKD protocol.

Theorem 9.2.3 *The DIQKD protocol, Protocol 7, satisfies the following properties. There is a $0 < \kappa \leq 1$ and $C \geq 1$ (depending on the tolerance parameter δ) such that the following hold for $\ell = \kappa n$ and $\varepsilon \leq 2^{-Cn}$.*

First, there is an implementation of the devices such that the protocol does not abort with probability at least $1 - \varepsilon$.

Second, for any implementation of the devices, either the protocol aborts with probability larger than $1 - \varepsilon$, or conditioned on not aborting Alice and Bob each produce a key K_A and K_B of length ℓ such that $\Pr(K_A \neq K_B) \leq \varepsilon$ and

$$(1 - \Pr(abort)) \left\| \rho_{KE} - \frac{\mathbb{I}_K}{2^\ell} \otimes \rho_E \right\|_1 \leq \varepsilon,$$

where ρ_{KE} is the joint state of the key K_A output by Alice and all the side information available to the eavesdropper at the end of the protocol, conditioned on the protocol not aborting.

QUIZ 9.2.1 *Suppose that Alice and Bob perform the DIQKD protocol described in the chapter and succeed in 850 out of 1000 CHSH test rounds. What can we say about p_{win}, the probability that the CHSH test is passed on a future test round?*

(a) $p_{\text{win}} = 0.85$ *with certainty*
(b) $p_{\text{win}} = 0.85$ *with high probability*
(c) $|p_{\text{win}} - 0.85|$ *is small with high probability*
(d) $|p_{\text{win}} - 0.85|$ *is small with certainty*

QUIZ 9.2.2 *In the chapter we proved security for the collective setting, i.e. when Eve attacks each round of the protocol independently. Which of the following parts of the proof break when we move to the coherent setting, i.e. we no longer demand that ρ_{ABE} is the tensor product of $N \approx 12n$ identical states?*

(a) *The winning probability of the CHSH game on a random subset of rounds no longer predicts the winning probability on the rest of the rounds.*
(b) *The entropy guarantees from individual rounds of the tripartite guessing game no longer give an entropy guarantee on the whole key.*
(c) *Classical correlation inequalities fail when applied to random variables coming from measurements on entangled states.*

9.3   Testing EPR Pairs

We end the chapter with an optional section on a phenomenon called "rigidity" of the CHSH game. Remember that in the previous chapter (see Section 8.3.1) we argued that the "matching outcomes" test has essentially the same effect as projecting the state ρ_{ABE} on an EPR pair between A and B. This, however, crucially relied on the fact that measurements made by Alice and Bob used in the protocol are fully characterized. To see where this played a role, see, for example, Exercise 8.3.1 and the formula for Π_1 and Π_2; these are only correct because we *know* which basis the users are measuring in.

Now we would like to argue that a similar effect is achieved by the test based on the CHSH game, *without* needing to assume anything about the user's measurements! To get started on this let's recall the standard version of the CHSH game (not the one with an extra input that we used for the DIQKD protocol). In the game the referee sends each of the two players, Alice and Bob, a uniformly random bit $x, y \in \{0, 1\}$ respectively. The players have to return outcomes $a, b \in \{0, 1\}$ such that the CHSH condition $a \oplus b = x \wedge y$ is satisfied. We saw that the maximum success probability of classical noncommunicating players in this game is $p_{\text{win}} = 3/4$, while if Alice and Bob are quantum there is a strategy that allows them to succeed with probability $p^*_{\text{win}} = \cos^2 \pi/8 \approx 0.85$.

In the strategy described in Box 4.2, Alice and Bob share an EPR pair $|\text{EPR}\rangle_{AB}$ and make the following measurements. When $x = 0$, Alice measures her qubit in the standard basis $\{|0\rangle, |1\rangle\}$, and when $x = 1$, she measures in the Hadamard basis $\{|+\rangle, |-\rangle\}$. When $y = 0$, Bob measures his qubit in the basis

$$\{\cos(\pi/8)|0\rangle + \sin(\pi/8)|1\rangle, -\sin(\pi/8)|0\rangle + \cos(\pi/8)|1\rangle\}$$

and when $y = 1$, he measures in the basis

$$\{\cos(\pi/8)|0\rangle - \sin(\pi/8)|1\rangle, \sin(\pi/8)|0\rangle + \cos(\pi/8)|1\rangle\}\,.$$

Since these measurements are binary projective measurements, with POVM elements of the form $\{\Pi, \mathbb{I} - \Pi\}$, we can equivalently describe them using the associated *observables* $O = 2\Pi - \mathbb{I}$. Note that O is a Hermitian operator which squares to identity. For Alice's measurements the observables are

$$A_0 = 2|0\rangle\langle 0| - \mathbb{I} = Z \ (x = 0) \quad \text{and} \quad A_1 = 2|+\rangle\langle +| - \mathbb{I} = X \ (x = 1)\,.$$

For Bob we have

$$B_0 = H \ (y = 0) \quad \text{and} \quad B_1 = \tilde{H} = \frac{1}{\sqrt{2}} \begin{pmatrix} 1 & -1 \\ -1 & -1 \end{pmatrix} (y = 1)\,.$$

We introduced this as a "good" strategy for the players: it certainly beats the classical bound $p_{\text{win}} = 3/4$, and achieves $p^*_{\text{win}} = \cos^2 \pi/8$. But could there be better strategies, achieving an even larger value? Or, even if they are not better, different strategies, based on using a different type of entangled state, for achieving the same success probability?

We're going to show that this is not the case: the maximum success probability of any quantum strategy in the CHSH game, as complicated as it may be, is p^*_{CHSH}. Moreover, any strategy achieving this value must be "equivalent" to the strategy described above. What do we mean by equivalent? We couldn't possibly hope to claim that the strategy is strictly unique. For example, if Alice and Bob were to rotate their basis choices by the same angle, then since the EPR pair is itself rotation invariant their success probability would remain unchanged. The next theorem shows that this local degree of freedom is essentially the only flexibility that the players have in designing an optimal strategy.

Theorem 9.3.1 (CHSH rigidity) *Suppose we are given an entangled state* $|\psi\rangle_{AB} \in \mathbb{C}^{d_A} \otimes \mathbb{C}^{d_B}$ *and observables* A_0, A_1 *for Alice and* B_0, B_1 *for Bob such that the corresponding strategy has a success probability* $p^*_{CHSH} = \cos^2 \pi/8$ *in the CHSH game. Then there exist isometries* $U_A : \mathbb{C}^{d_A} \to \mathbb{C}^2 \otimes \mathbb{C}^{d_{A'}}$ *and* $V_B : \mathbb{C}^{d_B} \to \mathbb{C}^2 \otimes \mathbb{C}^{d_{B'}}$ *such that*

$$U_A \otimes V_B |\psi\rangle_{AB} = |\text{EPR}\rangle \otimes |junk\rangle_{A'B'}$$

and

$$
\begin{aligned}
(U_A \otimes V_B)(A_0 \otimes \mathbb{I}_B)|\psi\rangle &= ((Z \otimes \mathbb{I})|\text{EPR}\rangle) \otimes |junk\rangle , \\
(U_A \otimes V_B)(A_1 \otimes \mathbb{I}_B)|\psi\rangle &= ((X \otimes \mathbb{I})|\text{EPR}\rangle) \otimes |junk\rangle , \\
(U_A \otimes V_B)(\mathbb{I}_A \otimes B_0)|\psi\rangle &= ((\mathbb{I} \otimes H)|\text{EPR}\rangle) \otimes |junk\rangle , \\
(U_A \otimes V_B)(\mathbb{I}_A \otimes B_1)|\psi\rangle &= ((\mathbb{I} \otimes \tilde{H})|\text{EPR}\rangle) \otimes |junk\rangle .
\end{aligned}
$$

In words, the theorem says that if a strategy achieves the optimal value in CHSH then up to some local rotations on Alice and Bob's spaces it looks exactly like the strategy described above. We called the rotations "isometries" because their range might not be the whole space; in particular, it is not necessarily the case that d_A or d_B are even.[4] The state $|junk\rangle$ can be any state: it does not matter for analyzing the strategy, because as the last equations show, the strategy only acts on the "EPR" part of the state. We had to include the $|junk\rangle$ state because any strategy can always be made more complicated by extending the entangled state arbitrarily, and making the players' measurements act as identity on the extended space.

Note that the theorem assumes that the players' strategy can be described by observables, or equivalently binary projective measurements. More generally we may consider players that apply a nonprojective POVM. However, as described in Box 3.2, a POVM can always be simulated with a projective measurement acting on a larger space, so the assumption is without loss of generality.

Remark 9.3.2 In practice we cannot expect to verify that some players achieve the optimal success probability in the CHSH game: at best, by repeatedly playing the game we can verify that they succeed with probability at least $p^*_{\text{win}} - \delta$, where $\delta > 0$ is a quantity depending on the quality of the players' strategy and on the accuracy of the verification (i.e. the number of repetitions of the game). To handle this scenario

---

4 isometry is a linear map that preserves distances, but need not be invertible. A unitary is an isometry that is also invertible.

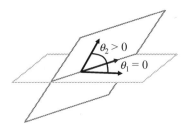

Fig. 9.1   Principal angles between two 2-dimensional subspaces in 3 dimensions. The subspaces intersect, and the smallest angle is $\theta_1 = 0$. The second principal angle is $\theta_2 > 0$.

we need "robust" analogues of Theorem 9.3.1 that have similar conclusions under the weaker assumption of near-optimal success. Such results are known, where the exact equalities in Theorem 9.3.1 are replaced by approximations in trace distance with an error scaling as $O(\sqrt{\delta})$.

Before we get to the proof of the theorem we make a small detour and explore the notion of angle between a pair of projection operators. This will be an important tool in the proof.

9.3.1   Principal Angles and Jordan's Lemma

Consider two lines through the origin in the complex plane $\mathbb{C}^2$. Each line is described by a unit vector $|u\rangle, |v\rangle$, and (ignoring any orientation) the angle between the two lines is the unique $\theta \in [0, \pi/2)$ such that $\cos^2 \theta = |\langle u|v\rangle|^2$. Up to a change of basis we can always consider that $|u\rangle = \begin{pmatrix} 1 \\ 0 \end{pmatrix}$ and (up to an irrelevant phase) $|v\rangle = \begin{pmatrix} \cos\theta \\ \sin\theta \end{pmatrix}$. A more convoluted way to describe the angle between the two lines is through the associated rank-1 projections $P = |u\rangle\langle u|$ and $Q = |v\rangle\langle v|$: there will always exist a choice of basis for $\mathbb{C}^2$ in which

$$P = \begin{pmatrix} 1 & 0 \\ 0 & 0 \end{pmatrix} \quad \text{and} \quad Q = \begin{pmatrix} \cos^2\theta & \cos\theta\sin\theta \\ \cos\theta\sin\theta & \sin^2\theta \end{pmatrix},$$

for some $\theta \in [0, \pi/2)$.

How do we generalize the notion of angle to higher-dimensional subspaces? The notion of principal angle gives an inductive definition. Suppose P and Q are two orthogonal projections in $\mathbb{C}^d$. (We identify the projections with the space on which they project.) The smallest principal angle between P and Q is defined as $\theta_1 \in [0, \pi/2)$ such that

$$\cos^2 \theta_1 = \sup_{|u\rangle\in P, |v\rangle\in Q} |\langle u|v\rangle|^2,$$

where by $|u\rangle \in P$ we mean any unit vector in the range of P, i.e. such that $P|u\rangle = |u\rangle$. This is a natural definition: we are finding the lines lying in P and Q that form the smallest possible angle. If P and Q intersect, then they share a vector and $\theta_1 = 0$.

We define principal angles $\theta_2, \ldots, \theta_d$, where $d = \min(\operatorname{rank} P, \operatorname{rank} Q)$, inductively via

$$\cos^2 \theta_i = \sup_{\substack{|u_i\rangle \in P. \, |u_i\rangle \perp \, \mathrm{Span}\{|u_1\rangle \ldots |u_{i-1}\rangle\} \\ |v_i\rangle \in Q. \, |v_i\rangle \perp \, \mathrm{Span}\{|v_1\rangle \ldots |v_{i-1}\rangle\}}} |\langle u_i | v_i \rangle|^2 ,$$

where $|u_1\rangle, \ldots, |u_{i-1}\rangle$ are unit vectors in P that achieve the optimum in the definition of $\theta_1, \ldots, \theta_{i-1}$ respectively, and similarly for the $|v_i\rangle$ and Q.

Jordan's lemma states that associated with the principal angles comes a very convenient simultaneous block decomposition of P and Q.

Lemma 9.3.3 (Jordan's lemma) *Let P and Q be two projection operators in $\mathbb{C}^d$. Then there exists a basis of $\mathbb{C}^d$ in which P and Q are simultaneously block diagonal, with blocks of size one or two such that either (for 1-dimensional blocks)*

$$P, Q \in \{(0), (1)\} ,$$

or (for 2-dimensional blocks)

$$P = \begin{pmatrix} 1 & 0 \\ 0 & 0 \end{pmatrix}, \qquad Q = \begin{pmatrix} \cos \theta_i^2 & \cos \theta_i \sin \theta_i \\ \cos \theta_i \sin \theta_i & \sin \theta_i^2 \end{pmatrix} ,$$

with $\theta_i \in (0, \pi/2]$.

The proof of the lemma is not very hard. It uses an alternative definition of the principal angles via the singular values of the operator PQ.

QUIZ 9.3.1 *Suppose we have three projectors P_0, P_1, P_2 over a d-dimensional complex vector space. What does Jordan's lemma guarantee about a common diagonalization?*

(a) *For any i, j, there is a basis in which P_i is diagonal and P_j is block diagonal with block size 2.*
(b) *There is a basis in which P_0, P_1, P_2 are all block diagonal with block size 2.*
(c) *There is a basis in which P_0, P_1, P_2 are all diagonal.*
(d) *If $\dim P_0 + \dim P_1 + \dim P_2 \leq d$, then there is a basis in which P_0, P_1, P_2 are all block diagonal with block size 2.*

9.3.2 Proof of the Rigidity Theorem

The proof of Theorem 9.3.1 has two steps. In the first step we use Jordan's lemma to reduce the case of general strategies to the case of "qubit strategies," for which the shared state is a two-qubit entangled state and the players' observables are single-qubit observables. In the second step we analyze qubit strategies in detail and show that they must take the form of Pauli measurements on an EPR pair.

A. Reduction to Qubit Strategies

Consider an arbitrary strategy $|\psi\rangle_{AB}, A_0, A_1, B_0, B_1$. Apply Jordan's lemma to the projections $P = \frac{1}{2}(\mathbb{I} + A_0)$ and $Q = \frac{1}{2}(\mathbb{I} + A_1)$. The lemma gives a basis for Alice's space $\mathbb{C}^{d_A}$ such that both P and Q are block diagonal in that basis, with blocks of size at most 2×2. Then $A_0 = 2P - \mathbb{I}$ and $A_1 = 2Q - \mathbb{I}$ are block diagonal in the same basis.

This block-diagonal decomposition lets us reformulate Alice's strategy as follows: each of her two-outcome projective measurements is equivalent to a measurement which (i) applies a multiple-outcome projective measurement that projects on the individual blocks of the decomposition, and (ii) depending on the block obtained as outcome performs the basis measurement associated with the restriction of A_0 (or A_1) to that block.

Exercise 9.3.1 Suppose that after application of Jordan's lemma we discover a basis

$$\{|u_1\rangle, |u_2\rangle, |u_3\rangle, |u_4\rangle, |u_5\rangle\} \tag{9.6}$$

of $\mathbb{C}^5$ in which

$$
A_0 = \begin{pmatrix} 1 & 0 & 0 & 0 & 0 \\ 0 & -1 & 0 & 0 & 0 \\ 0 & 0 & 1 & 0 & 0 \\ 0 & 0 & 0 & -1 & 0 \\ 0 & 0 & 0 & 0 & 1 \end{pmatrix} \quad \text{and} \quad A_1 = \begin{pmatrix} \frac{1}{2} & -\frac{1}{2} & 0 & 0 & 0 \\ -\frac{1}{2} & \frac{1}{2} & 0 & 0 & 0 \\ 0 & 0 & 1 & 0 & 0 \\ 0 & 0 & 0 & 1 & 0 \\ 0 & 0 & 0 & 0 & -1 \end{pmatrix}.
$$

Consider the two-outcome projective measurements associated with A_0 and A_1. Give an equivalent description of these measurements as the combination of a projective measurement $\{\Pi_0, \Pi_1, \Pi_2, \Pi_3\}$ followed by a basis measurement involving at most two basis elements. The projective measurement should be independent of Alice's input x, while the basis measurement should depend both on the outcome of the projective measurement and on Alice's input.

The same argument can be applied to Bob's observables. The key point is that, since the block decomposition is the same for A_0 and A_1 (resp. B_0 and B_1), step (i) associated with projection on the blocks does not depend on the player's question. Thus the step could be performed before the game even starts, without affecting their success probability! But then the players are really playing the game with a qubit strategy – whichever qubit strategy corresponds to the outcomes they obtained when applying the projective measurement from step (i).

This reformulation of an arbitrary strategy shows that it can always be reduced to a convex combination of qubit strategies, and it will be sufficient to analyze the latter.

B. Optimal Qubit Strategies

To prove the theorem we first express the success probability p^*_{win} of a given quantum strategy in terms of the observables A_x and B_y.

Exercise 9.3.2   Using the definition of the winning criterion $a \oplus b = x \wedge y$ and the relation between observables and binary measurements, show that

$$p^*_{\text{win}} = \frac{1}{2} + \frac{1}{8} \langle \psi | A_0 \otimes B_0 + A_0 \otimes B_1 + A_1 \otimes B_0 - A_1 \otimes B_1 | \psi \rangle . \qquad (9.7)$$

Let's call the operator appearing inside the bra-ket in (9.7) the CHSH operator,

$$\text{CHSH} = A_0 \otimes B_0 + A_0 \otimes B_1 + A_1 \otimes B_0 - A_1 \otimes B_1 .$$

The main trick in the proof is to consider the square of this operator. Using $A_0^2 = A_1^2 = B_0^2 = B_1^2 = \mathbb{I}$, we get

$$\begin{aligned}
\text{CHSH}^2 &= \big((A_0 + A_1) \otimes B_0 + (A_0 - A_1) \otimes B_1 \big)^2 \\
&= (A_0 + A_1)^2 \otimes \mathbb{I} + (A_0 - A_1)^2 \otimes \mathbb{I} + (A_0 + A_1)(A_0 - A_1) \otimes B_0 B_1 \\
&\quad + (A_0 - A_1)(A_0 + A_1) \otimes B_1 B_0 \\
&= 4\mathbb{I} + [A_0, A_1] \otimes [B_1, B_0] ,
\end{aligned} \qquad (9.8)$$

where $[A_0, A_1] = A_0 A_1 - A_1 A_0$ and $[B_1, B_0] = B_1 B_0 - B_0 B_1$ are the commutators. Since the operator norm (the largest singular value) of $[A_0, A_1]$ and $[B_1, B_0]$ is each at most 2, the norm of CHSH^2 is at most 8. Plugging back into (9.7), even an optimal choice of $|\psi\rangle$ (i.e. an eigenvector of CHSH associated with its largest singular value) will give a value at most $p^*_{\text{win}} \leq 1/2 + \sqrt{8}/8 = \cos^2 \pi/8$. Thus $\cos^2 \pi/8$ is indeed the maximum probability of success in the CHSH game.

Note that so far we have not used the reduction to qubit strategies discussed in the previous section, and the preceding argument is completely general. Let's now assume we are working with a qubit strategy that achieves the optimal p^*_{win}. Then all inequalities discussed above must be tight. In particular, $|\psi\rangle$ must be an eigenvector of CHSH with eigenvalue $2\sqrt{2}$, and as a consequence of (9.8) $|\psi\rangle$ must also be an eigenvector of $[A_0, A_1] \otimes [B_1, B_0]$ with associated eigenvalue 4. Squaring this operator,

$$\big([A_0, A_1]^2 \otimes [B_1, B_0]^2 \big) |\psi\rangle = 16 |\psi\rangle .$$

Using further that $[A_0, A_1]^2 \leq 4\mathbb{I}$ and $[B_1, B_0]^2 \leq 4\mathbb{I}$ we get that necessarily

$$\big([A_0, A_1]^2 \otimes \mathbb{I} \big) |\psi\rangle = \big(\mathbb{I} \otimes [B_1, B_0]^2 \big) |\psi\rangle = 4 |\psi\rangle , \qquad (9.9)$$

as neither operator can reduce the norm of $|\psi\rangle$. Assume $|\psi\rangle$ is not trivial, in the sense that its reduced density matrices on A and B have rank 2 (if this is not the case then it is easy to see that the strategy boils down to a classical strategy, which cannot achieve a success probability larger than $p_{\text{CHSH}} = 3/4$). Tracing out the A or B qubits in (9.9) and inverting the reduced density matrix of $|\psi\rangle$ on the remaining qubit gives us the operator equalities $A_0 A_1 = -A_1 A_0$ and $B_1 B_0 = -B_0 B_1$: Alice and Bob's observables pairwise anti-commute. It turns out that anti-commutation is a surprisingly strong constraint, as shown in the following exercise.

Exercise 9.3.3 Suppose that R and S are two observables on $\mathbb{C}^2$ such that $RS = -SR$. Then there exists a basis of $\mathbb{C}^2$ in which $R = Z$ and $S = X$. *[Hint: first show that we cannot have $R = \mathbb{I}$ or $R = -\mathbb{I}$, and deduce the eigenvalues of R. Use this to write R in a convenient form, and then use the anti-commutation relation to find the form of S.]*

Applying the results of the exercise to A_0 and A_1 we obtain a unitary U_A on Alice's qubit such that $U_A A_0 U_A^\dagger = Z$ and $U_A A_1 U_A^\dagger = X$. Similarly, for Bob's observables we may find a unitary U_B such that $U_B B_0 U_B^\dagger = H$ and $U_B B_1 U_B^\dagger = \tilde{H}$. Note that for Bob we are using H and $\tilde{H}$ in lieu of X and Z, but any pair of single-qubit observables will do. To conclude it remains to show the following.

Exercise 9.3.4 Show that the operator

$$Z \otimes H + X \otimes H + X \otimes \tilde{H} - Z \otimes \tilde{H}$$

has largest eigenvalue $2\sqrt{2}$, with a unique associated eigenvector equal to $|\text{EPR}\rangle$.

C. Putting Everything Together

We are almost done with the proof of Theorem 9.3.1. To summarize, we start with an arbitrary strategy $|\psi\rangle_{AB}$, A_0, A_1, B_0, B_1 with success probability $p^*_{\text{win}} = \cos^2 \pi/8$ in the CHSH game. Using part A this strategy can be decomposed in a convex combination of qubit strategies. More formally, there are projective measurements $\Pi^A = \{\Pi_1^A, \ldots, \Pi_{k_A}^A\}$ and $\Pi^B = \{\Pi_1^B, \ldots, \Pi_{k_B}^N\}$ for Alice and Bob, made of projectors with rank at most 2 each, such that $A_x = \sum_j \Pi_j^A A_x \Pi_j^A$ and $B_y = \sum_j \Pi_j^B B_y \Pi_j^B$. The associated block decomposition can be specified by unitary changes of basis U_A' and U_B' on Alice and Bob's systems respectively.

Using the first steps of part B, we know that any strategy can have success probability at most p^*_{win}, therefore all the qubit strategies, given by $(\Pi_j^A \otimes \Pi_\ell^B |\psi\rangle, \Pi_j^A A_x \Pi_j^A, \Pi_\ell^B B_y \Pi_\ell^B)$ for any $j \in \{1, \ldots, k_A\}$ and $\ell \in \{1, \ldots, k_B\}$, must have success probability p^*_{CHSH} (otherwise the overall strategy wouldn't achieve the optimal success probability).

By the remainder of part B, for all of these qubit strategies there exists a local change of basis U_j^A and U_ℓ^B in which it is equivalent to the canonical optimal strategy. By combining the unitaries U_A (resp. U_B), which specify the blocks, with the unitaries U_j^A (resp. U_ℓ^B), which identify a basis for each block in which $\Pi_j^A A_0 \Pi_j^A = Z$, $\Pi_j^A A_1 \Pi_j^A = X$, and similarly for Bob and $H, \tilde{H}$, we obtain the isometries claimed in the theorem: the proof is complete!

CHAPTER NOTES

The idea for the DIQKD protocol, which is that entanglement between Alice and Bob can be tested using the phenomenon of nonlocality, is due to A. K. Ekert (Quantum cryptography based on Bell's theorem. *Physical Review Letters*, **67**(6):661, 1991).

Example 9.1.1 is taken from a paper by S. Pironio, et al. (Device independent quantum key distribution secure against collective attacks. *New Journal of Physics*, **11**(4):045021, 2009).

Regarding Lemma 9.2.1, there are many ways it can be shown, yielding bounds of varying quality. The simplest analysis would consider a relaxation of the problem where the three players are allowed any kind of *nonsignaling strategy*: in this case a bound can be obtained via linear programming. The bound can then be strengthened by considering the fact that the players must be quantum, using a semidefinite relaxation of the problem. But the optimal bound can be obtained by a direct analytic calculation, using the fact that Alice only has two possible inputs to reduce to the 2-dimensional case via an application of Jordan's lemma. This is done in another paper by Pironio, et al. (Random numbers certified by Bell's theorem. *Nature*, **464**(7291):1021–1024, 2010), from which the bound given here, which is due to U. Vazirani and T. Vidick (Fully device-independent quantum key distribution. *Physical Review Letters*, **113**(14):140501, 2014), can be derived.

For a proof of Jordan's lemma, see, for example, Exercise VII.1.10 in the book by R. Bhatia, *Matrix Analysis* (Vol. 169, Springer Science & Business Media, 2013). A robust version of Theorem 9.3.1 on rigidity of the CHSH game is shown in the work of M. McKague, Tzyh Haur Yang, and V. Scarani (Robust selftesting of the singlet. *Journal of Physics A: Mathematical and Theoretical*, **45**(45):455304, 2012), with an earlier argument appearing in a paper by S. J. Summers and R. Werner (Maximal violation of Bell's inequalities for algebras of observables in tangent spacetime regions. *Annales de l'IHP Physique théorique*, **49**(2):215–243, 1988). The use of these results for DIQKD is explored in work by B. W. Reichardt, F. Unger, and U. Vazirani (Classical command of quantum systems. *Nature*, **496**(7446):456–460, 2013). In terms of parameters such as tolerance to errors and key rate, this technique yields weaker results than the approaches using uncertainty relations and guessing games; however, it also proves a stronger characterization of the devices that can be useful in other scenarios (see, for example, the problem of delegating computations in Chapter 13). For a quantitatively stronger approach, based on the "entropy accumulation theorem (EAT)," see the paper by R. Arnon-Friedman, R. Renner, and T. Vidick (Simple and tight device-independent security proofs. *SIAM Journal on Computing*, **48**(1):181–225, 2019).

PROBLEMS

9.1 BB'84 fails in the device-independent setting

Consider the purified variant of the BB'84 protocol. Suppose that Eve prepares the state ρ_{ABE} in the following form:

$$\rho_{ABE} = \frac{1}{2} \sum_{x,z=0}^{1} |xz\rangle\langle xz|_A \otimes |xz\rangle\langle xz|_B \otimes |xz\rangle\langle xz|_E \,,$$

where $|xz\rangle$ is short-hand notation for $|x\rangle \otimes |z\rangle$. Now suppose Alice and Bob's measurement device, instead of measuring a single qubit in the standard or Hadamard bases, as they think the device does, in fact performs the following:

- When the device is told to measure in the standard basis, it measures the first qubit of the two-qubit system associated with the device in the standard basis.
- When the device is told to measure in the Hadamard basis, it measures the *second* qubit of the two-qubit system associated with the device in the *standard* basis.

1. Alice and Bob put blind faith in their hardware and attempt to implement BB'84. They want to check that their state is an EPR pair, so Alice asks her box to measure in the standard basis. The box returns a measurement outcome of 0. What is the post-measurement state?

2. After Alice's measurement, Bob asks his box to measure in the Hadamard basis. What measurement outcome will Bob receive?

3. Suppose that instead Bob asks his box to measure in the standard basis. What measurement outcome will Bob receive?

4. After Bob asked his box to measure in the standard basis, Eve measures her first qubit in the standard basis. What measurement outcome does she receive?

5. As per the BB'84 protocol, Alice and Bob look at all the rounds on which they made the same measurement as each other. They pick a random subset of those rounds and test whether they received the same output on all the rounds. With what probability do they pass the test?

6. Let T' be a set of rounds on which Alice and Bob made the same measurement but didn't perform a test. Let $\{\theta_j\}_{j \in T'}$ be the measurements they made and $\{x_j\}_{j \in T'}$ be the results they received. The θ_j have been communicated over the public channel. Eve wishes to learn the x_j. Which measurements should she make?

7. Let X be the classical key generated by Alice and Bob. What is $H_{\min}(X \mid E)$, where E is Eve's system?

9.2 Commuting observables are compatible

In order to analyze the upcoming protocols we need to use the following fact: if A, B are commuting observables, then the product of the results of measuring A and then B is the same as the result of measuring AB.

In this problem, you'll verify a special case of the previous fact in a way that should illuminate the proof. Consider $X \otimes X$ and $Z \otimes Z$. Since they commute, they have a simultaneous eigenbasis. It happens to consist of the Bell states, which are

$$\left|\phi_{00}^+\right\rangle = \frac{1}{\sqrt{2}}(|00\rangle + |11\rangle) \,, \quad \left|\phi_{01}^-\right\rangle = \frac{1}{\sqrt{2}}(|00\rangle - |11\rangle) \,,$$

$$\left|\psi_{10}^+\right\rangle = \frac{1}{\sqrt{2}}(|01\rangle + |10\rangle) \,, \quad \left|\psi_{11}^-\right\rangle = \frac{1}{\sqrt{2}}(|01\rangle - |10\rangle) \,.$$

1. Suppose we measure the two-qubit state $|\phi\rangle$ using the observable $X \otimes X$ and receive the outcome -1. The post-measurement state belongs to which 2-dimensional eigenspace?

2. Next, we measure the observable $Z \otimes Z$ and receive the outcome 1. What is the post-measurement state $|\phi'\rangle$?

3. Suppose that instead we performed the measurement $-Y \otimes Y = (X \otimes X)$ $(Z \otimes Z)$ directly, and the post-measurement state had nonzero overlap with $|\phi'\rangle$. What measurement outcome would we have received? (In other words, what is the eigenvalue of the $-(Y \otimes Y)$-eigenspace in which $|\phi'\rangle$ lies?) Compare your answer to the product of the answers in the previous problems.

9.3 Another pseudo-telepathy game

Alice and Bob tell their friend Eve that they have a magic 3×3 square of numbers with the following properties:

- Every entry is either 1 or -1
- The product of each column is 1
- The product of each row is -1

1. What is the product of all of the entries of Alice and Bob's square?

As you may now see, Eve is not convinced by Alice and Bob's claim. Therefore, she asks them to play the following *magic square game*. First, Eve randomly generates two numbers $i, j \in \{0, 1, 2\}$. She gives i to Alice and j to Bob. Alice and Bob each produce a triple of ± 1 numbers $(a_0, a_1, a_2), (b_0, b_1, b_2)$. They win if $a_0 a_1 a_2 = 1$, $b_0 b_1 b_2 = -1$, and $a_j = b_i$. In other words, they win if Alice produces the ith column of the magic square and Bob produces the jth row of the magic square.

2. Suppose that Alice and Bob use a deterministic strategy in this game. What is the highest success probability they can achieve?

Consider the following 3×3 square of observables:

$$\begin{pmatrix} -\mathbb{I} \otimes Z & X \otimes \mathbb{I} & X \otimes Z \\ -Z \otimes \mathbb{I} & \mathbb{I} \otimes X & Z \otimes X \\ Z \otimes Z & X \otimes X & Y \otimes Y \end{pmatrix}.$$

Use the Pauli commutation relations, recalled here, to answer the following two questions.

$$X^2 = Y^2 = Z^2 = \mathbb{I}; \quad XYZ = -i\mathbb{I}; \quad XY = -YX, YZ = -ZY, ZX = -XZ.$$

3. Which tensor products of Pauli operators commute with $Y \otimes Y$?

4. Which tensor products of Pauli operators commute with $\mathbb{I} \otimes X$?

5. How would you describe the commutation pattern of the square?

6. The product of the first row of the magic square is $(\mathbb{I} \otimes Z)(Z \otimes \mathbb{I})(Z \otimes Z)$. What are the eigenvalues of this operator? Convince yourself that the product of each row is the same as the product of the first row.

7. The product of the third column of the magic square is $(X \otimes X)(Y \otimes Y)$ $(Z \otimes Z)$. What are the eigenvalues of this operator? Convince yourself that the product of each column is the same as the product of the third column.

This square of operators gives rise to a quantum strategy for Alice and Bob in the magic square game. Alice and Bob share two EPR states, with each of them holding one qubit of each pair. In other words, their overall state is $|\psi\rangle = |\mathrm{EPR}\rangle_{12} \otimes |\mathrm{EPR}\rangle_{34}$. When Eve distributes (i, j), Alice picks the ith column of the square, measures the three operators, and returns the three measurements in order. Similarly, Bob measures the three operators in the jth row of the square and returns their measurements in order. For example, if $(i, j) = (1, 2)$ then Alice will put (a_0, a_1, a_2) equal to the measurement results of $(-\mathbb{I}_1 \otimes Z_3, -Z_1 \otimes \mathbb{I}_3, Z_1 \otimes Z_3)$. Bob will put (b_0, b_1, b_2) equal to the measurement results of $(-Z_2 \otimes \mathbb{I}_4, \mathbb{I}_2 \otimes X_4, Z_2 \otimes X_4)$.

8. What does the previous question, along with the previous problem "Commuting observables are compatible," tell us about Alice and Bob's output in this game?

9.4 A nonlocal game

In this problem we study an example of a game such that repeating the game twice does not decrease the maximum winning probability. As we will see, this demonstrates an example of a "coherent attack," where two independent repetitions can be "attacked" better than a single one.

We begin by describing the game. The referee starts by generating a pair $(s, t) \in \{(0, 0), (0, 1), (1, 0)\}$ uniformly at random. She gives s to Alice and t to Bob. Alice and Bob generate output bits $a, b \in \{0, 1\}$, respectively. They win if $a \vee s \neq b \vee t$. (If a and b are bits, then $a \vee b = 0$ if a and b are both 0 and $a \vee b = 1$ if $a = 1$ or $b = 1$.) As a warm-up, consider the strategy in which $a = s$ and $b = t$.

1. What is the win probability? Which inputs cause Alice and Bob to lose?

In the two-parallel-repeated version $G^{(2)}$ of the game we just described, the referee picks two strings $(s_0, t_0), (s_1, t_1)$ from $\{(0, 0), (0, 1), (1, 0)\}$ independently and uniformly at random. She gives (s_0, s_1) to Alice and (t_0, t_1) to Bob, and demands outputs $(a_0, a_1), (b_0, b_1)$ from Alice and Bob. They win if $a_0 \vee s_0 \neq b_0 \vee t_0$ and $a_1 \vee s_1 \neq b_1 \vee t_1$.

2. Which of the following is a valid strategy giving Alice and Bob the highest win probability?

    I. $a_0 = s_0, a_1 = t_0; b_0 = s_1, b_1 = t_1$

    II. $a_0 = s_0, a_1 = s_1; b_0 = t_0, b_1 = t_1$

    III. $a_0 = s_0 = a_1; b_0 = t_0 = b_1$

    IV. $a_0 = s_1, a_1 = s_0; b_0 = t_1, b_1 = t_0$

Suppose Alice and Bob have a valid strategy for the two-parallel-repeated game which wins with probability ω_c.

3. Which of these protocols is a valid strategy in the single-repeated game guaranteeing that Alice and Bob win with probability at least ω_c?

   I. Alice and Bob receive their inputs (s,t), then run their two-parallel-repeated strategy on the inputs $s_1 = s = s_0, t_1 = t = t_0$, and output (a_0, b_0).

   II. Alice and Bob receive their inputs (s,t), then communicate their bits to each other and run their two-parallel-repeated strategy on the inputs $s_1 = s = t_0, t_1 = t = s_0$. They output (a_0, b_0).

   III. Alice and Bob agree on a shared string (s_1, t_1) uniformly at random from $\{(0,0),(0,1),(1,0)\}$. When they receive their inputs (s,t), they run their two-parallel-repeated strategy on the inputs $((s,t),(s_1,t_1))$, and output (a_0, b_0).

   IV. Alice and Bob independently generate random bits $s_1, t_1 \in \{0,1\}$. When they receive their inputs (s,t), they run their two-parallel-repeated strategy on the inputs $((s,t),(s_1,t_1))$, and output (a_0, b_0).

This proves that the optimal success probability in the one-shot game is an upper bound for the optimal success probability in the two-parallel game.

Now we will find an upper bound on the success probability of the single-repeated game, assuming that Alice and Bob may use shared entanglement in addition to classical resources.

The most general strategy that Alice and Bob can take is as follows. They each have two ± 1-eigenvalue-observables A_0, A_1, B_0, B_1. They share an entangled state $|\psi\rangle$. Alice measures her share of $|\psi\rangle$ using A_s, Bob measures his share using B_t, and they each output 0 if they measured a 1 and 1 if they measured a -1.

In general, if X is an observable, then $\langle \psi | X | \psi \rangle$ is equal to the probability of measuring a 1 minus the probability of measuring -1. In other words, the probability of measuring a 1 is $p_1 = \frac{1}{2} + \frac{1}{2} \langle \psi | X | \psi \rangle$.

4. For which of the following M is the probability that Alice and Bob win the game equal to $\frac{1}{2} + \frac{1}{2} \langle \psi | M | \psi \rangle$? *[Hint: consider the three possible inputs (s,t) separately. What must Alice and Bob's measurements be in each case to guarantee victory?]*

   I. $M = -\frac{1}{3}A_0 \otimes B_0 + \frac{1}{3}A_0 \otimes \mathbb{I} + \frac{1}{3}\mathbb{I} \otimes B_0$

   II. $M = -\frac{1}{3}A_0 \otimes B_0 + \frac{1}{3}A_0 \otimes B_1 + \frac{1}{3}A_1 \otimes B_0$

   III. $M = \frac{1}{3}A_0 \otimes B_0$

   IV. $M = -\frac{1}{3}(A_0 \otimes \mathbb{I} + \mathbb{I} \otimes B_0) + \frac{1}{3}(A_0 \otimes \mathbb{I} + A_0 \otimes B_1) + \frac{1}{3}(\mathbb{I} \otimes B_0 + A_1 \otimes B_0)$

This quantity $\langle \psi | M | \psi \rangle$ is bounded above by the maximum eigenvalue of M. With a bit of arithmetic, we can find the eigenvalues of M exactly, despite our ignorance about Alice and Bob's observables!

5. Which of the following equations is satisfied by M?

   I. $M^2 = \frac{1}{3}\mathbb{I} - \frac{2}{3}M$

   II. $M^2 = \frac{1}{9}\mathbb{I} - \frac{7}{9}M$

   III. $M^2 = \frac{1}{3}\mathbb{I} - \frac{1}{3}M$

IV. $M^2 = \mathbb{I} - 2M$

6. The answer to the last question gives the characteristic polynomial of M (indeed, it is the unique monic quadratic satisfied by M). Use it to solve for the largest eigenvalue $\lambda_{\max}$ of M.

7. Now use the facts that $p_{\text{win}} \leq \frac{1}{2} + \frac{1}{2} \langle \psi | M | \psi \rangle$ and $\langle \psi | M | \psi \rangle \leq \lambda_{\max}$ to find an upper bound on p_{win}.

10

Quantum Cryptography beyond Key Distribution

While quantum key distribution (QKD) is arguably the most celebrated cryptographic application of quantum communication, many others are known. In this chapter and the next we look at a variety of other settings where quantum information provides an advantage. As before, we focus on applications involving two parties, namely our usual suspects, Alice and Bob. In contrast to earlier chapters, however, in what follows Alice and Bob no longer trust each other. As such, our objective is not to protect Alice and Bob from a third entity – Eve – but instead we would like to ensure that an honest Alice (or Bob) will be protected against a dishonest Bob (or Alice).

10.1 Coin Flipping

As a warm-up let us start with a problem, introduced by Blum in the 1980s, that is deceptively simple: coin flipping. Imagine that Alice and Bob jointly won a single laptop as a prize in a quantum programming competition. They agree to flip a coin $c \in \{0, 1\}$ to determine which one of them can take home the laptop. Unfortunately for our protagonists, Alice is in Europe and Bob is in North America. So they need to perform this coin flip over the phone (or the internet), by exchanging classical (or quantum!) communication. Can we find a good protocol to help them solve this task?

To think through this, let us first try to define a bit more carefully what we mean by "their task," which is to produce a coin flip c (see Figure 10.1). As usual, we would first like the protocol to be correct: if Alice and Bob are both honest, then they should eventually agree on the outcome c of the coin flip in order to avoid further arguments. Ideally, when they are both honest, the coin should also be fair, in the sense that the probability of obtaining an outcome $c = 0$ (say, heads) is the same as the probability of producing the outcome $c = 1$ (say, tails).

Finding a protocol that works if both of them are guaranteed to be honest is indeed an easy task. For example, they could simply agree that Alice flips a coin locally to obtain c, and then she tells Bob the outcome c on the phone. But if the stakes are high (remember that the coin flip outcome will determine who gets the laptop!) then Bob may not trust Alice to perform the local coin flip. In this case the trivial protocol we just described no longer provides any guarantees: Alice has full control over the final outcome, and this is clearly not satisfactory.

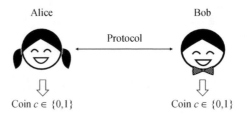

Fig. 10.1   Coin flipping: Alice and Bob engage in a protocol in order to produce a coin flip $c \in \{0,1\}$. After the protocol, Alice and Bob should both agree on the value of c, where ideally both outcomes of c should be equally likely. Moreover, neither Alice nor Bob should be able to bias the coin in a specific direction.

If Alice and Bob do not trust each other, then they would like the protocol to be not only correct, but also *secure*. What does security mean in the context of coin flipping? One possible definition of security demands that neither one of the two parties can bias the coin too much in either direction. That is, if Alice is honest, she is guaranteed that, whatever Bob does, the probability of obtaining either heads or tails is close to uniform, i.e. $p(c = 0) \approx p(c = 1) \approx 1/2$. This form of coin flipping is known as *strong coin flipping*, where the word strong refers to the fact that the dishonest party cannot bias the coin in either direction. This leads to the following (informal) definition.

Definition 10.1.1 (Strong coin flipping). Strong coin flipping *is a two-party task between Alice and Bob. The goal is for both parties to output the same value $c \in \{0,1\}$ such that the following properties hold.*

- Correctness: *If both Alice and Bob are honest, then c is uniformly distributed:* $p(c = 0) = p(c = 1) = 1/2$.
- ε-secure: *If Alice (or Bob) is honest, then Bob (or Alice) cannot bias the coin by more than ε:*
$$\frac{1}{2} - \varepsilon \le p(c = 0), \; p(c = 1) \le \frac{1}{2} + \varepsilon ,$$
where $p(c)$ denotes the probability that the honest party outputs the value c.

The smallest ε for which a protocol is ε-secure is called the (strong coin flipping) bias of the protocol.

What if both Alice and Bob are dishonest? In this case all bets are off, and there are no guarantees required for the protocol. Indeed, in any multiparty protocol in which some of the parties may be dishonest we never need to worry about writing down security guarantees protecting the dishonest party. Our objective is only to design protocols that protect the honest party(ies) from the dishonest one(s). This is because it is impossible to write down a security definition that makes guarantees for the dishonest parties: one strategy of the dishonest parties could always be to produce a random output, or any other output of their choice, just to make sure that our definition is not satisfied.

Thinking back to the laptop example that we gave earlier, we can see that we are overshooting the goal a little by asking for a strong coin flipping protocol. After all, we can reasonably assume that both Alice and Bob want to obtain the laptop, otherwise they could simply give up. So let's say that Alice and Bob agree that $c = 0$ means that Alice gets the laptop, and $c = 1$ indicates that Bob obtains the laptop. A protocol that would assure us that Alice cannot force $p(c = 0) > 1/2 + \varepsilon$ and Bob cannot force $p(c = 1) > 1/2 + \varepsilon$ for some (hopefully small!) error parameter $\varepsilon \geq 0$ would evidently be sufficient to solve our problem. This motivates the definition of a weaker cryptographic primitive.

Definition 10.1.2 (Weak coin flipping). Weak coin flipping *is a two-party task between Alice and Bob. Neither party has an input. The goal is for both parties to output the same value $c \in \{0, 1\}$ such that the following properties hold.*

- Correctness: *If both Alice and Bob are honest, then c is uniformly distributed:* $p(c = 0) = p(c = 1) = 1/2$.
- ε-secure: *If Alice is honest, then $p(c = 1) <= 1/2 + \varepsilon$. If Bob is honest, then* $p(c = 0) <= 1/2 + \varepsilon$.

As before, the smallest ε for which a protocol is ε-secure is called the (weak coin flipping) bias *of the protocol.*

While this second definition is clearly less demanding than the first, it is not immediately clear that either definition can be satisfied at all. Is secure coin flipping possible?

QUIZ 10.1.1 *A strong coin flipping protocol immediately implies a weak coin flipping protocol with the same bias. However, we can also derive a strong coin flipping protocol from a weak one. A simple way to do this is to make the modification that whoever wins the weak coin flip gets to flip their own 50-50 coin (if acting honestly) and announce the final outcome of the protocol. What is the bias of this strong coin flipping protocol, if the weak coin flipping protocol had bias ε?*

(a) $\frac{3}{8} + \frac{\varepsilon}{4}$
(b) $\frac{1}{4} + \frac{\varepsilon}{2}$
(c) $2(\varepsilon + \frac{1}{2}) - \frac{1}{2}$
(d) ε

10.1.1 Classical Coin Flipping

Let us first have a look at whether we might be able to construct a classical protocol for coin flipping. Classical means that the resource that Alice and Bob have at their disposal is a classical communication channel connecting them. No other assumptions are made: in particular, Alice and Bob have unlimited computational resources, and we have no guarantees, either about the locations of Alice and Bob or about the classical communication channel, other than that it will eventually deliver a message. Maybe a protocol a bit more sophisticated than the trivial protocol discussed earlier would work? Instead of just one party flipping a coin, a natural idea is to have both of them do it simultaneously, and then announce their choice via the classical communication channel.

Protocol 8 (Blum coin flipping)

1. Alice flips a random bit $a \in \{0,1\}$ and sends it to Bob.
2. Bob flips a random bit $b \in \{0,1\}$ and sends it to Alice.
3. Both parties output the coin flip $c = a \oplus b$.

As usual, our first goal is to check if the protocol is correct: indeed, it is easy to see that if both parties are honest, then c is uniformly distributed. In fact, it seems that it might be sufficient that only one party is honest: as long as a or b is random then $a \oplus b$ will be random. Is this right? Note that the protocol forces us to specify an order in which the parties exchange their messages (indeed, it is never wise to attempt to speak simultaneously over the phone). Here we made Alice go first, and Bob second. So Bob receives Alice's message a before he sends her his choice of b. But this makes it possible for him to cheat! Bob can easily force any outcome $c = b'$ of his choice by choosing b' first and then selecting $b = b' \oplus a$ at step 2. Thus this protocol does not even fulfill the security requirement of a weak coin flipping protocol.

At first glance you might hope that it will be possible to find a more sophisticated classical protocol. Unfortunately, it turns out that there exists *no* classical protocol for coin flipping that is secure without making any additional assumptions: no value of $\varepsilon < 1/2$ can be achieved for security. In other words, whenever one party cannot completely bias the outcome of the protocol to a certain value, then the other party can: there is always at least one of Alice or Bob who can cheat perfectly (in the protocol above, it is Bob). Very intuitively, the reason is precisely the same as what made the Blum protocol insecure: one can argue that, whatever the outcome c of the protocol will be, it has to be determined at *some* point in the protocol. By considering the messages during the course of the protocol (which can involve many rounds of interaction) one can determine the message before which the outcome c was not yet determined, but once the message is sent, the outcome becomes determined. In the case of the Blum protocol above, this message is Bob's message to Alice. However, given that c is not yet determined before that message, the next message sent effectively will determine a value for c. The party who sends that message thus has the ability to bias the coin as they desire.

If we allow ourselves to make a few more assumptions, then a slight variation of the Blum protocol *can* work. For example, if the channel connecting Alice and Bob features a guaranteed message delivery time t that cannot be influenced by either Alice or Bob (e.g. via special relativity if Alice and Bob's locations are fixed), and they have synchronized time slots (possibly by using the message delivery times), then we could modify the Blum protocol by asking that Alice and Bob both send a and b simultaneously. Any message that arrives after time t is immediately rejected by the recipient.

This way, we can be sure that none of the two parties can base their choice of which bit to send on their knowledge of the other's message, and the protocol becomes secure.

10.1.2 Quantum Strong Coin Flipping

Luckily, our impossibility argument does not apply to quantum protocols. Can you see why? Try to run the argument in your head – what part breaks down? This is a bit subtle: the main reason is that for a quantum protocol the notions of a transcript, and the outcome being "determined," are not well-defined, because everything can happen in superposition. So we can't really talk of a specific message in the protocol when the outcome becomes determined: the determining message could in fact be a superposition of messages exchanged at different times.

As is often the case, there is a good reason for a classical argument not to extend to the quantum setting: strong coin flipping actually *is* possible using quantum information, for some nontrivial values of the bias $\varepsilon < 1/2$. Let us see an example such protocol. To describe the protocol we introduce the term "qutrit," which is used to refer to a quantum state in the space $\mathbb{C}^3$, i.e. a state of the form $\alpha_0 |0\rangle + \alpha_1 |1\rangle + \alpha_2 |2\rangle$.[1]

Protocol 9 (Quantum strong coin flipping) For $a, x \in \{0, 1\}$ define the qutrit

$$|\phi_{a,x}\rangle = \frac{1}{\sqrt{2}} \left(|0\rangle + (-1)^x |a+1\rangle \right) .$$

1. Alice selects $x \in \{0, 1\}$ and $a \in \{0, 1\}$ uniformly at random and sends $|\phi_{a,x}\rangle$ to Bob.
2. Bob selects $b \in \{0, 1\}$ uniformly at random and sends b to Alice.
3. Alice sends a and x to Bob.
4. Bob verifies that the state he received from Alice in step 1 is $|\phi_{a,x}\rangle$ (e.g. by measuring in any orthonormal basis containing $|\phi_{a,x}\rangle$). If it is not the case then he declares that Alice has been cheating and aborts the protocol.
5. Both return the outcome $c = a \oplus b$.

Note the similarity between this protocol and the Blum protocol from the previous section. Here as well, Alice and Bob each choose "half" of the outcome c: Alice chooses a, Bob b, and they return $c = a \oplus b$. However, Alice does not fully reveal a to Bob in her first message: instead, she provides him with some form of "weak commitment" to a in the form of the state $|\phi_{a,x}\rangle$. Because the four states $|\phi_{a,x}\rangle$ are not orthogonal, it is impossible for Bob to completely discover the value of a without first learning x, which only happens after he had to make his choice of b.

Exercise 10.1.1 Compute the reduced density matrices $\rho^B_{|a=0}$ and $\rho^B_{|a=1}$ associated with Bob's view of the protocol after Alice's first message has been sent, for a uniformly random choice of $x \in \{0, 1\}$ and $a = 0$ or $a = 1$ respectively. Show that the trace distance between these two matrices is $1/2$. Conclude that the probability with which Bob can force an outcome c of his choice is at most $3/4$.

---

1 One way to realize qutrits is by using two qubits, e.g. by identifying $|0\rangle \leftarrow |00\rangle$, $|1\rangle \leftarrow |01\rangle$, and $|2\rangle \leftarrow |10\rangle$ (and we make sure that $|11\rangle$ always has zero amplitude).

The exercise shows that the maximum bias that a cheating Bob can force in the protocol is $\varepsilon = 1/4$. Security for cheating Alice is a bit harder to argue, because we have to consider the possibility for her to prepare an arbitrary state in the first step, which may be entangled with some information she keeps on the side. She could then subsequently use all of these, together with the value b received from Bob, to determine her message in the third step of the protocol. We will not give the details here; the main result one can show is the following.

Theorem 10.1.1 *The quantum coin flipping protocol, Protocol 9, is correct and ε-secure for $\varepsilon = 1/4$.*

Can we do even better? Unfortunately it turns out that perfectly secure strong coin flipping is also impossible for quantum protocols: Kitaev showed that the smallest bias any protocol could achieve is $\varepsilon = (\sqrt{2}-1)/2 \approx 0.207$. Kitaev's proof is an extension of the classical impossibility argument, based on an ingenuous representation of transcripts for quantum protocols. If you are interested, in Section 10.5 at the end of the chapter we sketch an argument that is in some sense "dual" to Kitaev's.

10.1.3 Quantum Weak Coin Flipping

If we relax our requirements to *weak* coin flipping now, it turns out that by using quantum protocols this is possible with *arbitrarily small* (but nonzero) bias ε. The best protocol known for achieving this, however, is very complex! In particular, the protocol requires a large number of rounds of interaction, which scales exponentially with $1/\varepsilon$. Using this protocol it is possible to show that there exist quantum strong coin flipping protocols with bias $(\sqrt{2}-1)/2 + \varepsilon$ for any $\varepsilon > 0$, nearly matching Kitaev's lower bound.

10.2 Two-Party Cryptography

Coin flipping is an example of a two-party task. In this task, the parties have no input at all, and they each produce a single bit of output (the coin flip). More generally, we can imagine two parties (Alice and Bob) who each have some classical input, x for Alice and y for Bob, and each would like to compute a certain function of both their inputs, $f_A(x,y)$ for Alice and $f_B(x,y)$ for Bob.[2] To do this they can communicate over a classical, or even quantum, channel. An easy solution would be for Alice and Bob to exchange their respective inputs and perform the computation locally. Unfortunately they don't trust each other: neither party wants to reveal more information about his or her input than is absolutely necessary. How do they do it?

A concrete example of such a task is "Yao's millionaires' problem." Here x and y represent Alice and Bob's respective fortunes. The functions $f_A(x,y) = 1_{x>y}$ and $f_B(x,y) = 1_{y>x}$, where $1_{x>y}$ is 1 if $x > y$ and 0 otherwise, tell Alice and Bob if their fortune is larger than the other's. Can they decide who is the richest without announcing

---

2 If we are precise, coin flipping requires a randomized output, so it does not strictly fall in the standard framework for two-party cryptography, which requires f_A and f_B to be deterministic functions. For this reason it is best to treat it separately.

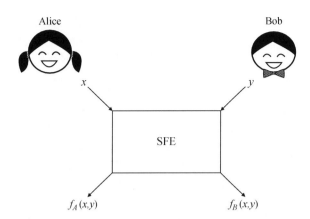

Fig. 10.2   Secure function evaluation (SFE): Alice has an input x, Bob has an input y. After the protocol is completed Alice should learn $f_A(x,y)$ for some function f_A, and Bob should learn $f_B(x,y)$. Neither of them should gain any further information.

their actual fortune? (It turns out they can, but it's not so easy: we'll see how to do it later.)

10.2.1 Secure Function Evaluation

 The general task we just introduced is called secure function evaluation (SFE). Let's formalize it (see also Figure 10.2).

> **Definition 10.2.1.** *Secure function evaluation (SFE) is a task involving two parties, Alice and Bob. Alice holds an input $x \in \mathcal{X}$ and Bob holds an input $y \in \mathcal{Y}$. Alice and Bob interact over a communication channel, and output an $a \in \mathcal{A}$ and $b \in \mathcal{B}$ respectively. We say that a given protocol is a secure protocol computing a pair of functions ($f_A : \mathcal{X} \times \mathcal{Y} \to \mathcal{A}$, $f_B : \mathcal{X} \times \mathcal{Y} \to \mathcal{B}$) if it satisfies the following properties:*
>
> - Correctness: *If both Alice and Bob follow the protocol (we say that they are honest) then $a = f_A(x,y)$ and $b = f_B(x,y)$.*
> - Security against cheating Bob: *If Alice is honest, then Bob cannot learn more about her input x than he can infer from $f_B(x,y)$.*
> - Security against cheating Alice: *If Bob is honest, then Alice cannot learn more about his input y than she can infer from $f_A(x,y)$.*

Note that the definition does not guarantee anything when Alice and Bob are both dishonest. In this case there is nothing we can do! The goal is only to protect the honest parties. The definition is intuitive, and it is rather informal – for example, what does it mean that "Bob cannot learn more about Alice's input x than he can infer from $f_B(x,y)$"? It turns out that this requirement is very delicate to make precise! We will return to it in a moment. First, let's consider some examples of SFE tasks.

Example 10.2.1 Alice and Bob are contemplating going to a movie. Here, $x, y \in \{0, 1\}$ where "0" denotes "no" and "1" denotes "yes." The function they wish to compute is

$$f(x, y) = f_A(x, y) = f_B(x, y) = x \ AND \ y.$$

Let us see what security means here. If $f(x, y) = 1$, then it must be that $x = y = 1$ and both parties learn the other's input. Alice and Bob go to the movies. If $f(x, y) = 0$, then it must be that either $x = 0$ or $y = 0$ (or both). If Alice's input is $x = 1$, i.e. she would like to go to a movie, and the output is $f(x, y) = 0$, then Alice can infer $y = 0$, but Bob will never learn whether $x = 0$ or $x = 1$. So a party only learns the other's input if they themselves declared that they wanted to go to a movie. ■

Example 10.2.2 Alice (a customer) wants to identify herself to Bob (an ATM). Here, x is the password honest Alice should know, and y the password (for Alice) that the honest ATM should have stored in its database. The function f is the equality, that is, $f(x, y) = 1$ if and only if $x = y$, and $f(x, y) = 0$ otherwise. Security means that if Alice is dishonest (she might not know x but is still trying to break through the ATM's authentication system), then Bob should have the guarantee that Alice will never learn anything more about his input y than she can infer from $f(x, y)$ – that is, whatever x she tries, that $x \neq y$! (Unless she happens to be lucky of course.) Similarly, if Bob is a fraudulent ATM who is out to steal passwords from the users, the best he can do is guess a y and see whether it worked. No more information is revealed. ■

Example 10.2.3 Alice wants to sell a book to Bob. Here, x is Alice's asking price, and y is Bob's bid. The function they wish to compute is $f(x, y) = (ok, y)$ if $y \geq x$, and $f(x, y) = (no, 0)$ if $y < x$. If $f(x, y) = (ok, y)$, Alice can proceed to sell the book to Bob. Bob pays what he offers, and Alice gets at least her asking price. Security means that dishonest Bob can never learn what the asking price actually was, only that it was less than or equal to his bid. If $f(x, y) = (no, 0)$, then Alice will not sell her book. Security means that Alice will never learn exactly what Bob's bid actually was, only that it was lower than her asking price. Similarly, Bob will only learn that Alice's asking price was higher than his bid. ■

QUIZ 10.2.1 *True or false? In an SFE protocol Bob learns nothing at all about Alice's input.*

QUIZ 10.2.2 *True or false? Consider an SFE protocol that outputs $f_A(x, y) = f_B(x, y) = x \oplus y$. Then a malicious Bob can learn with certainty Alice's input.*

10.2.2 The Simulation Paradigm for Security

The key ideas to formalize the intuition behind our informal definition of security for SFE are the concepts of an *ideal functionality* and a *simulator*.

Definition 10.2.2 (Ideal functionality). *Let $(f_A : \mathcal{X} \times \mathcal{Y} \to \mathcal{A}, f_B : \mathcal{X} \times \mathcal{Y} \to \mathcal{B})$ be a pair of functions corresponding to an SFE task. The* ideal functionality *is a trusted device which takes as input $x \in \mathcal{X}$ and $y \in \mathcal{Y}$, and returns $f_A(x,y)$ and $f_B(x,y)$.*

Thus the ideal functionality does precisely what a protocol solving the SFE task is supposed to achieve: it directly and honestly returns the values of each of the two functions. It is "ideal" in the sense that it does not require any interaction between the two parties: you should picture a "black box," similar to the box marked SFE in Figure 10.2, which takes the inputs and provides the outputs, no questions asked. Informally, we will then say that a protocol for SFE is secure if, provided one of the parties is honest, whatever the other party does there is nothing more they can obtain that they could not have obtained by interacting with the ideal functionality.

Example 10.2.4 Consider again the millionaires' problem. Here the ideal functionality takes as input x from Alice and y from Bob, and returns $f_A(x,y) = 1_{x>y}$ to Alice and $f_B(x,y) = 1_{y>x}$ to Bob. Suppose we are given a protocol for this problem, and suppose a malicious Bob was able to infer Alice's fortune x through his interaction with her. Then the simulation paradigm dictates that, if the protocol is secure, he should be able to do the same through an interaction with the ideal functionality. But the ideal functionality just takes any y' of Bob's choice and returns to him $f_B(x,y') = 1_{y'>x}$. Since only one interaction is allowed, the best Bob can do is find out if $x < y'$ for a single y' of his choice, which for this SFE task is unavoidable.   ∎

Now suppose that we have a candidate protocol for the millionaires' problem. This protocol states what Alice and Bob's actions should be in the protocol (if they are honest). How do we prove that the protocol is secure in the simulation paradigm? According to the discussion above we need to show that, whatever Bob (or Alice) can do in the real protocol, he "should be able to do the same through an interaction with the ideal functionality." So how do we show that this is the case? The idea behind this is to define a *simulator*. A simulator is simply an algorithm that interacts with the ideal functionality on the one hand and with Bob on the other. We use the simulator to show that, for any dishonest Bob that obtains some information in the real protocol, there is a "simulated Bob," obtained by inserting the simulator between the real Bob and the ideal functionality, that obtains the same information as the dishonest Bob. If the simulated Bob could obtain the information from the ideal functionality, without even involving Alice, then by definition this contains no more information about Alice's input than is already revealed in Bob's output. More formally, we make the following definition.

Definition 10.2.3 (Security against cheating Bob). *A protocol for an SFE task (f_A, f_B) is secure if, for any malicious Bob interacting with an honest Alice in the protocol, there exists a* simulator *which, by controlling Bob in an interaction with the ideal functionality, is able to generate a distribution on outputs that is indistinguishable from the distribution produced by malicious Bob in the real interaction.*

Of course, a symmetric definition can be given for security against cheating Alice. The definition refers to the output distributions being "indistinguishable" from one another. The strongest notion of indistinguishability is called "statistical indistinguishability." In this case it means that the two distributions, the one produced by Bob and the one produced by the simulator, cannot be differentiated by any adversary except with very small probability. As we know, this is equivalent to the distributions having small total variation distance.

A weaker notion is "computational indistinguishability." In this case it is only required that the distributions cannot be distinguished by any *efficient* algorithm. Here "efficient" means that the distinguisher has to run in time that is polynomial in some parameter of the problem, which is usually referred to as the "security parameter." Here we focus on finding certain tasks for which statistical indistinguishability can be achieved by quantum protocols.

If the notion of a simulator remains somewhat hazy to you it's ok, we only want you to be familiar with the terminology and get the flavor of it. We won't worry too much about giving formal proofs of security. Nevertheless, you should be aware that doing so can be quite tricky. In fact, many *wrong* proofs of security of quantum protocols have been given in the past by relying on "intuitive" arguments rather than the simulation paradigm. We will point out one such example later.

10.3 Oblivious Transfer

Let us discuss an important example of an SFE task, known as *oblivious transfer* (OT). Here $\mathcal{X} = \{0,1\}^\ell \times \{0,1\}^\ell$ and $\mathcal{Y} = \{0,1\}$. That is, Alice receives as input two ℓ-bit strings $x = (s_0, s_1)$, and Bob receives as input a single bit y. The goal is for Bob to obtain $f_B(x,y) = s_y$, while Alice will obtain nothing, $f_A(x,y) = \perp$. Thus a protocol will implement this task securely if, first of all, it is correct, and second, malicious Alice will not obtain any information at all about Bob's input bit (since the ideal functionality never returns anything to Alice), and malicious Bob will at best learn one of Alice's strings, but never more. Figure 10.3 gives the ideal functionality for this task.

As an example scenario where this task could be useful, imagine that Alice is a database that contains two entries s_0 and s_1. Bob would like to retrieve one of them, but does not want Alice to know which one he retrieved, preserving his privacy. Alice can also be sure that he does not retrieve her entire database.

What makes OT interesting is the fact that it is *universal* for two-party cryptography. That is, if we can build a secure protocol for 1-2 OT then we can solve *any* SFE problem by just using 1-2 OT multiple times. In this respect 1-2 OT is analogous to a universal gate in computing. This universality property can be shown in different ways; see the chapter notes for pointers to how this can be done.

Unfortunately it is also known that OT cannot be implemented securely without computational assumptions … in the classical world. What about quantum protocols? Given all that we know about encoding classical information in quantum states, the following protocol is a natural first attempt.

Protocol 10 (Quantum OT protocol) Alice has input $x = (s_0, s_1) \in \{0,1\}^\ell \times \{0,1\}^\ell$ and Bob has input $y \in \{0,1\}$. Their goal is to compute $(f_A(x,y) = \perp, f_B(x,y) = s_y)$.

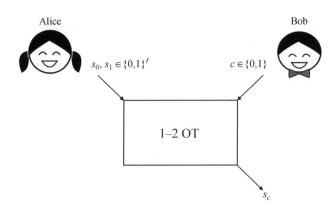

Fig. 10.3   1-2 oblivious transfer. Alice has two inputs $s_0, s_1 \in \{0,1\}^\ell$. Bob has a choice bit $c \in \{0,1\}$, according to which he receives the desired string s_c as output. Alice has no output, although it is often implicitly assumed that Alice obtains a notification that the 1-2 OT is completed (i.e. Bob made a choice for c, and received the corresponding output).

1. Alice selects uniformly random $x \in \{0,1\}^{2\ell}$ and $\theta \in \{0,1\}^{2\ell}$. She prepares BB'84 states $|x_j\rangle_{\theta_j}$ for $j = 1, \ldots, 2\ell$ and sends them to Bob.

2. Bob measures each of the qubits he received from Alice in a random basis $\tilde{\theta}_j$, obtaining outcomes $\tilde{x}_1, \ldots, \tilde{x}_{2\ell} \in \{0,1\}$. He notifies Alice that he is done with his measurements.

3. Alice reveals her choice of bases $\theta_1, \ldots, \theta_{2\ell}$ to Bob.

4. Bob sets $I = \{i : \theta_i = \tilde{\theta}_i\}$, $I_y = I$ and $I_{1-y} = \{1, \ldots, 2\ell\} \setminus I$. (For simplicity, assume that $|I_0| = |I_1| = \ell$.) Bob sends (I_0, I_1) to Alice.

5. Alice sends $t_0 = s_0 \oplus x_{I_0}$ and $t_1 = s_1 \oplus x_{I_1}$ to Bob.

6. Alice outputs $\perp$, and Bob outputs $t_y \oplus \tilde{x}_{I_y}$.

Let's first check that this protocol is correct. This is clear: whenever $j \in I_y$, by definition $\theta_j = \tilde{\theta}_j$, therefore $x_j = \tilde{x}_j$ and $(s_y \oplus x_{I_y}) \oplus \tilde{x}_{I_y} = s_y$.

Is it secure? Security against cheating Alice is not hard to verify. Indeed, the only information she gets from an honest Bob is two sets (I_0, I_1). If Bob is honest, even if Alice sent him arbitratry states in the first step, and misleading basis information in the third, since Bob's choice of $\tilde{\theta}_j$ is uniformly random the sets I_0, I_1 will be a uniformly random partition of $\{1, \ldots, 2\ell\}$ that contains no information at all about his input y. So anything a dishonest Alice could do in this protocol can be simulated by an interaction with the ideal functionality, where the simulator would replace Bob's message (I_0, I_1) (which is not provided by the ideal functionality) with a uniformly random choice.

How about security against cheating Bob? The idea is supposed to be that, given Bob's basis choices are random, he can at best learn roughly half of Alice's inputs $\tilde{x}_j$. Of course he could lie about which half he learned, but in any case he will only be able to recover about half of the bits of Alice's input $x = (s_0, s_1)$. Note he could still, for example, learn half of s_0 and half of s_1 (instead of the whole s_0 or s_1 and nothing about the other). This could be prevented by adding in a layer of privacy amplification (see Chapter 6) to the protocol, so let's assume it is not a serious issue.

You might already have noticed that there is a more worrisome hitch. The protocol requires Bob to "measure each qubit he received from Alice," and then to "notify Alice that he is done with his measurements." But what if Bob is malicious – what if he stores Alice's qubits in a large quantum memory, without performing any immediate measurement, and lies to her by declaring that he is done? Alice would then naively reveal her basis information, and Bob could measure all the qubits he stored using $\tilde{\theta}_j = \theta_j$. He would thus obtain outcomes $\tilde{x}_j = x_j$ for all j, and he could recover both $s_0 = t_0 \oplus \tilde{x}_{I_0}$ and $s_1 = t_1 \oplus \tilde{x}_{I_1}$.

So the protocol we gave is not at all secure! There are two ways to get around the problem. One possibility is to make certain physical assumptions on the capacities of cheating Bob. For example, that Bob has a bounded quantum memory, in which case he wouldn't be able to store all of Alice's qubits. We will explore this assumption in the next chapter. Another possibility would be to somehow force Bob to *commit* to a choice of basis $\tilde{\theta}_j$, and outcomes $\tilde{x}_j$ that he obtained, *before* Alice would accept to reveal her θ_j. Of course, to avoid reversing the difficulty it should be that Alice cannot learn any information about the $\tilde{\theta}_j$ just from Bob's commitments. The task we're trying to solve is called *bit commitment*, and it is another fundamental primitive of two-party cryptography. Let's explore it next.

QUIZ 10.3.1 *A cryptographic primitive related to the 1-2 OT protocol is Rabin's OT. In Rabin's OT, Alice's input is a message. She sends it to Bob who receives it with probability $1/2$, while Alice remains oblivious as to whether the message was received or not.*

True or false? Rabin's OT can be constructed from 1-2 OT.

10.4   Bit Commitment

Let's move on to our second fundamental example of a two-party task, *bit commitment*. The idea of a bit commitment protocol (Figure 10.4) is that it should allow Alice to *commit* to an unbreakable promise, without revealing any information about the promise to Bob until Alice decides to *open* her promise – without being allowed to change her mind in-between the "commit" and "open" phases. You can think of a commitment as a *cryptographic safe*: in the commit phase, Alice places her bit inside the safe and locks the door; in the open phase she reveals the key to the safe.

Definition 10.4.1 (Bit commitment). *Bit commitment is a task involving two parties, Alice (the committer) and Bob (the receiver). The input to Alice is a single bit $b \in \{0,1\}$, and she has no output. Bob has no input, and his output is a bit b'. A protocol for bit commitment has two phases, the* commit phase *and the* open phase, *and it should satisfy the following properties:*

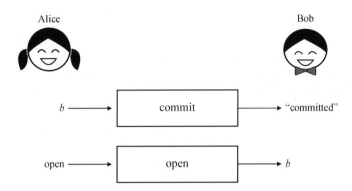

Fig. 10.4 Bit commitment is a two-party primitive with two phases: commit and open. During the commit phase, Bob should learn that Alice is indeed committed, but gain no further information about b. When Alice decides to initiate the open phase, Bob learns b, where Alice cannot change her mind about the bit b defined by the open phase. Protocols typically also allow abort.

1. Correctness: *If both Alice and Bob are honest then at the end of the protocol Bob outputs a bit $b' = b$.*
2. Hiding: *For any malicious Bob, the state of Bob at the end of the commit phase (including all his prior information and information received from Alice during the commit phase, classical or quantum) is independent of b.*
3. Binding: *For any three possible malicious behaviors of Alice, A, A_0, and A_1, the probabilities p_b that Bob outputs $b' = b$ after interacting with A in the commit phase and A_b in the open phase satisfy $p_0 + p_1 \leq 1$.*

The hiding property is clear: it states that, after the commit phase, Bob still has no information at all about the bit b that honest Alice committed to. From his point of view, it could really go either way. The binding property is more subtle. Intuitively, what it is trying to capture is that once Alice has committed to a specific value b (this is the role of A in the definition), then she shouldn't be able to come up with two possible different behaviors (A_0 and A_1) such that she has a strictly higher than $1/2$ chance of being able to convince Bob that $b = 0$ (she would run A_0) *or* that $b = 1$ (she would run A_1).

Bit commitment is a good example of a cryptographic task for which it is crucial to define security as precisely as possible, especially in the quantum setting. Consider the following "intuitive" definition of the binding property: "It should be impossible for malicious Alice to convince honest Bob that $b = 0$ *and* $b = 1$ with probability strictly larger than 1." Do you see the difference? I wouldn't blame you if you didn't – the pioneers of quantum information and cryptography didn't either! In 1991 Brassard et al. famously proposed an "unconditionally secure" quantum protocol for bit commitment that satisfied the above intuitive notion of security. However, their protocol was later completely broken! (Indeed, as we will soon see, perfectly secure bit commitment is impossible in both the classical and the quantum world.) Their "mistake" is that they interpreted the italicized "and" in the intuitive definition above in a strong sense: they show that, in their protocol, it wouldn't be possible for a malicious Alice to *simultaneously* convince Bob that $b = 0$ and $b = 1$, by assuming that, if this were the case, the two final quantum states of the protocol associated with the outcomes "Bob returns $b' = 0$" and "Bob returns $b' = 1$" would exist simultaneously. However, as we know very well by now, quantum information is subtle, and the fact that Alice can "change her mind" after the commit phase does *not* imply that she can generate *both* the $b' = 0$ and $b' = 1$ states for Bob from the same state at the end of the open phase; only that she can generate either of them.

> **QUIZ 10.4.1** *Consider the following protocol for bit commitment: Alice prepares* $|\psi_{00}\rangle_{AB} = \frac{1}{\sqrt{2}}(|00\rangle + |11\rangle)$ *if she commits to $x = 0$, or she prepares* $|\psi_{01}\rangle_{AB} = \frac{1}{\sqrt{2}}(|00\rangle - |11\rangle)$ *if she commits to $x = 1$. Then she sends the register B to Bob. Finally, in the open phase, Alice sends Bob her register A, so that Bob can perform a measurement in the Bell basis on the two qubits in registers A and B to learn Alice's bit. Is this protocol correct and secure?*
>
> **(a)** *Yes*
> **(b)** *No*

10.4.1 Universality of Bit Commitment

Bit commitment is an important task in quantum multiparty cryptography, because – just as 1-2 OT – it is known to be universal. This is demonstrated by a small modification to the protocol for OT that we gave in the previous section: as we discussed, the protocol by itself is not secure; however, if one has access to a secure protocol for bit commitment then it can be turned into a secure OT protocol as well. The essential idea is to enhance step 2 of the protocol by asking Bob to commit to the measurement outcomes. This allows Alice to gain confidence that Bob has measured the qubits after step 2 of the protocol, avoiding the attack described above.

Since OT itself is universal for multiparty computation, we deduce that bit commitment is universal. However, note that the protocol for OT based on bit commitment we gave is quantum, even if bit commitment is implemented using a classical protocol. Interestingly, this is unavoidable: indeed, bit commitment is *not* universal for *classical* multiparty computation! Nevertheless, it suffices as a building block for many useful protocols – such as the millionaires' problem, as the next exercise asks you to show.

Exercise 10.4.1   Give a secure protocol for Yao's millionaires' problem, assuming you have access to a protocol securely implementing bit commitment.

Let's see how the reverse can be accomplished, using OT as a building block to achieve bit commitment. For this, we will consider an approximate version of bit commitment, in which Alice can change her mind with some small error probability ε. That is, we say that the protocol is ε-binding if the requirement $p_0 + p_1 \le 1$ is relaxed to $p_0 + p_1 \le 1 + \varepsilon$.

The following protocol takes 1-2 OT and turns it into bit commitment. In the protocol we invert the use of 1-2 OT: Bob is now the sender, and Alice the receiver.

Protocol 11 (Bit commitment from 1-2 OT) Alice's input is $b \in \{0,1\}$. Bob has no input.

1. Commit phase: Bob chooses two strings $s_0, s_1 \in \{0,1\}^\ell$ uniformly at random. Bob and Alice execute a protocol for OT, with the role of the parties reversed: OT-Alice's input is (s_0, s_1) (provided by Bob), and OT-Bob's input is b (provided by Alice). Thus Alice receives s_b, and Bob receives $\perp$.

2. Open phase: Alice sends $\hat{b}$ and $\hat{s} = s_b$ to Bob. If $\hat{s} = s_{\hat{b}}$, then Bob accepts and concludes that Alice committed herself to $b = \hat{b}$. If $\hat{s} \neq s_{\hat{b}}$, then Bob rejects.

Why does this give bit commitment? First of all, if both parties behave honestly the protocol is clearly correct. Now let's consider the hiding property. We need to show that, at the end of the commit phase, Bob has no information about b. This follows right away from the definition of OT, which guarantees that the sender never receives any information about the receiver's input.

It remains to show that the protocol is ε-binding. This again follows from the security of OT, for $\varepsilon = 2^{-\ell}$. Indeed, the ideal functionality for OT is such that the receiver can learn only *one* of the two strings. Suppose Alice has two possible strategies, one to open $\hat{b} = 0$ and the other to open $\hat{b} = 1$. Let p_0 be the probability that the first strategy succeeds, and p_1 the probability that the second succeeds. As a consequence, Alice can recover both of s_0 and s_1 with probability at least $p_0 + p_1 - 1$. By the security of the OT primitive, this can happen with probability at most the probability that a random guess of the nonreceived string would succeed, i.e. $2^{-\ell}$. By taking ℓ large enough we can achieve any desired ε-security for the binding property.

If you have been reading carefully you may have noticed that in the argument we made a jump from "Alice can recover s_0 with probability p_0, and s_1 with probability p_1" to "Alice can recover both of s_0 and s_1 with probability at least $p_0 + p_1 - 1$." This is correct if Alice is classical, but if her strategies involve incompatible quantum measurements then the implication might no longer be true. Hence, in the case when we allow the protocol implementing OT to be a quantum protocol, we have to be extra careful to show that the resulting protocol for bit commitment satisfies the required definition. This is possible (so the protocol described above *is* secure provided the implementation of OT is, whether classical or quantum), but one must take even greater care in making the right security definitions to ensure that they satisfy the stringent criteria of "universal composability."

10.4.2 Impossibility of Bit Commitment

Since, as we argued, bit commitment implies OT (in the quantum world), but perfect OT is impossible, it must be that bit commitment is impossible as well! Let's see why. We give an informal argument, and refer you to the chapter notes for pointers.

Consider a protocol for bit commitment that is perfectly hiding, i.e. at the end of the commit phase Bob has absolutely no information about Alice's bit b. Let's show that in this case a malicious Alice can cheat *arbitrarily*: she is able to open any bit that she likes.

Suppose that the initial state of Alice and Bob is a pure state $|\psi\rangle_{AB}$, which in particular contains Alice's input. We can always assume that this is the case by considering a purification and giving the purifying system to Alice.

Now suppose Alice executes the bit commitment protocol with input b. At the end of the commit phase, the joint state of Alice and Bob can be described by some pure state $|\psi(b)\rangle_{AB}$. Since the bit commitment protocol is perfectly hiding, it must be the case that

$$\rho_B(0) = \mathrm{tr}_A(|\psi(0)\rangle\langle\psi(0)|_{AB}) \quad \text{and} \quad \rho_B(1) = \mathrm{tr}_A(|\psi(1)\rangle\langle\psi(1)|_{AB})$$

are absolutely identical, as otherwise there would be a measurement that Bob can make on his system to distinguish (even partially) between the two states, giving him some information about b.

Now it is time to take out our quantum information theorist's toolbox and extract one of its magic tools: Uhlmann's theorem! The theorem implies that, if $\rho_B(0) = \rho_B(1)$, then necessarily there exists a unitary U_A on Alice's system such that $U_A \otimes \mathbb{I}_B |\psi(0)\rangle_{AB} = |\psi(1)\rangle_{AB}$. But this means that Alice can perfectly change her mind, thereby completely breaking the binding property for the protocol.

Rather unfortunately for the fate of quantum multiparty cryptography, it is possible to generalize this argument to show that any protocol for *any* task in multiparty quantum cryptography must be "totally insecure" in the following sense: if the protocol is perfectly secure against a malicious Bob, then it must be that a malicious Alice (interacting with honest Bob) can recover the value $f_A(x,y)$ associated with Bob's input y for *all* possible values of x, simultaneously! (To see why this indeed makes the protocol totally insecure, consider, for example, the millionaires' problem: Alice would learn if $x > y$ for any x, and could thus perform a quick binary search to learn Bob's fortune y exactly.)

Given such a strong impossibility result, we are left with two possibilities. The first possibility is that we could place limiting assumptions on any malicious party's abilities. In classical cryptography these are mostly computational assumptions, and we give an example in the next section. In quantum cryptography we can also consider placing physical assumptions on the adversary, such as it having limited storage capabilities. We will explore such assumptions in detail in the next chapter.

The second possibility is to accept the impossibility of *perfect* protocols for multiparty cryptography and instead settle for protocols with a relaxed notion of security, where, for instance, Bob can learn "some" information about both Alice's input strings s_0, s_1 in oblivious transfer, but not all. This is indeed possible, and can be quite useful in spite of the relaxed security condition; we saw an example of this when discussing coin flipping in Section 10.1.

10.4.3 Computationally Secure Commitments

We have seen that it is impossible to perfectly implement bit commitment, whether we use quantum information or not. The fact that it is such a useful primitive, however, should encourage us to be creative. In many contexts we would be willing to put up with our usual requirement for perfect, information-theoretic security, and start making assumptions – of course, the fewer the better! For example, we can assume that the malicious party has bounded computational power. This is a very standard assumption in classical cryptography, as indeed very little can be achieved without it (in contrast to quantum cryptography). Of course, here we would only want to make assumptions that hold even if the malicious party has bounded *quantum* computational power. The weakest such assumption under which any interesting cryptographic task is made possible is the existence of *one-way functions*. Informally, a function is one-way if it is easy to evaluate the function on any input (there is an efficient algorithm to compute it), but it is hard to invert the function (given a point in the range of the function, find a pre-image).

There are many candidate constructions of one-way functions, including some that are believed to be hard to invert even for quantum computers. And it turns out that, assuming one-way functions exist, there is a simple protocol for bit commitment that is statistically binding ($p_0 + p_1$ can be made as close to 1 as desired by increasing the amount of communication required in the scheme) and computationally hiding (the hiding property holds as long as it can be assumed that the malicious party cannot invert the one-way function).

Since our focus in this book is on information-theoretic, not computational, security, we won't explain all the details of the protocol. Just for fun, though, let's see the flavor of it. What the protocol actually needs is a special object that can be constructed from a one-way function and is called a *pseudorandom generator* (PRG). For us a PRG is a family of functions $G_n : \{0,1\}^n \to \{0,1\}^{3n}$. These functions have two important properties: (i) for any $x \in \{0,1\}^n$, it is easy for anyone to compute $G_n(x)$, and (ii) it is "computationally infeasible" for anyone to demonstrate the difference between the following two distributions U_1 and U_2: U_1 is obtained by computing $G_n(x)$ for a uniformly random $x \in \{0,1\}^n$, and U_2 is obtained by directly returning a uniformly random $y \in \{0,1\}^{3n}$. Note that since U_1 has support size $2^n \ll 2^{3n}$, these distributions are very different; however, the assumption is that, if we are only given a number of samples and an amount of time that scales polynomially with n, then it is impossible to tell the difference between the two distributions.

The PRGs are all we need; let's see the protocol! Recall that Alice has as input a bit $b \in \{0,1\}$, and Bob has no input at all. To commit,

1. Bob selects a uniformly random $r \in \{0,1\}^{3n}$ and sends it to Alice.
2. Alice selects a uniformly random $s \in \{0,1\}^n$ and sends $\sigma = (b \cdot r) \oplus G(s)$ to Bob, where $b \cdot r$ is the string 0^{3n} if $b = 0$ and the string r if $b = 1$.

We can already see that the protocol is *computationally* hiding, because whatever Bob does, Alice sends him a sample from the distribution U_1 or the distribution $r \oplus U_1$. These distributions are indistinguishable because both are indistinguishable from U_2, by the PRG assumption.

Now, to reveal her bit, Alice simply sends s to Bob, and Bob checks if $G(s) \oplus \sigma$ is 0^{3n} or r. Using the fact that the range of G has size 2^n, it is not too hard to check that, except with probability 2^{-n} over Bob's choice of r, Alice will not be able to reveal to a value $b' \neq b$. So the protocol is *statistically* binding, meaning that the chance that Alice violates the binding condition is exponentially small, independently of the amount of time or computational power that she has.

We won't go into the protocol in more detail here, but we hope that it gives you the flavor of how certain tasks can be securely implemented assuming that some problem is computationally hard. (Here, the hard problem is to distinguish U_1 from U_2.)

10.5 Kitaev's Lower Bound on Strong Coin Flipping

In this section we give a proof of Kitaev's lower bound on the bias of secure protocols for strong coin flipping. The argument relies on simple notions of linear and semidefinite programming with which you may not already be familiar. If so, we encourage you to read a bit about these techniques before proceeding. But if you don't have the time then you can skip this section: it is independent of the remainder of the book.

For any coin flipping protocol, define p_{1*} as the probability that Alice outputs a 1, maximized over all possible (cheating) strategies for Bob. Define p_{*1} symmetrically. Then the condition for the protocol to be a secure strong coin flipping protocol with bias ε is that both $p_{1*}, p_{*1} \in [1/2 - \varepsilon, 1/2 + \varepsilon]$. For example, for Blum's protocol we have $p_{1*} = p_{*1} = 1$ (make sure you understand why this is the case), and so the protocol is *not* secure.

Theorem 10.5.1 (Kitaev) *For any strong coin flipping protocol, we have* $p_{1*}p_{*1} \geq 1/2$.

Note that the condition in the theorem immediately implies that any strong coin flipping protocol has bias at least $(\sqrt{2} - 1)/2$, as claimed. This is because if $x, y \in [0, 1]$ are such that $(1/2 + x)(1/2 + y) \geq 1/2$ then $\max(x, y) \geq (\sqrt{2} - 1)/2$; this is easy to show by contrapositive (assume both $x, y < (\sqrt{2} - 1)/2$ and reach a contradiction).

Kitaev's theorem applies to both classical and quantum protocols. We'll see the proof for classical protocols first, and then move to the quantum setting. Both proofs have the same structure: Bob's maximum cheating probability can be expressed as the optimum of a linear program (LP) (or semidefinite program (SDP) in the quantum case), and similarly for Alice's. We will show that any feasible solution to the duals of each LP provides an upper bound on the probability of success of the cheating strategy. The crucial insight is that the cheating probabilities need to be considered *together*, through the quantity $p_{1*}p_{*1}$: a good upper bound on this quantity expresses the fact that either Alice can force Bob to output a 1 or, if she can't, then it must be that Bob can force her to produce a 1. We will obtain a bound on this quantity by taking the product of some of the LP (or SDP) constraints. Let's proceed with the details.

10.5.1 The Bound on Classical Protocols

Fix a classical protocol for strong coin flipping. Since we are proving a lower bound, i.e. an impossibility result, we need to find a way to represent the most general protocol possible. In general, a classical protocol can be thought of as a tree. Each node in the tree is labeled by a variable u, which represents the transcript that led to this node. So if we are in a node labeled by u, and Alice plays by sending a message a, then we arrive at node (u, a). The root of the tree, before any message has been sent, is simply labeled as $\emptyset$.

The honest protocol is then given by probabilities $p_A(a|u)$, $p_B(b|u)$, which are Alice's (resp. Bob's) transition probabilities. Suppose first that Alice is honest. Then Bob's maximum cheating probability can be expressed as a *linear program* LP_B. This can be seen as follows. For each node u in the tree introduce a variable $p_B(u)$, which represents the probability of reaching node u when Alice is honest and Bob cheats using some cheating strategy of his choice. Note that each cheating strategy of Bob leads to a $p_B(u)$, but not all $p_B(u)$ may be achievable, given that Alice is being honest. Recall that since we are considering p_{*1} we consider that Bob's goal is to maximize the probability of reaching a leaf labeled with a 1. Denote the set of all such leaves as L_1. We introduce constraints to express the fact that Bob can choose any distribution on edges of the tree when it is his turn to play, but he still has to follow Alice's distribution when it is her turn.

$$(\text{LP}_B, \text{primal}) \qquad \max \quad \sum_{u \in L_1} p_B(u)$$

$$p_B(u)p(a|u) = p_B(u,a) \qquad \forall a, \forall u \text{ node for Alice}$$
$$p_B(u) = \sum_b p_B(u,b) \qquad \forall u \text{ node for Bob}$$
$$p_B(0) = 1$$
$$p_B(u) \geq 0 \qquad\qquad\qquad \forall u$$

If we could solve this linear program we would obtain the value of p_{*1}. However, without knowing the protocol, which determines all the $p(a|u)$, this seems hard. To get information about a linear program we know that it is always wise to compute the dual linear program. In order to do so, introduce variables $Z_A(u, a)$ for the first set of constraints and $Z_A(u)$ for the second set. With a little work the dual can be written in the form

$$(\text{LP}_B, \text{dual}) \qquad \min \quad Z_A(0)$$
$$Z_A(u) \geq \sum_a p(a|u)Z_A(u,a) \qquad \forall u \text{ node for Alice}$$
$$Z_A(u) \geq Z_A(u,b) \qquad \forall b, \forall u \text{ node for Bob}$$
$$Z_A(u) \geq 1 \qquad\qquad \forall u \in L_1$$

What to make of this? For each node u, we can interpret $Z_A(u)$ as the maximum probability with which Bob can cheat, starting at node u.

Now we can proceed in an exactly symmetric manner to introduce another linear program LP_A, this time for a cheating Alice. Taking the dual, we get variables $Z_B(u)$. If we accept to interpret these as maximum cheating probabilities from a given node, then we are motivated to introduce the following quantity:

$$F_\ell = E_{u \sim \ell}\left[Z_A(u)Z_B(u)\right], \qquad (10.1)$$

where $u \sim \ell$ is shorthand for u being taken according to the probability distribution on nodes at depth ℓ which arises from the honest protocol (with both parties playing honestly). In this expression, $Z_A(u)Z_B(u)$ should be interpreted as the bias that cheating parties can achieve, if any of them starts cheating at node u.

Let Z_A, Z_B be optimal solutions to the duals of LP_B and LP_A respectively. The last constraint of the dual implies that without loss of generality we can assume that Z_A and Z_B are both exactly 1 at all leaves labeled with a 1 (as if they were larger, a better solution to the LP could be obtained by scaling). Hence if n is the last level of the tree, then

$$F_n = p_{1,1} = 1/2 . \tag{10.2}$$

This is because we assume that the protocol is correct, and so for a random leaf (taken according to the distribution on leaves obtained by having both parties play honestly) the leaf is labeled with a 1 with probability $1/2$ and a 0 with probability $1/2$.

Now let's use strong duality: the optimums of the primal and dual forms of LP_B are equal, and similarly for LP_A. This means that F_0, which is defined as the product of the two dual optimums, is also equal to the product of the two primal optimums, i.e.

$$F_0 = p_{1*}p_{*1} . \tag{10.3}$$

To conclude, we multiply out the constraints of the two duals to show that for any $\ell \geq 0, F_\ell \geq F_{\ell+1}$. So then $F_0 \geq F_n$, which using (10.2) and (10.3) proves Theorem 10.5.1 for the case of classical protocols.

10.5.2 The Bound on Quantum Protocols

The lower bound for quantum protocols is very similar to the lower bound for classical protocols, and if you understood how that works then you already have all of the main ideas in place. The main technical hurdle that we need to overcome is the following: how do we model a general quantum protocol? For a classical protocol we could list all the honest probabilities of sending a certain message m given that the parties are in a certain stage u of the protocol. However, for quantum protocols, there is not such a simple notion of sending a given message. In particular, for a fully general protocol the messages will be in a quantum state, and that quantum state could be entangled with the party's memory! It seems that we are facing an almost impossible mess.

Luckily there is a very clean and principled way to go about modeling a general quantum protocol, and this is useful even outside of the present context of coin flipping. First, let us make the assumption that, at each step of the protocol, either Alice or Bob applies a unitary operation on the message space, as well as their private space. This is without loss of generality, because any measurement can be pushed to the end of the protocol using the principle of deferred measurement. We also assume that the parties start in the $|0...0\rangle$ state: if Alice or Bob wants to prepare a different state then they can do this as part of their first action. At the first step, Alice (or Bob, whoever plays first) applies a unitary A_1 on her private space and the message space, and she sends the message to Bob. Then Bob applies B_1 to his private space and the message space, which he then returns to Alice, etc. See Figure 10.5 for a pictorial representation of Alice's i-th and $(i+1)$-st actions, and Bob's i-th action, in the protocol. The density matrices ρ_i and ρ_{i+1} represent the joint state of Alice's private space, and the message space, right before she performs her i-th and $(i+1)$-st actions respectively.

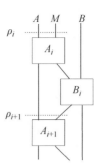

Fig. 10.5 Round i in a quantum coin flipping protocol. A and B denote registers associated with Alice's and Bob's private space respectively, and M with the quantum messages they exchange. At round i Alice applies operation A_i on her local view ρ_i. Then Bob applies B_i, leading to an updated local view ρ_{i+1} for Alice.

At the end of the interaction, Alice measures her entire space using a projective measurement $\{\pi_A, \mathbb{I} - \pi_A\}$, and similarly Bob with $\{\pi_B, \mathbb{I} - \pi_B\}$. Each of them obtains a binary outcome, and this is what they return as their value for the coin flip.

Using this notation, let's first write what it means for the protocol to be correct. Simply writing out the party's actions in turn we see that

$$p_{1.1} = \|(\pi_A \otimes \mathbb{I}_M \otimes \pi_B)B_n A_n \cdots B_1 A_1 |0 \ldots 0\rangle\|^2 \, ,$$

and we can compute $p_{0.0}$ symmetrically using $(\mathbb{I} - \pi_A), (\mathbb{I} - \pi_B)$ as the measurements. Correctness requires both of these quantities to equal $1/2$.

Proceeding analogously to the classical case, we now need to write an optimization problem that captures the maximum cheating probability for Bob, assuming that Alice plays honestly. The key idea is that instead of using probabilities $p_B(b|u)$ as the variables, we use density matrices ρ_i, where i indexes the messages from Bob to Alice in the protocol, that are supported on Alice's private space and the message space (see Figure 10.5). Since Alice is playing honestly, and Bob can act on the messages, but not on Alice private's space, we know that for any i, $\mathrm{tr}_M(\rho_{i+1}) = \mathrm{tr}_M(A_i \rho_i A_{i+1}^{\dagger})$. Here we traced out the message register M because it can be affected by cheating Bob, and in general we do not know how. Putting everything together we get the following optimization problem, which places a limit on Bob's probability to cheat in the protocol:

$$\text{(SDP}_B\text{, primal)} \qquad \max \quad \langle \pi_B \otimes \mathbb{I}, \rho_n \rangle$$
$$\mathrm{tr}_M(\rho_{i+1}) = \mathrm{tr}_M(A_i \rho_i A_{i+1}^{\dagger}) \qquad \forall i$$
$$\rho_0 = |0\rangle\langle 0|_{A \times M}$$
$$\rho_i \geq 0 \qquad \forall i$$

This is not a linear program because the last condition is a matrix positivity condition: it says that density matrices must be positive semidefinite. Instead, it is a semidefinite program. Semidefinite programs are a generalization of linear programs, and they keep many of the same advantages, including that they can be solved in polynomial time and have a rich duality theory.

The dual of the semidefinite program can be computed mechanically, and we obtain the following formulation:

$$(\text{SDP}_B, \text{dual}) \qquad \min \quad \langle 0|Z_A(0)|0\rangle$$
$$Z_A(i) \otimes \mathbb{I}_M \geq A_{i+1}^\dagger(Z_A(i+1) \otimes \mathbb{I}_M)A_{i+1} \qquad \forall i$$
$$Z_A(n) = \pi_A$$
$$Z_A(i) = (Z_A(i))^\dagger \qquad \forall i$$

Now let $|\psi_\ell\rangle$ be the state of all registers, Alice's private space, the message register, *and* Bob's private register, assuming both parties play honestly. Then the analogue of (10.1) is to define

$$F_\ell = \langle \psi_\ell | Z_A(\ell) \otimes \mathbb{I}_M \otimes Z_B(\ell) | \psi_\ell \rangle. \tag{10.4}$$

Correctness then implies the condition $F_n = 1/2$, where n is the total number of rounds, and strong duality implies the condition $F_0 = p_{1*}p_{*0}$, exactly as before. As before, it is not hard to show that the relation $F_\ell \geq F_{\ell+1}$ follows from the dual constraints. We then get that $F_0 \geq 1/2$, which is precisely Kitaev's lower bound.

CHAPTER NOTES

The quantum coin flipping protocol discussed in Section 10.1.2 is due to D. Aharonov, et al. (Quantum bit escrow. In *Proceedings of the Thirty-Second Annual ACM Symposium on Theory of Computing*, pp. 705–714. ACM, 2000). A proof of Theorem 10.1.1 can be found in the paper by A. Ambainis (A new protocol and lower bounds for quantum coin flipping. In *Proceedings of the Thirty-Third Annual ACM Symposium on Theory of Computing*, pp. 134–142. ACM, 2001).

Quantum weak coin flipping with arbitrarily small bias was first discovered by C. Mochon (Quantum weak coin flipping with arbitrarily small bias. arXiv:0711.4114, 2007). The proof was rewritten and simplified by D. Aharonov, et al. (A simpler proof of the existence of quantum weak coin flipping with arbitrarily small bias. *SIAM Journal on Computing*, **45**(3):633–679, 2016), who showed the existence of a protocol with a number of rounds that is exponential in $1/\varepsilon$, where ε is the bias. Subsequently an explicit description of the protocols was given by A. Singh Arora, J. Roland, and S. Weis (Quantum weak coin flipping. In *Proceedings of the 51st Annual ACM SIGACT Symposium on Theory of Computing*, pp. 205–216, 2019). C. A. Miller (The impossibility of efficient quantum weak coin flipping. In *Proceedings of the 52nd Annual ACM SIGACT Symposium on Theory of Computing*, pp. 916–929, 2020) showed that a polynomial dependence of the number of rounds on $1/\varepsilon$ is necessary. It remains an open question what is the optimal round complexity of quantum weak coin flipping.

A. Chailloux and I. Kerenidis (Optimal quantum strong coin flipping. In *50th Annual IEEE Symposium on Foundations of Computer Science*, pp. 527–533. IEEE, 2009) showed that any weak coin flipping protocol with bias ε could be used to build a strong coin flipping protocol with bias $(\sqrt{2}-1)/2 + O(\varepsilon)$, thereby matching Kitaev's lower bound.

Good references to learn more on the topic of SFE and more generally multi-party cryptography include a survey by O. Goldreich, *Foundations of Cryptography: A Primer, Volume 1* (Now Publishers, 2005) (especially Section 7), and one by Y. Lindell and B. Pinkas (Secure multiparty computation for privacy-preserving data mining. *Journal of Privacy and Confidentiality*, **1**(1):5, 2009). For the case of quantum protocols, it is only recently that a satisfactory definition has been introduced; see the paper by D. Unruh (Universally composable quantum multi-party computation. In *Annual International Conference on the Theory and Applications of Cryptographic Techniques*, pp. 486–505. Springer, 2010) for the notion of universal composability or another by S. Fehr and C. Schaffner (Composing quantum protocols in a classical environment. In *Theory of Cryptography Conference*, pp. 350–367. Springer, 2009) for a weaker but perhaps more approachable definition.

The "unconditionally secure" bit commitment protocol of G. Brassard, et al. appeared in a conference paper (A quantum bit commitment scheme provably unbreakable by both parties. In *Proceedings 34th Annual Symposium on Foundations of Computer Science*, pp. 362–371. IEEE, 1993). The impossibility of (quantum) bit commitment is due to D. Mayers (Unconditionally secure quantum bit commitment is impossible. *Physical Review Letters*, **78**(17):3414, 1997) and Hoi-Kwong Lo and Hoi Fung Chau (Is quantum bit commitment really possible? *Physical Review*

Letters, **78**(17):3410, 1997). For an approachable, self-contained proof we recommend the detailed notes by J. Watrous (Impossibility of quantum bit commitment. Lecture 19, CPSC 519/619 University of Waterloo: https://cs.uwaterloo.ca/watrous/QC-notes/QC-notes.19.pdf). The impossibility of perfectly secure two-party cryptography, even using quantum information, is shown in H. Buhrman, M. Christandl, and C. Schaffner (Complete insecurity of quantum protocols for classical two-party computation. *Physical Review Letters*, **109**(16):160501, 2012). For much more on computational security and how to implement various cryptographic tasks classically under computational assumptions, we recommend the freely available lecture notes by R. Pass and A. Shelat (*A Course in Cryptography*, 2010. Lecture notes available at `www.cs.cornell.edu/courses/cs4830/2010fa/lecnotes`); see in particular Section 4.7 for a discussion of bit commitment.

The proof of Kitaev's lower bound on strong quantum coin flipping given in Section 10.5 is due to G. Gutoski and J. Watrous (Toward a general theory of quantum games. In *Proceedings of the Thirty-Ninth Annual ACM Symposium on Theory of Computing*, pp. 565–574. ACM, 2007).

For a survey of other directions for quantum cryptography "beyond key distribution," we point to the survey by Broadbent et al. (Broadbent, Anne, and Christian Schaffner. Quantum cryptography beyond quantum key distribution. *Designs, Codes and Cryptography*, **78**:351–382, 2016).

PROBLEMS

10.1 A weak coin flipping protocol

In the chapter, we studied a strong quantum coin flipping protocol with bias $1/4$. In this problem you'll see how a variation of that same protocol allows us to construct a weak coin flipping protocol with bias smaller than $1/4$.

Recall that in a weak coin flipping protocol, we define Alice's cheating probability as $P_A^* = \Pr[\text{Alice wins}]$, maximized over Alice's (cheating) strategies, and similarly P_B^* for Bob, and we say that the cheating probability of the protocol is $\max\{P_A^*, P_B^*\}$. The protocol in this problem is parametrized by $\alpha \in [0, \pi]$, over which you'll optimize later on.

For $a, x \in \{0, 1\}$, define the qutrit state $|\psi_{a,x}\rangle$ in the space $\mathcal{H}_t = \mathbb{C}^3$ as

$$|\psi_{a,x}\rangle = \cos\left(\frac{\alpha}{2}\right)|0\rangle + \sin\left(\frac{\alpha}{2}\right)(-1)^x |a+1\rangle$$

and $|\psi_a\rangle \in \mathcal{H}_s \otimes \mathcal{H}_t = \mathbb{C}^2 \otimes \mathbb{C}^3$ as

$$|\psi_a\rangle = \frac{1}{\sqrt{2}}(|0\rangle |\psi_{a,0}\rangle + |1\rangle |\psi_{a,1}\rangle)$$

The protocol is as follows.

Step 1. Alice picks $a \in_R \{0, 1\}$, prepares the state $|\psi_a\rangle \in \mathcal{H}_s \otimes \mathcal{H}_t$ (i.e. a state of one qubit and one qutrit), and sends to Bob the second half of the state (the qutrit).

Step 2. Bob picks $b \in_R \{0, 1\}$ and sends it to Alice.

Step 3. Alice reveals the bit a to Bob. Let $c = a \oplus b$. If $c = 0$, then Alice sets $c_A = 0$ and sends to Bob the other part of the state $|\psi_a\rangle$ (the qubit). Bob checks that the qutrit-qubit pair he received is indeed in the state $|\psi_a\rangle$ (by making a measurement with respect to any orthonormal basis of $\mathcal{H}_s \otimes \mathcal{H}_t$ containing

$|\psi_a\rangle$). If the test is passed, Bob sets $c_B = 0$, and so Alice wins the game. Otherwise Bob concludes that Alice has deviated from the protocol, and aborts.

Step 4. If, on the other hand, $c = a \oplus b = 1$, then Bob sets $c_B = 1$, and returns the qutrit he received in round 1. Alice checks that her qubit-qutrit pair is in state $|\psi_a\rangle$. If the test is passed, she sets $c_A = 1$, so Bob wins the game. Otherwise Alice concludes that Bob has tampered with her qutrit to bias the game, and aborts.

1. Verify that this protocol satisfies correctness.

2. What is Bob's reduced density matrix ρ_a after step 1, in the case that Alice has prepared the honest state $|\psi_a\rangle$? (Note that the subscript a refers to the classical bit and not the system of Alice or Bob.)

Now, suppose Bob is honest while Alice may cheat. We aim to obtain a (tight) upper bound on Alice's winning probability. The most general strategy is for Alice to prepare a pure state $|\phi\rangle \in \mathcal{H} \otimes \mathcal{H}_s \otimes \mathcal{H}_t$, where $\mathcal{H}$ is an ancillary space (one can always purify the state via $\mathcal{H}$). Then she sends the qutrit part in $\mathcal{H}_t$ to Bob, and keeps the part of the state in $\mathcal{H} \otimes \mathcal{H}_s$.

We can assume without loss of generality that in step 3 of the protocol Alice always replies with $a = b$ (so that $c = 0$), and consequently tries to pass Bob's check. For this, she performs a unitary U_b on her part of $|\phi\rangle$, so that she gets $|\phi_b\rangle = (U_b \otimes I)|\phi\rangle$, and then sends the qubit in $\mathcal{H}_s$ to Bob. The final joint state can then be written as $|\phi_b\rangle = \sum_i \sqrt{p_i}|i\rangle|\phi_{i,b}\rangle$ for some $\{p_i\}$ and Schmidt bases $\{|i\rangle\}$ of $\mathcal{H}$ and $\{|\phi_{i,b}\rangle\}$ of $\mathcal{H}_s \otimes \mathcal{H}_t$.

Now, recall the interpretation of the fidelity between two density matrices as the square root of the probability that Alice can convince Bob that one is the other. Let σ_b be the density matrix of Bob's qubit-qutrit pair at the end of the protocol. And let σ be Bob's reduced density matrix after the first step of the protocol (i.e. just the qutrit).

3. Show an upper bound on the probability that Alice wins given that Bob sent b (here ρ_a and $|\psi_a\rangle$ are defined as in the previous problem). *[Hint: express it first in terms of the fidelity of two density matrices and then use the fact that fidelity is nondecreasing under taking partial trace.]*

4. Use the above to bound the probability that Alice wins. *[Hint: you might find useful the fact that for any three density matrices σ, ρ_0, ρ_1, it holds that $F^2(\sigma, \rho_0) + F^2(\sigma, \rho_1) \leq 1 + F(\rho_0, \rho_1)$.]*

Now we turn to Bob's winning probability when he is potentially cheating and Alice is honest. He will be trying to infer as much as he can about the value of the bit a, so that he can send back a bit b such that $a \oplus b = 1$, at the same trying to cause as little disturbance as possible to the joint state $|\psi_a\rangle$, so as to pass Alice's final check. The most general strategy that he can employ is to perform a unitary U on the space $\mathcal{H}_t \otimes \mathcal{H} \otimes \mathbb{C}^2$ of the qutrit he received from Alice, some ancillary qubits, and a qubit reserved for his reply. He then measures the last qubit and sends the outcome as b to Alice.

Suppose without loss of generality that the unitary is such that

$$U : |i\rangle |\bar{0}\rangle |0\rangle \mapsto |\xi_{i,0}\rangle |0\rangle + |\xi_{i,1}\rangle |1\rangle$$

where $|\bar{0}\rangle$ is the initial state of the ancilla qubits, and for some states $|\xi_{i,0}\rangle, |\xi_{i,1}\rangle$, not necessarily orthogonal, such that $\|\xi_{i,0}\|^2 + \|\xi_{i,1}\|^2 = 1$.

5. Calculate the probability that Bob wins given that Alice sent a. Simplify the expression you find using the definitons of $|\psi_{a,0}\rangle$ and $|\psi_{a,1}\rangle$.

6. Is the expression found in the previous question at most

$$\left(\cos^2 \left(\frac{\alpha}{2} \right) \|\xi_{0,\bar{a}}\| + \sin^2 \left(\frac{\alpha}{2} \right) \right)^2 ?$$

7. Use the abovementioned bound to calculate an upper bound on the probability that Bob wins, and maximize it over the choice of $|\xi_{0,0}\rangle$ and $|\xi_{0,1}\rangle$.

8. Determine the value of the parameter α that minimizes the overall bias of the protocol. What is the bias?

10.2 A simple quantum bit commitment protocol

As you know, perfectly secure quantum bit commitment is impossible. Nonetheless, it is possible to construct protocols in which Alice and Bob can cheat to some extent, but not completely.

For a cheating Alice and honest Bob, we define Alice's cheating probability as

$$P_A^* = \frac{1}{2} \left(\Pr(\text{Alice opens } b = 0 \text{ successfully}) + \Pr(\text{Alice opens } b = 1 \text{ successfully}) \right),$$

maximized over Alice's (cheating) strategies. For a cheating Bob and an honest Alice, instead, we let Bob's cheating probability be

$$P_B^* = \Pr(\text{Bob guesses } b \text{ after the commit phase}),$$

maximized over Bob's (cheating) strategies. The cheating probability of the protocol as a whole is defined as $\max\{P_A^*, P_B^*\}$. In this problem, we introduce a simple example of such a protocol.

Step 1. *Commit phase*: Alice commits to bit b by preparing the state

$$|\psi_b\rangle = \sqrt{a} |bb\rangle + \sqrt{1-\alpha} |22\rangle$$

and Alice sends the second qutrit to Bob.

Step 2. *Open phase*: Alice reveals the classical bit b and sends the first qutrit over to Bob, who checks that the pure state is the correct one by making a measurement with respect to any orthogonal basis containing $|\psi_b\rangle$.

1. What is the density matrix ρ_b that Bob has after the *commit phase* if Alice has committed to bit b and honestly prepared state $|\psi_b\rangle$?

2. Compute Bob's cheating probability P_B^* by recalling the operational interpretation of the trace distance.

Next, let's calculate Alice's cheating probability. Let the underlying Hilbert space be $\mathcal{H} \otimes \mathcal{H}_s \otimes \mathcal{H}_t$, where $\mathcal{H}_t$ corresponds to the qutrit that is sent to Bob in the commit phase, $\mathcal{H}_s$ to the qutrit that is sent during the open phase, and $\mathcal{H}$ is any auxiliary system that Alice might use. For the most general strategy, we can assume that she prepares the pure state $|\phi\rangle$, as it can always be purified on $\mathcal{H}$.

We can write $|\phi\rangle = \sum_i \sqrt{p_i} |i\rangle |\tilde{\psi}_{i,b}\rangle$ where $\{|i\rangle\}$ and $\{|\tilde{\psi}_{i,b}\rangle\}_{i,b}$ are Schmidt bases of $\mathcal{H}$ and $\mathcal{H}_s \otimes \mathcal{H}_t$ respectively. So, the reduced density matrix on $\mathcal{H}_s \otimes \mathcal{H}_t$ is $\sigma_b = \sum_i p_i |\tilde{\psi}_{i,b}\rangle\langle\tilde{\psi}_{i,b}|$. Moreover, let σ be Bob's reduced density matrix after the commit phase, i.e. just a qutrit.

3. Compute the probability of dishonest Alice successfully opening bit b in terms of the fidelity of two density matrices, and hence give an upper bound on Alice's cheating probability. *[Hint: use the fact that the fidelity is nondecreasing under taking partial trace, in particular tracing out system $\mathcal{H}_s$.]*

4. Give an upper bound to Alice's cheating probability in terms of α. *[Hint: you might find useful the inequality $F^2(\rho_1, \rho_2) + F^2(\rho_1, \rho_3) \leq 1 + F(\rho_2, \rho_3)$ for arbitrary density matrices ρ_1, ρ_2, ρ_3.]*

Note that the bound on Bob's cheating probability that you obtained in question 2 of this problem is tight, since it is the best possible probability of distinguishing between two known states, and he knows what the two states are when Alice is honest.

Importantly, the bound above on Alice's cheating probability that we just obtained is also tight. There is a simple cheating strategy that allows Alice to achieve this bound, without even making use of the ancillary system $\mathcal{H}$.

5. Which of the following states of two qutrits can Alice prepare?

   I. $|\psi_0\rangle + |\psi_1\rangle$ normalized

  II. $|\psi_0\rangle - |\psi_1\rangle$ normalized

 III. $|\psi_0\rangle + \frac{\sqrt{3}}{2}|\psi_1\rangle$ normalized

6. Finally, by combining the calculations so far on Alice and Bob's cheating probabilities, determine the α that minimizes the overall cheating probability of the protocol.

10.3 From coin flipping to bit commitment

In this chapter you learned that in the quantum world it is possible to construct a weak coin flipping protocol with arbitrarily small bias, i.e with a cheating probability of $1/2 + \varepsilon$ for any $\varepsilon > 0$, something that is not possible in the classical world. We refer to ε as the bias.

In this question, you'll explore how such a weak coin flipping protocol can be used to construct a quantum bit commitment protocol. This protocol is inspired by that of the previous problem, and improves on it (so we recommend that you go through the previous problem before attempting this). It will in fact be optimal, in the sense that no lower cheating probability can be achieved. Specifically, we will be using an unbalanced weak coin flipping protocol with ε bias (unbalanced just means that the honest winning probabilities are different than $1/2$). The main idea is to reduce Bob's cheating probability by increasing slightly the amplitude of the term $|22\rangle$ in $|\psi_b\rangle$ from the previous problem.

1. Just to make sure we are all on the same page, why would doing so decrease Bob's cheating probability?

However, this modification might allow Alice to cheat even more. We take care of this by introducing a weak coin flipping procedure between Alice and Bob so that they jointly create the initial state, as opposed to Alice creating it all by herself. We describe in detail the new bit commitment protocol.

Step 1. *Commit phase, part 1:* Alice and Bob perform an ε-bias unbalanced weak coin flipping protocol with winning probabilities $1 - p$ and p for Alice and Bob respectively. Assume that the final part of that coin flipping protocol would require Alice and Bob to read out their respective outcomes by measuring an output one-qubit register. But suppose that they don't carry out the final measurements, and instead Alice just sends to Bob all her qubits, but keeps her output register.

After this, Alice and Bob share the following state:

$$|\Omega\rangle = \sqrt{p}\,|L\rangle_A \otimes |L, G_L\rangle_B + \sqrt{1-p}\,|W\rangle_A \otimes |W, G_W\rangle_B \,,$$

where L corresponds to Alice loses and W to Alice wins. $|G_L\rangle$ and $|G_W\rangle$ are ancilla states.

Step 2. *Commit phase, part 2:* Alice performs the following operation. Conditioned on her qubit being $|W\rangle$ she creates two qutrits in the state $|22\rangle$, and sends the second to Bob. Conditioned on her qubit being $|L\rangle$, she creates two qutrits in the state $|bb\rangle$, and sends the second to Bob, where b is the classical bit she wants to commit to. Then, if Alice and Bob behave honestly, they share the state

$$|\Omega_b\rangle = \sqrt{p}\,|L, b\rangle_A \otimes |L, b, G_L\rangle_B + \sqrt{1-p}\,|W, 2\rangle_A \otimes |W, 2, G_W\rangle_B \,.$$

Step 3. *Open phase:* Alice reveals b, and sends all of system A to Bob, who checks that he has the correct state $|\Omega_b\rangle$, by making a measurement in the basis $\{|\Omega_b\rangle, |\Omega_b\rangle^\perp\}$.

It is clear that if both Alice and Bob are honest, then Alice always succesfully reveals the bit b she had committed to. Now, if Alice is honest and Bob tries to cheat, he can make it so that they instead prepare, after part 1 of the commit phase, the state

$$|\Omega^*\rangle = \sqrt{p'}\,|L\rangle_A \otimes |L, G_L'\rangle_B + \sqrt{1-p'}\,|W\rangle_A \otimes |W, G_W'\rangle_B \,,$$

where p' is constrained by the fact that the weak coin flipping protocol has ε-bias.

2. Compute a tight bound on Bob's cheating probability P_B^*, i.e. the probability that he can guess b after part 2 of the commit phase.

Let ρ_b be Bob's reduced state after the commit phase with an honest Alice that commits to classical bit b, and let σ be Bob's state after the commit phase with a cheating Alice.

Then $\rho_b = p|\bar{b}\rangle\langle\bar{b}| + (1-p)|\bar{2}\rangle\langle\bar{2}|$, where $|\bar{b}\rangle = |L, b, G_L\rangle$ for $b \in \{0,1\}$ and $|\bar{2}\rangle = |W, 2, G_W\rangle$.

3. Let $r_i = \langle \bar{i}|\sigma|\bar{i}\rangle$, $i \in \{0,1,2\}$. What bounds do the r_i satisfy? *[Hint: recall that the underlying weak coin flipping protocol has ε-bias.]*

   I. $r_0, r_1 \le p + \varepsilon$

  II. $r_2 \le (1 - p + \varepsilon)$

 III. $r_0 + r_1 + r_2 \le 1$

Now, we turn to Alice's cheating probability.

4. Bound the fidelity $F(\sigma, \rho_b)$ in terms of p, r_b, and r_2, making use of the fact that the fidelity is nondecreasing under operations, for the particular operation that carries out a measurement on Bob's output qutrit in the computational basis.

5. Use the result of the previous question to obtain a tight bound on Alice's cheating probability. *[Hint: recall that in Problem 9.2 you have already shown how $P_A^* \le \frac{1}{2}\left(F^2(\sigma, \rho_0) + F^2(\sigma, \rho_1)\right)$, and the same holds here analogously.]*

6. You are told that, for $\varepsilon < p(1 - \frac{1}{2-p})$, the bound you obtained in question 4 is maximal when r_2 is maximal and $r_0 = r_1$. Compute the maximal value of said bound in terms of p and ε.

7. Finally, assuming the bound you found above is tight, determine the cheating probability of the overall bit commitment protocol. *[Hint: you can ignore the $O(\varepsilon)$ terms, since we have a weak coin flipping protocol for any bias $\varepsilon > 0$.]*

11

Security from Physical Assumptions

 As we saw in the previous chapter, security for two-party cryptography is difficult to achieve. Even given the ability to use quantum communication, we still cannot hope to achieve the security conditions of general two-party cryptography, including 1-2 oblivious transfer (OT) and bit commitment. Yet, two-party cryptography includes many interesting challenges we would like to solve on a daily basis! Due to the interest in solving this challenge, we may be willing to bend our principles somewhat and limit ourselves to obtain security guarantees under some assumption about the adversary. The idea is that we'd show security only for "reasonable" adversaries, where "reasonable" is still a very large class of adversaries that we expect might arise in real life.

Classically, the most commonly used kind of assumption on the adversary is a *computational assumption*. This assumption is two-fold: First, we assume that a specific problem, such as factoring a large integer, requires a large amount of computational resources to be solved. Second, we assume that the adversary has a relatively limited amount of computational resources available – namely, an amount that is insufficient to solve the difficult problem within a practically interesting time frame. An example of a computational assumption is the pseudorandom generator assumption, which we used to construct a secure bit commitment protocol in Section 10.4.3.

An important limitation of computational assumptions is that they tend not to stand the test of time. For example, once we build a quantum computer, any assumption based on the hardness of factoring will become vacuous, because the Shor algorithm provides an efficient way to factor large numbers on a quantum computer. Moreover, security can often be broken retroactively: if we build a quantum computer tomorrow, most two-party protocols that have been executed to date can lose their security, as long as the adversary has stored a copy of the protocol transcript in their classical memory, such as a flash drive. Clearly, for high-security applications this is quite undesirable.

In this chapter we present an alternative path to base security in challenging settings. The main idea is to make physical, rather than computational, assumptions. Importantly, we would like that security requires the assumption of interest to only be valid during the execution of the protocol. If someone builds better equipment tomorrow, good for them, but we don't want that to compromise our cryptographic interactions from today! This guarantee allows us to use physical assumptions that are motivated by technological challenges that may be overcome in the future, but may be reasonable to assume today given humankind's present technological abilities.

How is it possible to obtain long-lasting security under an assumption that is valid today, but may become invalid tomorrow? Intuitively, the idea is that even temporary physical assumptions can lead to a permanent lack of information for the adversary that prevents them from ever breaking the protocol in the future. It is interesting to note that technologically motivated physical assumptions can also enable key exchange or two-party cryptography using only classical communication (see the chapter notes). Indeed, we already saw an example of such an assumption in Section 7.3. There, Alice and Bob were able to generate secure key using only classical communication, as long as the eavesdropper is limited in their ability to listen in to their communication channel.

When looking for technologically motivated assumptions, it is useful to keep in mind the perpetual conflict we face when designing cryptographic protocols: On the one hand, the protocol should be secure; that is, whatever assumptions we impose on the adversary should be sufficient to protect the honest parties. On the other hand, we of course want the protocol to be correct. That is, the honest parties should be able to execute it correctly. When considering good technological assumptions to make, this conflict translates into a desire to design a protocol that is technologically *easy* to execute for the honest parties, but at the same time technologically *infeasible* to break for the adversary. The gap between the resources needed to execute the protocol and the resources needed to break the protocol should be as large as possible; if it is too small then we may need to look for a better security assumption.

11.1 The Noisy Storage Model

At present, our abilities to store large amounts of quantum information without errors and for a long time are extremely limited. In quantum systems with a quantum memory capable of accessing quantum communication, i.e. quantum memory systems with an optical interface, the state of the art at the time of writing is that storage times of the order of seconds have been observed for a small number of qubits. Moreover, the transfer of qubits into such a memory is typically itself already lossy.

These technological limitations motivate the assumption that the abilities of the adversary to store quantum information are limited. The *noisy storage model* assumes that during finite waiting times Δt introduced in the protocol, the adversary is limited in their ability to store quantum information (see Figure 11.1). Specifically, to keep information during such waiting times, the adversary is limited to using a quantum memory whose action on the quantum data is modeled by some quantum channel $\mathcal{F}$. The fact that $\mathcal{F}$ is not necessarily the identity represents the fact that there may be losses in the memory. Other than this limitation the adversary remains all powerful: they may perform arbitrary quantum operations, including arbitrary encoding and decoding procedures before and after the waiting time, and store an unlimited amount of classical information. In particular, before and after the waiting time, the adversary is allowed to have an arbitrary faithful quantum memory. Since proofs of security in

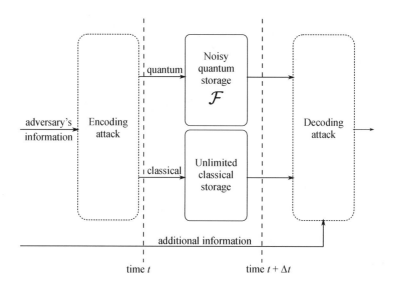

Fig. 11.1 Noisy storage model: During waiting times Δt introduced in the protocol, the adversary is required to store any quantum information using a quantum memory modeled by the channel $\mathcal{F}$. Other than this requirement the adversary may be arbitrary, including having access to a perfect large-scale quantum computer to encode and decode the quantum information before and after using the memory $\mathcal{F}$.

the noisy storage model only require the adversary's memory to be limited during the waiting time Δt, even if tomorrow we can build better quantum memories then security can nevertheless not be broken retroactively.

A simple example of a channel $\mathcal{F}$ we might use to model the adversary's memory corresponds to a quantum memory that is error free, but limited in size to only q qubits for some integer q. That is, $\mathcal{F} = \mathbb{I}^{\otimes q}$. This special case is also known as the *bounded quantum storage* model. In general, one typically assumes $\mathcal{F}$ to be of the form $\mathcal{F} = \mathcal{N}^{\otimes q}$; that is, the quantum memory is bounded in size as well as noisy. Here, $\mathcal{N}$ may be a noisy one-qubit channel, such as depolarizing noise. Taking $\mathcal{F}$ to be of this form is appealing to analyze since one understands general properties of such channels to store quantum information from quantum information theory.

11.2 1-2 Oblivious Transfer in the Noisy Storage Model

From the previous chapter we know that any two-party protocol can be constructed by combining multiple copies of a 1-2 OT protocol, making this primitive a fundamental building block for two-party cryptography. We have also seen that, despite the attempt to construct OT protocols that make use of quantum communication, no unconditionally secure protocols can exist. Nevertheless, we made a valiant attempt, Protocol 10 from Section 10.3, which we unfortunately observed could easily be broken by a malicious Bob by using, for example, a large quantum memory. We thus know that our earlier protocol cannot be secure without some form of assumption on the power of the adversary. Could the protocol be secure against a memory-bounded malicious Bob? Let's give it a shot.

Recall that in 1-2 OT, Alice has two inputs $s_0, s_1 \in \{0,1\}^\ell$ and Bob has a single input $y \in \{0,1\}$. At the end of the protocol, Alice should return no output and Bob should obtain the string s_y. Here we recap essentially the same 1-2 OT protocol as in the previous chapter, except for two small modifications. First, we make explicit a small waiting time Δt during which the noisy storage assumption is applied. Second, for reasons that will soon become clear, we introduce a secure strong extractor Ext : $\{0,1\}^{n/2} \times \{0,1\}^t \to \{0,1\}^\ell$ whose exact parameters we discuss later. First, let's see the protocol.

Protocol 12 Protocol for 1-2 OT in the noisy storage model. Alice has inputs $s_0, s_1 \in \{0,1\}^\ell$ and Bob has input $y \in \{0,1\}$.

1. Alice chooses a uniformly random string $x = x_1, \ldots, x_n \in \{0,1\}^n$ and $\theta = \theta_1, \ldots, \theta_n \in \{0,1\}^n$. She prepares BB'84 states $|x_j\rangle_{\theta_j}$ for $j = 1, \ldots, n$ and sends them to Bob.
2. If $y = 0$ then Bob measures all of the qubits in the standard basis. If $y = 1$, he measures in the Hadamard basis. He records the resulting outcome string $\tilde{x} = \tilde{x}_1, \ldots, \tilde{x}_n$.
3. Both parties wait time Δt. (Storage assumption is applied!)
4. Alice sends to Bob the string $\theta_1, \ldots, \theta_n$. Bob computes the set of indices $I = \{j \mid \theta_j = y\}$ where he measured in the same basis as Alice.
5. Alice chooses two independent random seeds $r_0, r_1 \in \{0,1\}^t$. She computes $k_0 = \mathrm{Ext}(x_+, r_0)$ and $k_1 = \mathrm{Ext}(x_\times, r_1)$, where x_+ is the substring of x that Alice encoded in the standard basis, and $x_\times$ is the substring where she used the Hadamard basis. Alice sends r_0 and r_1 to Bob.
6. Alice sends to Bob $m_0 = s_0 \oplus k_0$ and $m_1 = s_1 \oplus k_1$, where $\oplus$ denotes the bitwise XOR.
7. Bob computes $k = \mathrm{Ext}(x_I, r_y)$ and $s_y = k_y \oplus r_y$.

We see that the honest parties need *no* quantum memory to execute this protocol. While we have for ease of explanation described the protocol in a way that may suggest that Alice first prepares all n qubits, and only then sends all of them to Bob, Alice and Bob can also execute the protocol with Alice preparing and transmitting only one qubit at a time, which Bob immediately measures upon receipt. This makes the comparison of the resources needed to execute the protocol (no quantum memory at all) to the resources needed to break the protocol (a large amount of quantum memory, as we will see) especially appealing.

Why does this protocol work? Let us first double check that the protocol is correct, that is, Bob actually obtains s_y in accordance with his choice bit $y \in \{0,1\}$. Note that if there is no noise, then whenever $\theta_j = y$, we have $x_j = \tilde{x}_j$. That is, whenever Alice encoded in the basis in which Bob measures, then Bob learns the corresponding element of Alice's bit string. This means that if Alice applies Ext to hash down the elements of the strings corresponding to the standard and the Hadamard basis respectively, then Bob knows one of them perfectly. Since Alice sends him r_0 and r_1, he learns the correct k_y. In the protocol the string k_y is used as a key that encrypts s_y using one-time pad encryption, and so Bob can recover Alice's string s_y as well.

QUIZ 11.2.1 *In the protocol for oblivious transfer, honest Alice and Bob do not need any quantum memory. Does that mean that if Alice possessed an unbounded and noise-free quantum memory, then the protocol would no longer be secure against dishonest Alice?*

(a) *Yes, the protocol would not be secure against cheating Alice in this case; that is, she could then easily learn Bob's bit y.*

(b) *No, even with such a quantum memory Alice would not be able to learn Bob's bit y because nowhere in the protocol is there any communication from Bob to Alice taking place.*

11.3 Security from Quantum Uncertainty

Is Protocol 12 secure? Let us first check security against dishonest Alice. This is virtually identical to the argument from Chapter 10. Recall that we only need to show that Alice cannot learn Bob's choice bit y. If you stare at the protocol description, it is clear that this is definitely the case: Bob never sends any information at all to Alice, from which she could learn anything about y. This idea can be formalized into a complete security proof (against dishonest Alice), which we omit.

The main difficulty is to show security against dishonest Bob. Recall from the definition that we need to show that, while Bob might learn one of Alice's two strings s_0 and s_1, there must always exist some $\bar{y}$ such that Bob learns (almost) nothing about $s_{\bar{y}}$.

As you can see, at step 5 of the protocol we have used an old "trick" which we already encountered in the analysis of quantum key distribution (QKD), namely the use of privacy amplification to reduce the amount of information that a potential adversary has about a certain string. Based on what we know about privacy amplification and extractors, to achieve our goal it will be sufficient to somehow ensure that the conditional min-entropy of either X_+ or $X_\times$, conditioned on all information available to Bob, is high. As long as we can establish a good lower bound on this, by selecting the parameters of the extractor appropriately we can make sure that from Bob's point of view the string k_0 or k_1 is ε-close to uniformly distributed, making him unable to recover the corresponding s_0 or s_1.

To summarize, our goal is now to show that there exists a $\bar{Y}$ such that the conditional min-entropy of X_+ for $\bar{Y} = 0$ or $X_\times$ for $\bar{Y} = 1$, conditioned on Bob's information, is high. Observe how here we started using a capital letter for $\bar{Y}$. You know that we use capital letters for random variables. We did this because, if you think about it, for a given malicious Bob we may not be able to identify a single string X_+ or $X_\times$ that is unknown to Bob. This is simply because Bob, being quantum, could use a quantum coin flip to govern his behavior in the protocol, and end in a superposition of a Bob who knows a little bit about X_+ with a Bob who knows a little bit about $X_\times$. Therefore, it is inevitable that the answer to the question "which string is unknown to Bob" not only depends on Bob, but may in fact be a "quantum" answer. We warned you that proofs of security for quantum protocols can be subtle! Let's see how to make it work.

11.3.1 Security in the Bounded Quantum Storage Model

For security we will focus on the special case of the bounded quantum storage model, where $\mathcal{F} = \mathcal{N}^{\times q}$ with $\mathcal{N} = \mathbb{I}$. The argument in this case is simpler; we will sketch how it can be generalized in the following subsection.

In order to examine the conditional min-entropy of the strings X_+ and $X_\times$, let us first consider if we can say anything at all about Bob's min-entropy about *the entire* string x. Clearly, if Bob could store all of Alice's qubits, then he could just measure the entire string in the correct basis after having learned it at step 4 and obtain the entire string x.

This means that security against Bob can never be achieved if the number q of qubits that Bob can store is $q \geq n$. Let us thus assume that $q < n$. In practice this means that if we assume that the adversary can store at most q qubits – for example, we might take q to be many times the size of the largest quantum memory known to date – we will choose the parameter n in the protocol to be large enough to achieve security. Note that, since we allow Bob to have an arbitrary quantum memory and computer *before* the waiting time, he can first store all of Alice's qubits and perform an arbitrary quantum operation on all of them. For example, he might measure some of them, resulting in some classical information K. However, he can then keep only q qubits in his quantum memory, which we denote by Q.

Since we are interested in Bob's min-entropy about the entire string X, we want to show a lower bound on the quantity

$$\mathrm{H}_{\min}(X_+ X_\times | \Theta, K, Q) ,$$

where we have written $X_+ X_\times$ for the string X to remind ourselves that we are ultimately interested in the two portions corresponding to the two different bases. To make his guess, Bob can use the classical information K, his quantum memory Q, and the basis information Θ that Alice sends to him after the waiting time.

When we studied QKD we noticed that is it often much easier to show security against an adversary who is entirely classical. Here, the adversary, malicious Bob, has information that is partially classical (Θ and K) and partially quantum (Q). How can we get rid of Q? The intuition of course is that Q, being made of only q qubits, should contribute "only q" to Bob's information. How do we make this formal? If you remember our work in previous chapters you might notice that we already faced a similar issue a couple times, and we handled it using the powerful *chain rule* for the conditional min-entropy. To recall the chain rule, see Box 5.2. In our context the chain rule gives us that

$$\mathrm{H}_{\min}(X_+ X_\times | \Theta, K, Q) \geq \mathrm{H}_{\min}(X_+ X_\times | \Theta, K) - \log |Q|$$
$$= \mathrm{H}_{\min}(X_+ X_\times | \Theta, K) - q .$$

What this shows is that Bob's limited quantum memory can indeed only contribute an additive q bits to his conditional entropy. Now, it remains to bound the information aquired by a Bob, who has only Θ and K. How could we possibly analyze this?

Once more, let us think back to the ideas that we learned while studying QKD. We know that the conditional min-entropy has an interpretation as a guessing probability, and that guessing probabilities can be bounded by studying an associated guessing game. How did we reduce the analysis of the BB'84 QKD protocol to a guessing game? By purifying the protocol and introducing an equivalent formulation that indeed looked like a game. So let's do the same here.

Consider a purified version of the 1-2 OT protocol where at the first step, instead of Alice sending BB'84 states to Bob, Bob prepares n EPR pairs and sends the first qubit of each pair to Alice. As for QKD, let's give even more power to Bob and let him prepare any state that he wants, as long as there are n qubits that are sent to Alice. Having received the qubits from Bob, Alice chooses one of two random bases to measure each qubit and announces the basis choice to Bob. Bob, given his side information (any qubits that he kept to himself, as well as the strings Θ and K), has to guess Alice's string of outcomes.

This is precisely the bipartite guessing game from Chapter 5! The only difference is that in Chapter 5 we called Bob "Eve," and only looked at the version where Bob/Eve sends a single qubit to Alice, not n as here. Also, in Chapter 5 we didn't explicitly write out the classical side information K – but because we allow Bob/Eve to prepare any bipartite quantum state he wants, he can as a special case keep an arbitrary string K of information about the state he prepared.

Applying the bound shown in Chapter 5,[1] we obtain the familiar

$$H_{\min}(X_+ X_\times | \Theta, K) = n\left(-\log\left(\frac{1}{2} + \frac{1}{2\sqrt{2}}\right)\right) \approx 0.22n .$$

Of course, what we really need is to make a statement about the different parts X_+ and $X_\times$. That is, we would like to show that there exists a $\bar{Y} \in \{+, \times\}$ such that Bob's entropy about $X_{\bar{Y}}$ is high. This is some form of *min-entropy splitting*: if the uncertainty about a string is high, there must be some half of it that is unknown. And indeed, this is true (but a bit harder to prove): there exists some register $\bar{Y}$ such that

$$H_{\min}(X_{\bar{Y}} | \Theta, K, \bar{Y}) \geq \frac{H_{\min}(X_+ X_\times | \Theta, K)}{2} - 1 .$$

Note that here, as we expected, $\bar{Y}$ is a classical random variable that can be correlated with the side information held by Bob. It is also noteworthy that min-entropy splitting only works if K really is classical, which is why we first have to get rid of Q. Putting all the steps together we obtain

$$\begin{aligned} H_{\min}(X_{\bar{Y}} | \Theta, K, \bar{Y}, Q) &\geq H_{\min}(X_{\bar{Y}} | \Theta, K, \bar{Y}) - q \\ &\geq \frac{H_{\min}(X_+ X_\times | \Theta, K)}{2} - 1 - q . \end{aligned}$$

---

1 In fact we need an n-qubit version of the bound from Section 5.4. This can be done using the same ideas as those introduced to study the tripartite guessing game, and it leads to the same bound $p_{\text{succ}}^{n \text{ rounds}} \leq \left(\frac{1}{2} + \frac{1}{2\sqrt{2}}\right)^n$.

To conclude, we can employ the properties of randomness extraction to claim that Bob is ε-close to being ignorant about $X_{\bar{Y}}$ whenever

$$\ell < H_{\min}(X_{\bar{Y}}|\Theta, K, \bar{Y}, Q) - O(\log 1/\varepsilon) - 1$$
$$\approx 0.11n - q - O(\log 1/\varepsilon) - 2 \,.$$

This means that whenever $q \lesssim 0.11n$, we can have security for some $\ell > 0$! Or, reading it the other way around, assuming a maximum q for the adversary tells us that we need to send roughly $n \approx q/0.11$ qubits in order to achieve security.

> **QUIZ 11.3.1** *In the security proof against cheating Bob we encountered again the uncertainty game discussed in Chapter 5. However, here we want to evaluate $P_{\mathrm{guess}}(X|\Theta K)$ rather than just $P_{\mathrm{guess}}(X|\Theta)$. That is, Bob can have some classical coin that determines what state he sends to Alice. Consider two states ρ_1 and ρ_2, such that in a deterministic game the guessing probability of Bob corresponding to each of those states is $P_{\mathrm{guess}}(X|\Theta)_{\rho_1} = P_{\mathrm{guess}}(X|\Theta_A)_{\rho_2} = p$. Depending on the outcome of the random coin K, Bob sends to Alice the first or the second state. Would the resulting guessing probability $P_{\mathrm{guess}}(X|\Theta K)$ be smaller than, equal to, or bigger than p?*
>
> (a) $P_{\mathrm{guess}}(X|\Theta K) = p$
> (b) $P_{\mathrm{guess}}(X|\Theta K) > p$
> (c) $P_{\mathrm{guess}}(X|\Theta K) < p$

11.3.2 General Channels

Using a more sophisticated analysis it is possible to show that security can be achieved as long as $q \le n - O(\log^2 n)$, which is essentially optimal. We thus see that security becomes possible by sending just a few more qubits than Bob can store. This result was obtained following a series of advanced ideas in quantum information theory that were motivated by the noisy storage model. While outside the scope of our book, we briefly explain the notion of *capacity* of a quantum channel which underpins these ideas through the simple example of the classical capacity of a perfect one-qubit channel.

Earlier we directly assumed that $\mathcal{F} = \mathcal{N}^{\times q}$, where $\mathcal{N} = \mathbb{I}$ is a perfect – noise-free – one-qubit channel. Looking at Figure 11.1, it is intuitive that the problem of proving security in the noisy storage model, is directly related to a problem at the heart of quantum information theory, namely the study of the *capacity* of the channel $\mathcal{F}$. Roughly speaking, the capacity of a channel quantifies how much information we may send through it (with some limited amount of error), provided that we can use the best possible error-correcting code imaginable. That is, the best possible encoding procedure to protect the information from the noise in the channel, and the best possible decoding procedure to recover it.

For sending some classical information $i \in \{0,1\}^m$ through the channel $\mathcal{F}$, an encoding scheme consists of a mapping $i \to \rho_i$ of the classical string i to some quantum state ρ_i. A decoding scheme corresponds to a measurement on the channel output $\mathcal{F}(\rho_i)$ with outcome $\hat{i}$ that corresponds to a guess for i. The probability that a string i is correctly decoded may thus be written as $q_i = \mathrm{tr}(M_i \mathcal{F}(\rho_i))$, where $\{M_i\}_i$ are the POVM operators corresponding to the decoding measurement. The quantity of interest when studying the classical capacity of a channel $\mathcal{F}$ is the average probability of recovering an unknown string i. This can be expressed as

$$P_{\text{succ}}(\mathcal{F}, m) = \max \frac{1}{2^m} \sum_{i \in \{0,1\}^m} \text{tr}(M_i \mathcal{F}(\rho_i)) \ ,$$

where the maximization is taken over all possible encodings and all possible decodings. In general, even for sending classical information, the capacity of most quantum channels is difficult to understand!

Luckily, for our simple example of $\mathcal{F} = \mathbb{I}^{\otimes q}$ it is easy to examine "how much" information we can convey. We call $R = m/q$ the *rate* of sending classical information though $\mathcal{F}$. The *classical capacity* C is defined such that for $R \leq C$ there exists an error-correcting code to send information with $P_{\text{succ}}(\mathcal{F}, qR) \to 1$ and for $R > C$ we have $P_{\text{succ}}(\mathcal{F}, qR) \to 0$ as $q \to \infty$. That is, the capacity forms a sharp threshold for sending information! For our example with $m = qR$ we have

$$P_{\text{succ}}(\mathcal{F}, qR) = \max \frac{1}{2^{qR}} \sum_{i \in \{0,1\}^{qR}} \text{tr}(M_i \rho_i)$$

$$\leq \frac{1}{2^{qR}} \sum_{i \in \{0,1\}^{qR}} \text{tr}(M_i)$$

$$= \frac{1}{2^{qR}} \text{tr}(\mathbb{I})$$

$$= 2^{-qR} 2^q = 2^{-q(R-1)} \ ,$$

where the second line follows from $0 \leq \rho_i \leq \mathbb{I}$, and the third from $\sum_i M_i = \mathbb{I}$. Since we are considering a space of q qubits, we have $\text{tr}(\mathbb{I}) = 2^q$. We thus see that in our case $C = 1$ forms a sharp threshold, since for $R > C$ we get that $P_{\text{succ}} \to_{q \to \infty} 0$. That is, we can transmit no more than one classical bit per qubit of channel use. Moreover, clearly for $R \leq C$ we may indeed send one classical bit per qubit, for example by encoding it into the standard basis.

QUIZ 11.3.2 *Consider the setting described in this section. How reliably can we reconstruct the input state at the output of the channel $\mathcal{F}$ in the regime where $R > 1$?*

(a) *There is no difference between $R > 1$ and $R \leq 1$; that is, suitable encodings can allow us to reliably decode the input state at the output.*

(b) *In the regime where $R > 1$, the set of encoding procedures that allow for reliable transmission of all the N qubits becomes restricted to encodings that satisfy certain specific conditions.*

(c) *In this regime there exist no reliable encodings; that is, for all encoding procedures, the trace distance between the input and the output state will be exponentially (in N, the number of input qubits that we want to store) close to 1.*

11.3.3 Working with Noisy Devices

Our protocol above assumes that honest Alice and Bob can execute the protocol without errors. What if even the honest users are subject to errors? This is important for practical applications. Adapting the protocol to this situation can be done using the techniques we already learned in our study of QKD. Indeed, similar to the case of QKD when a small amount of errors occur on the channel, information reconciliation can be performed in order to ensure that Bob is able to correct for errors if he only receives noisy versions $\tilde{X}_+$ or $\tilde{X}_\times$ of Alice's string. Alice can then employ a classical error-correcting code, and she will need to send Bob the error syndromes $e_+ = s_H(x_+)$ and $e_\times = s_H(x_\times)$ computed according to the error-correcting code used, with parity-check matrix H. Bob then proceeds to use the relevant syndrome to correct his noisy string $\tilde{x}_+$ to x_+ for $y = 0$, and $\tilde{x}_\times$ to $x_\times$ for $y = 1$. Correctness of the protocol then follows by the properties of the classical error-correcting code used.

How about security? The security argument for the case that Alice is dishonest remains the same as before, so we only have to worry about dishonest Bob. If Bob is dishonest, we make a worst-case assumption and assume that all noise is in fact due to Bob's attack and no other noise occurs in the transmission. In other words, we assume that if Bob is dishonest then he may also eliminate all other errors – for example, those occurring during transmission – and he is only limited by the assumption on his noisy storage device. Evidently, this means that Bob could now also use the error-correcting information that Alice sends in order to correct errors in his noisy memory $\mathcal{F}$! Our task is hence to understand the min-entropy *conditioned* also on the syndrome information S:

$$H_{\min}(X|\Theta, K, Q, S) \,.$$

Luckily, we can again employ the chain rule to bound the reduction in min-entropy due to conveying this additional information to Bob, just as we did in the case of QKD. That is,

$$H_{\min}(X|\Theta, K, Q, S) \geq H_{\min}(X|\Theta, K, Q) - \log|S| \,,$$

where $\log|S|$ corresponds to the number of bits of syndrome information sent. Remember from the case of QKD that in the limit of large n there exist error-correcting codes such that $\log|S| \approx nh(p)$, where p is the bit-flip error rate of each bit in x and $h(p) = -p\log p - (1-p)\log(1-p)$ is the binary entropy.

To conclude, note that just as in the case of QKD it may no longer be possible to achieve security for a particular storage assumption $\mathcal{F}$ once p gets too large. This is indeed very intuitive, since if p is too large, Alice has to send so much error-correcting information that dishonest Bob (whose only noise comes from $\mathcal{F}$) can use this information to correct the errors in this quantum memory $\mathcal{F}$ and therefore break the security of the protocol.

QUIZ 11.3.3 *Consider a scenario where Alice's device is noisy, such that whenever she wants to prepare the states $\{|0\rangle, |1\rangle\}$, she actually prepares states $|0'\rangle = \cos\varepsilon\,|0\rangle + \sin\varepsilon\,|1\rangle, |1'\rangle = \sin\varepsilon\,|0\rangle - \cos\varepsilon\,|1\rangle$ for some small $\varepsilon > 0$. Suppose that Bob is aware of this imperfection. In such a scenario is $H_{\min}(X|Bob)$ larger or smaller than with the perfect device?*

(a) *Larger*
(b) *Smaller*

CHAPTER NOTES

The classical bounded storage model was defined by U. M. Maurer (Conditionally-perfect secrecy and a provably-secure randomized cipher. *Journal of Cryptology*, **5**(1):53–66, 1992). A good introduction to the use of physical assumptions in classical communication can be found in a paper by A. Koch (The landscape of security from physical assumptions. In *IEEE Information Theory Workshop (ITW)*, pp. 1–6, 2021). Yet, not only is classical memory cheap and plentiful, the small gap between what classical parties need in order to implement the protocol ($\Omega(n)$ bits of classical memory) versus what the adversary needs to break the protocol (typically, $O(n^2)$ bits of classical memory) is too small to make this assumption useful in general. Inspired by these classical results, the bounded quantum storage model was put forward by I. B. Damgård, et al. (Cryptography in the bounded-quantum-storage model. In *46th IEEE Symposium on Foundations of Computer Science*, pp. 449–458. IEEE, 2005), and the more general noisy storage model by S. Wehner, C. Schaffner, and B. M. Terhal (Cryptography from noisy storage. *Physical Review Letters*, **100**:220502, 2008) and R. Konig, S. Wehner, and J. Wullschleger (Unconditional security from noisy quantum storage. *IEEE Transactions on Information Theory*, **58**(3):1962–1984, 2012).

To learn more about protocols and their properties in the noisy storage model, we refer to the review article by P. J. Coles, et al. (Entropic uncertainty relations and their applications. *Reviews of Modern Physics*, **89**:015002, 2017) and references therein. The best bounds for protocols in the noisy storage model mentioned in this chapter were obtained by F. Dupuis, O. Fawzi, and S. Wehner (Entanglement sampling and applications. *IEEE Transactions on Information Theory*, **61**(2):1093–1112, 2015). The min-entropy splitting lemma is due to J. Wullschleger (Oblivious-transfer amplification. In *Annual International Conference on the Theory and Applications of Cryptographic Techniques*, pp. 555–572. Springer, 2007).

One can in principle implement two-party protocols using the same equipment used to realize BB'84 QKD. Implementations have been reported for bit commitment by N. H. Y. Ng, et al. (Experimental implementation of bit commitment in the noisy-storage model. *Nature Communications*, **3**(1):1–7, 2012) and oblivious transfer by C. Erven, et al. (An experimental implementation of oblivious transfer in the noisy storage model. *Nature Communications*, **5**(1):1–11, 2014). If you want to learn more about the analysis of quantum cryptography protocols in the presence of imperfections encountered in real-world systems, we encourage you to take a look at these papers.

PROBLEMS

11.1 Security from special resources: The PR-box

In this problem we will investigate side-stepping the impossibility of perfectly secure 1-2 OT in some fun ways! To do this we'll imagine that Alice and Bob are given some special resources that they can use during their protocol. We start our investigations with Alice and Bob having access to a very special box – also known as a PR-box or nonlocal box – used in the study of quantum nonlocality.

Imagine thus that Alice and Bob get access to the following shared box. This box takes two inputs: one input bit x from Alice, and one input bit y from Bob. Once both inputs have been given, the box generates a random bit r with $p(r = 0) = p(r = 1) = 1/2$. The box then outputs $b = r + x \cdot y \mod 2$ to Bob and $a = r$ to Alice.

Alice and Bob want to find a protocol to solve the following task: Alice holds a database of two bits x_0 and x_1 and Bob holds some bit y. Bob would like to retrieve the bit x_y from Alice's database. To achieve this, they can use the box above and in addition Alice is allowed to send one bit to Bob. Bob is not allowed to send anything to Alice.

Given these constraints, Alice and Bob have come up with three different protocols:

- **Protocol 1**

Step 1. Alice inputs the sum $x = x_0 + x_1 \mod 2$ into the box.
Step 2. Alice sends the message $m = x_1$ to Bob through her one-bit classical channel.
Step 3. Bob inputs y into the box and obtains b.
Step 4. Bob can now recover $x_y = b + m \mod 2$.

- **Protocol 2**

Step 1. Alice inputs the sum $x = x_0 + x_1 \mod 2$ into the box.
Step 2. Alice sends the message $m = r$ to Bob through her one-bit classical channel.
Step 3. Bob inputs y into the box and obtains b.
Step 4. Bob can now recover $x_y = b + m \mod 2$.

- **Protocol 3**

Step 1. Alice inputs the sum $x = x_0 + x_1 \mod 2$ into the box.
Step 2. Alice sends $m = x_0 + r$ to Bob through her one-bit classical channel.
Step 3. Bob inputs y into the box and obtains b.
Step 4. Bob can now recover $x_y = b + m \mod 2$.

1. Which of these protocols allows Bob to obtain the bit x_y for any value of the bit y?

A box of this type is called a PR-box, named after Popescu and Rohrlich, who invented it in 1993 to investigate nonlocality. It has some very interesting properties! Consider, for example, a situation where Alice and Bob are in possession of a PR-box and they decide to play the CHSH game using this box. That is, as a reminder, Alice receives a question bit x at random such that $p(x = 0) = p(x = 1) = 1/2$ and Bob receives a question bit y at random such that $p(y = 0) = p(y = 1) = 1/2$. Alice and Bob then come up with answer bits a and b respectively (without communicating) and they win the game if $a + b = x \cdot y$ (where addition and multiplication are modulo 2). They will follow any one of the following strategies.

- **Protocol 1**

Step 1. Alice puts her bit x in the box and receives a random bit r.
Step 2. Bob puts a bit $c = 1$ in the box and hence receives $x + r$.
Step 3. Alice sets her answer bit a to $x + r$.

Step 4. Bob sets his answer bit b to $b = y \cdot (x + r)$.

- **Protocol 2**

Step 1. Alice puts a bit $c = 1$ in the box and receives a random bit r.
Step 2. Bob puts a bit y in the box and hence receives $y + r$.
Step 3. Alice sets her answer bit a to $x + r$.
Step 4. Bob sets his answer bit to $b = y \cdot r$.

- **Protocol 3**

Step 1. Alice puts her bit x in the box and receives a random bit r.
Step 2. Bob puts his bit y in the box and hence receives $x \cdot y + r$.
Step 3. Alice sets her answer bit a to r.
Step 4. Bob sets his answer bit to $b = x \cdot y + r$.

Their winning probability p_{win} is given by

$$p_{\text{win}} = \frac{1}{4} \sum_{x,y \in \{0,1\}} \sum_{\substack{a,b \in \{0,1\} \\ a+b=x \cdot y}} p(a,b|y,x) \, .$$

2. For each of the three protocols, calculate p_{win} and conclude which protocol has the highest winning probability.

Slightly shocked by this outcome, Alice and Bob begin to suspect that their box is cheating. Namely, they suspect their box of actually communicating between themselves. Therefore, they resolve to test what are called the *nonsignaling conditions*. The nonsignaling conditions intuitively say that if Alice and Bob input something into the box and receive an output, Alice's output should not depend on Bob's input and Bob's output should not depend on Alice's input. Formally what they will do is the following: Alice generates a bit x at random such that $p(x = 0) = p(x = 1) = \frac{1}{2}$ and inputs it in the box on her side. Similarly, Bob generates a bit y at random such that $p(y = 0) = p(y = 1) = \frac{1}{2}$ and inputs it on his side. They receive output bits a and b from the box, which they will use to check the following conditions:

$$\sum_{b \in \{0,1\}} p(a,b|x,y) = \sum_{b \in \{0,1\}} p(a,b|x,\hat{y}), \quad \forall a,x,y,\hat{y} \in \{0,1\}$$

$$\sum_{a \in \{0,1\}} p(a,b|x,y) = \sum_{a \in \{0,1\}} p(a,b|\hat{x},y), \quad \forall a,y,x,\hat{x} \in \{0,1\}$$

where $p(a,b|x,y)$ is the probability that the box, given input bits x,y, will produce output bits a,b.

3. Does the box violate the nonsignaling conditions?

Now imagine that we get an upgraded version of the same box. This box takes a bit string x of length n as input on Alice's side and a single bit y on Bob's side. It outputs a string $a = r$ to Alice such that $p(r_i = 0) = p(r = 1) = 1/2$ for all $i \in \{1, \ldots, n\}$ and outputs the string b such that $b_i = r_i + x_i \cdot y \mod 2$ for all $i \in \{1, \ldots, n\}$. You can think of this as the "string" version of the PR-box from before.

4. Alice and Bob would like to use this box, and classical communication from Alice to Bob, to design a protocol for some form of 1-2 oblivious transfer. That is, Alice has two strings s_0, s_1, Bob has a bit b, and at the end of the protocol we would like Bob to hold the string s_b while having no knowledge of the other string. Below, you see a list of possible steps in the protocol. Your job is to pick the right steps and put them in the correct order.

For consistency, assume that if Alice and Bob input something in the box, Alice always does so first. Note also that we are looking for a protocol different from the one given in the first part of this problem.

(a) Alice creates two keys $k_0 = r$ and $k_1 = x + r$.

(b) Alice encodes the strings s_0, s_1 as $e_0 = s_0 + k_0$ and $e_1 = s_1 + k_1$.

(c) Alice generates a random n-bit string x, enters it into the box and receives an output string a.

(d) Bob uses his output string to decode the encoded message e_y.

(e) Alice sends the encoded strings to Bob.

(f) Alice inputs the string $s_0 + s_1$ into the box.

(g) Alice generates a random bit string x and encodes the messages as $e_0 = s_0 + x$ and $e_1 = s_1 + r$.

(h) Alice sends $s_0 + s_1$ to Bob.

(i) Bob inputs his bit y into the box and receives an output bit b.

(j) Alice sends $s_0 + s_1 + r$ to Bob.

(k) Alice sends the random string x to Bob.

5. Is this really 1-2 oblivious transfer? *[Hint: can we build bit commitment out of this form of 1-2 oblivious transfer? Why, or why not?]*

12

Further Topics around Encryption

One of the main goals of cryptography is to enable secure communication. In Chapter 2 we discovered the quantum one-time pad, which Alice can use to perfectly hide an n-qubit quantum state ρ using a $2n$-bit classical key k. This scheme has perfect correctness (given k it is possible to perfectly recover ρ from its encryption) and perfect secrecy (without knowledge of k no information at all can be obtained about ρ from its encryption). The only drawback is the necessity to share a large classical secret key. In Chapter 8 and following we developed quantum protocols for secure key distribution which precisely address the question of generating such a key. Problem solved?

Not quite. First of all, there are many topics of further interest in quantum cryptography. Some of these were explored in the previous chapters; moreover, every day researchers identify new, creative ideas of using quantum information to achieve certain tasks securely. Second, the question of encryption itself has many facets, most of which we have not had a chance to explore in depth. For example, what about this long key, is there any chance that we could make it shorter? After all, two rounds of quantum key distribution (QKD) (and more considering all the extra rounds for testing) for each qubit that we want to encrypt is a lot of effort. Could we not re-use the key, or extend it in some way? And what about the requirement that Alice and Bob need to share the same key: What if I want to communicate with a group of friends – how do we all get the same key? In this chapter we start by looking at the possibility of shortening the key used for quantum encryption. We will see an impossibility result, and then open the door to the fascinating world of computational security. Finally, we will discuss a new possibility for quantum encryption, which is known as *certified deletion*: this is the possibility for the encrypter of a secret to request that the ciphertext is provably and irrevocably *erased*!

12.1 The Key Length Requirements for Secure Quantum Encryption

We start by investigating the need for a large classical key to achieve secure quantum encryption. In Chapter 2 we discovered Shannon's theorem, which states that a perfectly secure classical encryption scheme requires keys of length at least as long as the messages. Let's first see how we can extend this result to the quantum case. For this we need a definition of perfect security for quantum encryption.

Definition 12.1.1. *A quantum encryption scheme for n-qubit messages is specified by two families of quantum maps* $\mathrm{Enc}_k : (\mathbb{C}^2)^{\times n} \mapsto (\mathbb{C}^2)^{\times m}$ *and* $\mathrm{Dec}_k : (\mathbb{C}^2)^{\times m} \mapsto (\mathbb{C}^2)^{\times n}$, *where the index k ranges over some set of keys* $\mathcal{K}$. *The scheme is said to be*

- *Perfectly correct if for any n-qubit state* ρ *and any key* $k \in \mathcal{K}$, $\mathrm{Dec}_k \circ \mathrm{Enc}_k(\rho) = \rho$.
- *Perfectly secure if there exists an m-qubit state* σ_0 *such that for any n-qubit state* ρ,

$$\frac{1}{|\mathcal{K}|} \sum_{k \in \mathcal{K}} \mathrm{Enc}_k(\rho) = \sigma_0 . \tag{12.1}$$

In this definition the meaning of the correctness requirement is clear: encryption followed by decryption with the same key should return the initial quantum state. Observe that, for our definition to be as general as possible, we allow encryption schemes that increase the number of qubits of the message; in contrast, the quantum one-time pad has $m = n$.

For security, the requirement is that when the key k is chosen uniformly at random in $\mathcal{K}$ and hidden from the adversary, then the encrypted state is independent of the message: the state σ_0 is some fixed state that may depend on the scheme but not on ρ. Intuitively a scheme satisfying this condition would indeed deserve the name "perfectly secure," and we will justify this intuition later. Note that the quantum one-time pad does satisfy the condition (12.1), with $\sigma_0 = \frac{1}{2^n}\mathbb{I}$.

Exercise 12.1.1 Call a quantum encryption scheme *super-perfectly secure* if there exists a state σ_0 such that for all quantum states ρ_{AE}, where A is n qubits and E is arbitrary,

$$\frac{1}{|\mathcal{K}|} \sum_{k \in \mathcal{K}} \left(\mathrm{Enc}_k \otimes \mathbb{I}_E\right)(\rho_{AE}) = \sigma_0 \otimes \rho_E .$$

Clearly this definition is at least as strong, and in particular if E is empty we recover Definition 12.1.1. Intuitively "super-perfect security" is meant to guard against a situation in which the malicious Eve has some information correlated with Alice's message. In this case we want to make sure that the ciphertext no longer contains any such correlation.

Show carefully that any perfectly secure scheme is automatically super-perfectly secure.

Let's show that the definition of perfect security in Definition 12.1.1 indeed requires long keys.

Theorem 12.1.1 *Let* $(\mathrm{Enc}, \mathrm{Dec})$ *be a perfectly correct and secure quantum encryption scheme for n qubits. Then* $|\mathcal{K}| \geq 2^{2n}$.

Proof For simplicity in the proof we consider a scheme that encrypts n qubits into n qubits (i.e. $m = n$) and is such that $\sigma_0 = \frac{1}{2^n}\mathbb{I}$. The general argument is a little more technical but leads to the same bound. Let $k = 1, \ldots, N$ index the possible keys, let

p_k be the probability that the k-th key is chosen, and let U_k be the encoding unitary on key k, i.e. $\text{Enc}_k(\rho) = U_k \rho U_k^\dagger$. (This is where we use that the scheme encrypts n qubits into n qubits; in general we would have to consider arbitrary channels.) We also define

$$\text{Enc}(\rho) = \frac{1}{N} \sum_{k=1}^{N} \text{Enc}_k(\rho) = \frac{1}{N} \sum_{k=1}^{N} U_k \rho U_k^\dagger .$$

Our goal is to show that necessarily $N \geq 2^{2n}$. So let's suppose for contradiction that $N < 2^{2n}$. The main observation is that when averaged over a random key the encoding map has exactly the same behavior as the one-time pad:

$$\mathcal{E}(\rho) = \frac{1}{2^{2n}} \sum_{(k_1', k_2')} X^{k_1'} Z^{k_2'} \rho \left(X^{k_1'} Z^{k_2'} \right)^\dagger = \frac{1}{2^n} \mathbb{I} ,$$

where k_1', k_2' range over n-bit strings and $X^{k_1'}$ denotes an X operator on the qubits associated with entries of k_1' that are equal to 1; similarly for $Z^{k_2'}$. We see that perfect security (together with our simplifying assumption that $\sigma_0 = \frac{1}{2^n}\mathbb{I}$) requires that $\text{Enc}(\rho) = \mathcal{E}(\rho)$ for all ρ. From this it is possible to show, using a notion of unicity of the Kraus decomposition of a quantum channel (see Box 3.1 – we omit the details; to show it, consider the effect of encrypting one half of a maximally entangled state using either scheme), that there must exist a $2^{2n} \times 2^{2n}$ unitary matrix A such that for each k,

$$\sqrt{p_k} U_k = \sum_{(k_1', k_2')} A_{k,(k_1', k_2')} \frac{1}{\sqrt{2^{2n}}} X^{k_1'} Z^{k_2'} .$$

Here the indexing for the rows of A makes sense because we assumed that $N \leq 2^{2n}$. Using that the matrices $X^{k_1'} Z^{k_2'}$ are orthonormal with respect to the normalized trace inner product $\langle A, B \rangle \mapsto \frac{1}{2^n} \text{tr}(A^\dagger B)$ we can compute

$$p_k = \frac{1}{2^n} \langle \sqrt{p_k} U_k, \sqrt{p_k} U_k \rangle$$

$$= \frac{1}{2^{2n}} \sum_{(k_1', k_2')} \left| A_{k,(k_1', k_2')} \right|^2$$

$$\leq \frac{1}{2^{2n}} ,$$

where the second line uses orthonormality and the last line uses that the rows of A have euclidean norm at most 1 since A is unitary. So

$$1 = \sum_{k=1}^{N} p_k \leq \frac{N}{2^{2n}} ,$$

from which we deduce that $N \geq 2^{2n}$, as desired. ∎

So, perfectly secure quantum encryption schemes require long keys! It seems like Alice and Bob will have to bite the bullet, and exchange long DVDs full of random bits to encrypt their quantum message. Or do they? Given Theorem 12.1.1, the only option is to modify our definition of perfectly secure encryption. Can we relax it in such a way that it remains meaningful, but allows shorter keys? In the next subsections we describe two ideas on how this can be done. First, we can relax the notion that encryptions of distinct messages are perfectly indistinguishable. We discuss this possibility in the next section. Second, we can weaken our security requirement to only require that encryptions of distinct messages look similar to "reasonably powerful" adversaries. We discuss this in Section 12.1.2. Finally, we conclude by giving a quick overview of the notion of *public-key* encryption.

12.1.1 Approximate Encryption

As we saw, the requirement of perfect security inevitably imposes long keys. But do we really need perfect security? Suppose, for example, that we could construct a scheme that satisfies the weaker condition that for any ρ,

$$\left\| \frac{1}{|\mathcal{K}|} \sum_{k \in \mathcal{K}} \mathrm{Enc}_k(\rho) - \sigma_0 \right\|_1 \leq \varepsilon , \tag{12.2}$$

where ε is some very small quantity. By the interpretation of the trace distance, this would immediately imply that no adversary, given either $\mathrm{Enc}_k(\rho_0)$ or $\mathrm{Enc}_k(\rho_1)$, would be able to distinguish these two states with an advantage larger than ε. If ε is, say, 2^{-80}, this seems pretty safe; the adversary would have to see 2^{80} copies of our encryptions to reliably distinguish the messages that they encrypt. A scheme that satisfies the weaker requirement (12.2) instead of (12.1) is called an *ε-approximate encryption scheme*. Do there exist ε-approximate encryption schemes with short keys?

A simple idea for constructing an approximate encryption scheme is to start with the quantum one-time pad but only use a subset of all the possible keys. Let $\mathcal{K} \subseteq \{0,1\}^{2n}$ denote a subset. How small can we find a $\mathcal{K}$ such that the equation (12.2) holds, where Enc_k is the quantum one-time pad? It turns out that the answer is, roughly, $\log |\mathcal{K}| = n + O(\log n) + O(\log(1/\varepsilon))$. That is, by considering approximate encryption and even asking for an ε that is almost exponentially small in n, we can get a saving of a factor 2 in the key length, bringing it down to almost the same length as the *classical* one-time pad.

So how do we choose the keys? Without going into details, it is possible to show that a randomly chosen set $|\mathcal{K}|$ of this size will work. Moreover, there also are explicit constructions of small sets of keys that will work, using techniques from the area of classical error-correcting codes. Finally, the size $|\mathcal{K}| \approx 2^n$ is optimal, i.e. no smaller set will give security.

QUIZ 12.1.1 *Imagine applying the quantum one-time pad to a single qubit $|\psi\rangle$, such that we choose only one of three possible keys, corresponding to doing nothing ($k = 1$), applying an X operation ($k = 2$), or a Z operation ($k = 3$). In other words, we never apply XZ. Under this scheme, an encryption of $|0\rangle\langle 0|$ and an encryption of $|1\rangle\langle 1|$ have trace distance*

(a) 0
(b) $\frac{1}{3}$
(c) $\frac{1}{2}$
(d) 1

There is an important caveat to keep in mind with approximate encryption, which is explored in the next exercise.

Exercise 12.1.2 Suppose we are given a message $|m\rangle$ that is chosen uniformly at random in $\{0,1\}^n$, but such that an adversary, Eve, holds a copy of $|m\rangle$. That is, we imagine that the initial state of Alice and the eavesdropper is $|\phi\rangle = 2^{-n}\sum_{m\in\{0,1\}^n}|m\rangle_A\,|m\rangle_E$. Show that if Alice encrypts her message using the quantum one-time pad, then the ciphertext is completely unknown to Eve, i.e.

$$\frac{1}{2^{2n}}\sum_{(k_1,k_2)}\left(\mathrm{Enc}_{(k_1,k_2)}\otimes\mathbb{I}_E\right)\left(|\phi\rangle\langle\phi|\right)=\rho_A\otimes\rho_E\,,\qquad(12.3)$$

for some density matrices ρ_A and ρ_E that you will determine.

Exercise 12.1.3 Imagine now that Alice decides to encrypt her message using an *approximate* encryption scheme, such that the total number of keys is $K\leq\frac{1}{2}2^{2n}$. Show that in this case it must be that the trace distance between the left-hand side and the right-hand side in (12.3) is at least some constant.

The exercises show that while approximate encryption may be good enough to encode messages that are in tensor product with the environment (the adversary), one has to be careful that it is *not* sufficient to destroy *correlations*, as an adversary that has some quantum correlation with Alice's message before encryption may retain some quantum correlation with the ciphertext after encryption. This is in contrast to perfect encryption, which as you showed in Exercise 12.1.2 always perfectly destroys all correlations.

A saving of a factor 2 in the key length might not seem worth the trouble. To save more, we consider a further relaxation of the security definition, to *computational* security.

12.1.2 Computational Security

We already encountered the idea that security can be based on computational, as opposed to physical, assumptions twice in this book – first in Chapter 7 when discussing authenticated channels, and then in Chapter 10 when discussing computationally secure commitments based on the notion of a pseudorandom generator. Formally, computational security can be defined in a similar way as physical security, through an appropriate *security game*. Let's see in more detail how this can be done for the case of encryption. Towards this we introduce a game between a *challenger*, Charlie, and an *adversary*, Eve. In the game, the challenger always plays honestly, while the adversary tries to win as best she can by optimizing her strategy.

1. Charlie generates parameters for the encryption scheme, i.e. he selects a key $k\in\mathcal{K}$ uniformly at random.
2. Eve prepares a quantum state ρ_{ME} of her choice, where M is a register the size of a plaintext message, and E is a quantum register that Eve keeps to herself. Eve sends the part of the quantum state in register M to Charlie.

3. Charlie selects a uniformly random $c \in \{0, 1\}$. If $c = 0$ then Charlie encrypts M:

$$\rho'_{CE} = (\text{Enc}_k \otimes \mathbb{I}_E)(\rho_{ME}),$$

and sends register C, now containing the ciphertext, to Eve. If $c = 1$ then Charlie first replaces the contents of M by a "dummy" message $|0\rangle\langle 0|_M$, encrypts the dummy message into register C, and sends C back to Eve. In this case,

$$\rho'_{CE} = (\text{Enc}_k \otimes \mathbb{I}_E)(|0\rangle\langle 0|_M \otimes \rho_E).$$

4. Eve produces a guess $d \in \{0, 1\}$ and sends it to Charlie.
5. Charlie declares that the adversary has won if and only if $d = c$.

To understand this "game" let's first go through it in the case where Eve chooses the register E to be empty. Let ρ_M be the state prepared by Eve at step 2. Then, in the case $c = 0$ the state sent back to the adversary at step 3 is precisely $\frac{1}{|\mathcal{K}|} \sum_k \text{Enc}_k(\rho)$, while if $c = 1$ it is $\frac{1}{|\mathcal{K}|} \sum_k \text{Enc}_k(|0\rangle\langle 0|)$. As we saw in Chapter 5, by definition of the trace distance Eve's maximum success probability to distinguish these two states is exactly

$$\frac{1}{2} + \frac{1}{2} \left\| \frac{1}{|\mathcal{K}|} \sum_{k \in \mathcal{K}} \text{Enc}_k(\rho) - \frac{1}{|\mathcal{K}|} \sum_{k \in \mathcal{K}} \text{Enc}_k(|0\rangle\langle 0|) \right\|_1.$$

We see that if Enc is ε-approximate secure in the sense of (12.2), then by the triangle inequality the adversary's success probability is at most $1/2 + \varepsilon$. Conversely, if the adversary's success probability is at most $1/2 + \varepsilon$, then by defining $\sigma_0 = \frac{1}{|\mathcal{K}|} \sum_{k \in \mathcal{K}} \text{Enc}_k(|0\rangle\langle 0|)$ we get that (12.2) holds with right-hand side 2ε. In other words, the two statements

(a) The scheme Enc is ε-approximate secure,
(b) Eve's maximum success probability in the game defined above, when she is restricted to not use any E and Charlie plays honestly, is at most ε,

are equivalent (up to a factor 2 in the ε's). Using such a "security game," as opposed to an equation that needs to be satisfied, is an intuitive yet mathematically rigorous (and fun!) way of making a security definition.

The same reasoning applies when E is included, and corresponds to a strengthening of (12.2) that takes into account correlations; this strengthening is motivated by the discussion at the end of the previous section, which shows that the definition is indeed strictly stronger (see Exercise 12.1.3).

 Remark 12.1.2 The choice of encrypting $|0\rangle\langle 0|_M$ in the case $c = 1$ in the definition of the security game is arbitrary. The point is that, whatever message the adversary chooses to give to the challenger, she should not be able to distinguish an encryption of it from an encryption of some fixed message, such as $|0\rangle\langle 0|_M$. While we could have directly required that encryptions of 0 are indistinguishable from encryptions of 1, by letting the adversary choose the message we give her more power and hence obtain a potentially stronger notion of security (for example, when there are more than two possible plaintexts).

Exercise 12.1.4 Show that a quantum encryption scheme (Enc, Dec) is super-perfectly secure (in the sense of Exercise 12.1.1) if and only if the maximum success probability for the adversary in the security game is exactly $1/2$. *[Don't forget to show both directions of the "if and only if"! The "if" part requires more work, so start with "only if."]*

The security game gives us a different way to think about the security definition for quantum encryption. Now we can ask the following question: What if we only care about certain types of adversaries, and we only need to be secure against them? We can then specialize the security game, and only ask about the success probability of the class of adversaries we care about. In a way we already did this when we considered two versions of the security definition, when E is required to be empty or not. And in the previous section we saw that if E is empty, and we allow success probabilities up to $1/2 + \varepsilon$, then we can use shorter keys. So, placing restrictions on the adversary can help us create more efficient schemes! Of course, we always have to remember that the scheme is only secure up to the security definition, not any further.

A restriction we could consider on the adversary is to have a bounded quantum memory, similar to the model considered in Chapter 11. Another alternative would be to imagine that the adversary only has a certain amount of computational time to invest in breaking the scheme. This amount of time should of course be allowed to be much bigger than the space or time it takes to honestly encrypt and decrypt, but perhaps it is not infinite either (because who has infinite time?).

Investigating these questions can lead to very large improvements in the performance of encryption schemes as well as other cryptographic primitives. For example, it is possible to show that secure encryption against computationally bounded adversaries is possible with a key length that is only polylogarithmic, as opposed to linear, in the length of the message (for long enough messages)! Unfortunately, introducing such schemes would take us far beyond the scope of this book. We give a very informal description here and refer you to the chapter notes for pointers on where to learn more.

The main idea behind computational security is to postulate that a certain computational problem is *hard on the average* and construct an encryption scheme such that the only case when an adversary can win in the security game is if the adversary also has the ability to solve an instance of the computational problem that is related to the key. A typical computational problem used in classical cryptography is the problem of factoring. However, since this problem is not hard for quantum computers, it is not a good problem on which to base quantum encryption schemes. Instead, some other computational problems have been used, including problems related to error-correcting codes (finding a minimum-weight codeword) and integer lattices (finding a closest lattice vector). Security of the scheme is proven by *reduction*: we show that if Eve is an adversary who succeeds in the security game with too large probability, say more than $1/2 + \varepsilon$, then the same Eve could be used to break the computational problem. So if we assume that the latter is hard, then the scheme is secure.[1]

---

1 Sometimes we think of this as a "win-win" notion: *either* the scheme is secure, *or* someone has found a new algorithm for a hard problem!

Beyond efficiency savings, computational security can also lead to conceptually different schemes. The most important family of schemes beyond private-key encryption is called *public-key* encryption, and we briefly discuss it in the next section.

12.1.3 Public-Key Encryption

The encryption schemes we have considered so far all work under the same natural premise: that Alice and Bob should both share the same key, and that this key can be used to either encrypt or decrypt. This is natural because, since decryption is the inverse of encryption, it makes sense to define both operations from the same piece of information, the key. Hence such schemes are usually called "private-key" quantum encryption schemes, where the term "private" refers to the fact that the scheme is only secure as long as the same key k remains private to the sender and receiver.

A striking observation that revolutionized cryptography in the 1990s is that encryption and decryption do not need to be treated symmetrically. From the security standpoint, it is crucial that only the authorized party is able to decrypt a ciphertext, but it is a priori not a problem that anyone would be able to encrypt.[2] From the mathematical standpoint there is also a natural asymmetry: some functions are easy to compute in one direction, but hard in the other. A prime example is multiplication (easy), whose inverse is factoring (hard). The idea of public-key cryptography is to leverage this mathematical asymmetry to implement an encryption scheme such that anyone can encrypt, but only the trusted user can decrypt.

More precisely, in *public-key encryption* the keys come in pairs (sk, pk) where sk is a *secret key* and pk is a *public key*. As their name indicates, the secret key is meant to be kept private (e.g. only Bob has it) while the public key can be made publicly available (Bob can publish it on his website, or include it as part of his email signature). Encryption can be done using only pk, while decryption requires sk. Therefore, the security game is exactly the same as the one in the previous section, except that the challenger generates a pair (sk, pk), keeps sk to themselves, and gives pk to the adversary. Public-key encryption schemes necessarily rely on computational security (because from an information-theoretic point of view, the public key pk uniquely specifies a private key sk, and so such a scheme can always be broken in principle), and can be implemented using similar assumptions to the ones discussed in the previous section.[3]

Since we are able to use QKD to distribute private keys, do we really need public-key cryptography? There are many reasons why we do. First of all, implementing QKD remains technologically challenging. It is already hard when there are two parties. However, a major benefit of public-key cryptography is the ability for *any* user to publicly send a message to Bob by encrypting using his public key, while having the guarantee that only Bob will be able to decrypt the message; in this sense the asymmetry helps. This use case is also relevant for quantum cryptography, so we expect that

---

2 One might nevertheless worry about the possibility of "spoofing" a message. This problem is solved by *authentication*, which is a separate technique that can be combined with encryption (see Section 7.5).

3 In general, the mathematical assumptions required to implement a public-key encryption scheme tend to be more demanding than the assumptions required to implement private-key encryption, because public-key encryption requires more structure.

both private-key schemes, for which encryption and decryption of a single message is typically very fast, and public-key schemes, which require longer encryption and decryption times but are much more efficient in settings where there are many users, since ciphertexts can be "re-used," can coexist.

12.2 Encryption with Certified Deletion

Even when Alice only wants to transmit classical messages to Bob, there may be a use for encryption using quantum ciphertexts. Suppose, for example, that Alice has some private information, such as the details of a future financial transaction that she wishes to have executed. Alice could share a key with her bank, encrypt the transaction details using a classical one-time pad, and send the resulting ciphertext to the bank. The bank can then use its own copy of the key to decrypt the information and execute the transaction. However, imagine now that this is a timed transaction: Alice would like to be able to send the ciphertext to the bank, and then tell the bank to store the ciphertext and wait for her next signal. A few days later Alice can decide that she wants to proceed with her transaction, in which case she would tell the bank to "go ahead," i.e. decrypt and execute. But what if Alice has changed her mind? Wouldn't it be nice if there was a special "delete" signal that she could send to request that the bank *delete* her ciphertext, without having learned any information at all about Alice's aborted transaction?

Another example is with data protection. Nowadays, we are constantly asked to submit personal information, such as our address or birth date, to some websites. In certain countries the websites are legally required to delete this information shortly after the transaction has been completed. But do they actually do this? How could one check that they have not kept a copy in some secret vault of theirs?

Classically both of these tasks are impossible to realize securely. If the bank can either decrypt or delete then it can also do both: simply copy the ciphertext, decrypt one copy and use the other copy to "prove" deletion. Similarly, there is no way to prevent a website from storing a copy of any classical information we send to it.

However, if the information is quantum then by the no-cloning principle it is no longer clear that this strategy can be applied. So, can certified deletion be realized by using quantum information? Let's start by introducing a security definition that captures this new property. Afterward we'll investigate a scheme that satisfies the definition.

12.2.1 Security Definition

For simplicity let's focus on defining security for schemes that encrypt a single classical bit at a time. As for a standard encryption scheme, there should be a key generation procedure, together with encryption and decryption procedures. In addition to these we now include two additional procedures, the procedure Del that given a ciphertext "deletes" it and obtains a "proof of deletion" π, and a "verification of deletion" procedure VerDel that checks that deletion has been produced correctly. Here is a more formal definition. (The first time you read the definition you can ignore the role of the "security parameter" λ. This quantifies the security of the scheme with respect to deletion, and we'll explain it later.)

Definition 12.2.1. *Let* $\lambda \geq 1$ *be an integer. A* quantum encryption scheme with certified deletion *with security* $2^{-\lambda}$ *is specified by the following procedures:*

1. *A classical probabilistic key generation procedure* $k \leftarrow \mathrm{Gen}(1^\lambda)$ *that generates a secret key* k.[4]
2. *An encryption procedure* $(c, dk) \leftarrow \mathrm{Enc}_k(m)$ *that given a plaintext* $m \in \{0, 1\}$ *returns a ciphertext* c *together with a "deletion key"* dk *(which may depend on* m*). The ciphertext* c *may contain a quantum component, that is, it can be a cq-state.*
3. *A decryption procedure* $m \leftarrow \mathrm{Dec}_k(c)$.
4. *A deletion procedure* $\pi \leftarrow \mathrm{Del}(c)$ *that given a ciphertext* c *produces a classical proof of deletion* π. *Note that* Del *does not require the key* k.
5. *A verification of deletion procedure* $v \leftarrow \mathrm{VerDel}_k(dk, \pi)$ *that takes as input a deletion key* dk *and a deletion proof* π *and returns a bit* $v \in \{0, 1\}$, *where* 1 *stands for "accept" and* 0 *stands for "reject."*

Let's now discuss security requirements. First, as usual the scheme is called *perfectly correct* if for every key k and plaintext m, $\mathrm{Enc}_k(m) = (c, dk)$ implies that $\mathrm{Dec}_k(c) = m$. In addition, we add the requirement that for any proof of deletion that is generated by the correct deletion procedure, $\pi = \mathrm{Del}(c)$, it holds that $\mathrm{VerDel}_k(dk, \pi) = 1$.

For security, we first require perfect security for the encryption scheme, i.e. condition (12.1).[5] What about deletion? Informally, we would like that for any "adversary" holding a ciphertext c, if the adversary successfully "proves deletion" then it becomes impossible for them to recover the plaintext m associated with c, even if they are later given the key (of course, if they don't have the key, then encryption security guarantees that they can't recover m). There are many quotes in this sentence! To formalize the intuition we introduce a security game. This game is of the same type as the one in Section 12.1.2, and once again it is played between an honest *challenger*, Charlie, and a possibly malicious *adversary*, Eve. The idea is that a scheme will be called a *certified deletion* (encryption scheme) with security λ if and only if no adversary can win in the game with probability much larger than $2^{-\lambda}$.

1. Charlie selects a key $k \leftarrow \mathrm{Gen}(1^\lambda)$.
2. Eve prepares an arbitrary quantum state ρ_{ME}, where M is a classical register the size of a plaintext message and E is a quantum register that the adversary keeps to herself. Eve sends register M to Charlie.

---

4 Recall that the notation $X \leftarrow \mathrm{PROC}(Y)$ means that we use the variable X to denote the outcome of running the procedure PROC on input Y.
5 More generally, we could consider a weaker security notion of the kind considered in the first part of this chapter. For simplicity, and because we can, we focus on the stronger notion.

3. Charlie selects a $c \in \{0,1\}$ uniformly at random.

- If $c = 0$ then Charlie encrypts M:

$$\rho'_{CDE} = (\mathrm{Enc}_k \otimes \mathbb{I}_E)(\rho_{ME}) \, ,$$

and sends register C, now containing the (possibly quantum) ciphertext, to Eve. Charlie keeps register D, which contains the (classical) deletion key dk.
- If $c = 1$ then Charlie first replaces the contents of M by a "dummy" message $|0\rangle\langle 0|_M$, encrypts the dummy message into register C, and sends C back to Eve. As before, Charlie keeps register D.

4. Eve sends a "proof of deletion" $\pi \in \{0,1\}^\lambda$ to Charlie.
5. Charlie sends the secret key k to Eve.
6. Eve produces a guess $d \in \{0,1\}$.
7. Charlie declares that Eve has won if and only if $d = c$ *and* $\mathrm{VerDel}_k(dk, \pi) = 1$.

We should convince ourselves that this game captures the intuition of the "deletion security" that we want for our encryption scheme. Compared to the game in Section 12.1.2, there are two key differences. First, Eve is asked for some additional information: at step 4, she has to return a "proof of deletion" π. What we imagine here is that Charlie has asked for his ciphertext to be deleted, and Eve is supposed to comply by sending the proof π, which is checked in the last step. Now, the validity of this proof is *supposed* to guarantee that Eve has deleted the ciphertext. How do we check this? Here comes the second difference: at the next step, Charlie *reveals* the secret key k to the adversary! We say that Eve wins the game if, first of all, her "proof of deletion" is accepted, and second, she is able to discover which plaintext was encoded by Charlie. Note that if the deletion that is supposed to have happened at step 4 does not affect the ciphertext (or the ciphertext can be copied) then by correctness of the encryption scheme it is easy to win in this game, simply by decrypting c once k is given (and, for example, choosing $\rho_M = |1\rangle\langle 1|$ in step 2, so that the challenger can indeed distinguish between the cases $c = 0$ and $c = 1$). So, for any scheme that satisfies this definition, clearly there must be something interesting going on: decryption is possible before π is produced, but no longer after; which is exactly what we want.

The next exercise shows that in the security game it is essential that the key is revealed to Eve only *after* the proof of deletion has been obtained. Otherwise, there will always be an adversary that is able, given the key, to produce both a valid deletion certificate and a correct guess for the bit c.

Exercise 12.2.1 Show that if steps 4 and 5 are inverted then for any perfectly correct certified deletion scheme there is an adversary that succeeds with probability 1 in the security game.

12.2.2 A Construction

As we hinted earlier, the task of encryption with certified deletion is closely related to the notion of no-cloning. This is because if the ciphertext is cloneable then it will always be possible to win in the security game; hence, the existence of an encryption scheme with certified deletion implies that there exist quantum states that cannot be cloned. Based on this observation, a natural idea for implementing a certified deletion scheme would be to include an "uncloneable" component in our ciphertexts. For example, we could add a randomly generated Wiesner quantum money state (see Chapter 3) to each ciphertext. This is a good idea but by itself it is unlikely to work, as we must somehow tie the part of the ciphertext that contains information about the plaintext to the "uncloneable" part.

We now introduce a scheme that does just that. To describe the scheme, we identify a string $\mathcal{I} \in \{0,1\}^\lambda$ with the subset $\mathcal{I} \subseteq \{1,\ldots,\lambda\}$ which is the list of positions at which $\mathcal{I} = 1$. We also recall the notation $|x\rangle_\theta = H^\theta |x\rangle$ for the BB'84 states, where $x, \theta \in \{0,1\}$.

1. The key space is $\mathcal{K} = \{0,1\} \times \{0,1\}^\lambda$. The key generation procedure returns a uniformly random $k = (u, \mathcal{I})$ such that $u \in \{0,1\}$ and $\mathcal{I}$ is a subset of $\{1,\ldots,\lambda\}$.
2. Given a message $m \in \{0,1\}$ and a key $k = (u, \mathcal{I})$, $\mathrm{Enc}_k(m)$ generates $x \leftarrow \{0,1\}^\lambda$ uniformly at random and returns the ciphertext $c = (c', |\phi\rangle)$, where $c' = m \oplus u \oplus_{i \in \overline{\mathcal{I}}} x_i$ and $|\phi\rangle = |x_1\rangle_{\mathcal{I}_1} \cdots |x_\lambda\rangle_{\mathcal{I}_\lambda}$, together with the deletion key $dk = x$.
3. Given a ciphertext $c = (c', |\phi\rangle)$ and a key $k = (u, \mathcal{I})$, Dec_k measures $|\phi\rangle$ in the standard basis to obtain a string y and returns $m = c' \oplus u \oplus_{i \in \overline{\mathcal{I}}} y_i$.
4. Given a ciphertext $c = (c', |\phi\rangle)$, Del measures $|\phi\rangle$ in the Hadamard basis to obtain a string z and returns $\pi = z$.
5. Given $k = (u, \mathcal{I})$, $\mathrm{VerDel}_k(\pi, dk)$ returns 1 if and only if $\pi_i = dk_i$ for all $i \in \mathcal{I}$.

To understand a scheme it is always useful to start by ignoring all the parts included to guarantee security and focus on checking that the scheme is correct. Let's do this. For any message $m \in \{0,1\}$, according to item 2 the associated ciphertext takes the form $c = (c', |\phi\rangle)$ where $c' = m \oplus u \oplus_{i \in \overline{\mathcal{I}}} x_i$ and $|\phi\rangle = |x_1\rangle_{\mathcal{I}_1} \cdots |x_\lambda\rangle_{\mathcal{I}_\lambda}$. When the decryption procedure measures $|\phi\rangle$ in the standard basis to obtain y, by definition we have that $y_i = x_i$ whenever $\mathcal{I}_i = 0$, because then $|x_i\rangle_{\mathcal{I}_i} = |x_i\rangle$. Since according to our notation $\mathcal{I}_i = 0$ is equivalent to $i \notin \mathcal{I}$, we get that

$$c' \oplus u \oplus_{i \in \overline{\mathcal{I}}} y_i = c' \oplus u \oplus_{i \in \overline{\mathcal{I}}} x_i = m .$$

So the scheme is perfectly correct. This correctness follows from the fact that, for decryption, the "$\mathcal{I}$" part of the private key tells us exactly which qubits of $|\phi\rangle$ were encoded in the standard basis and contain information that should be used for decryption.

The next step is to argue that the scheme is perfectly secure as an encryption scheme. To see this we can think of encryption as taking place in two steps. First, the encrypter chooses a random x and $\mathcal{I}$ and returns the state $|\phi\rangle$. Clearly this is completely independent of the message and leaks no information whatsoever about it. Second, the encrypter privately computes $m' = m \oplus_{i \in \overline{\mathcal{I}}} x_i$, which does depend on the message, and returns $m' \oplus u$ for a uniformly random u. Since adding u acts like a classical one-time pad, this part of the ciphertext is also perfectly secure and independent of $|\phi\rangle$. Hence, for any single-bit message m it holds that

$$\frac{1}{|\mathcal{K}|} \sum_{k \in \mathcal{K}} \mathrm{Enc}_k(m) = \frac{1}{2} \mathbb{I} \otimes \sigma_0 ,$$

where the first $\frac{1}{2}\mathbb{I}$ is the one-time padded m' and σ_0 represents a uniform mixture over all possible $|\phi\rangle$ (which you can check equals $2^{-\lambda}\mathbb{I}$, where the identity is over λ qubits). Therefore the scheme is perfectly secure.

It remains to show the certified deletion property! This requires more work, and we devote the next section to it.

> **QUIZ 12.2.1** *Suppose that in the construction we choose $\lambda = 1$. Is the scheme still a perfectly secure encryption scheme?*
>
> **(a)** *Yes*
> **(b)** *No*

12.2.3 Proof of Certified Deletion Property

We first consider the case where $\lambda = 1$. Let's rewrite the security game from Section 12.2.1, when specialized to our construction from the previous section, and the choice $\lambda = 1$. For reasons that will become clear later, we relabel $\mathcal{I}$ as $\theta \in \{0,1\}$. We obtain the following game.

1. Charlie selects $u \in \{0,1\}$ and $\theta \in \{0,1\}$ uniformly at random.
2. Charlie selects $c \in \{0,1\}$ uniformly at random. He selects $x \in \{0,1\}$ uniformly at random and creates $|\phi\rangle = |x\rangle_\theta$. If $\theta = 0$ Charlie sets $c' = c \oplus u \oplus x$. If $\theta = 1$ Charlie sets $c' = c \oplus u$. Charlie returns $c = (c', |\phi\rangle)$ to Eve.
3. Eve sends a deletion proof $\pi \in \{0,1\}$ to Charlie.
4. Charlie sends $k = (u, \theta)$ to Eve.
5. Eve produces a guess $d \in \{0,1\}$.
6. Charlie declares that Eve has won if and only if $d = c$ *and* (if $\theta = 1$ then $\pi = x$).

Remark 12.2.1 In this description we did not let Eve choose the plaintext m and we also ignored the possibility for her to prepare a plaintext m that is correlated with some quantum information in register E. It is a good exercise to convince yourselves that both changes are without loss of generality, i.e. they do not reduce Eve's power. More formally, if any adversary can succeed in the earlier security game with probability ε then they can also succeed in this new security game with probability ε. Note that this simplification relies on the fact that we are considering a scheme that encrypts a single classical bit only.

In the next step we are going to give more power to Eve. First of all, we will only check the condition that $d = c$ in the case when $\theta = 0$. Note that when $\theta = 0$, then Eve receives $c' = c \oplus u \oplus x$, and she also receives u at step 4. So the probability that she guesses c correctly is exactly the same as the probability that she guesses x (since she can convert from one to the other using $x = c \oplus (c' \oplus u)$, where she always has both c' and u). So, in this step we replace item 6 by

6. Charlie declares that Eve has won if and only (if $\theta = 0$ then $d = x$) *and* (if $\theta = 1$ then $\pi = x$).

This new version of the game can only be easier for Eve. Finally, we observe that in this new version u no longer plays any role at all, so we can simply remove it. Slightly reorganizing the description of the steps we arrive at the following game.

1. Charlie selects $x \in \{0, 1\}$ and $\theta \in \{0, 1\}$ uniformly at random. He creates $|\phi\rangle = |x\rangle_\theta$ and sends $|\phi\rangle$ to Eve.
2. Eve sends a deletion proof $\pi \in \{0, 1\}$ to Charlie.
3. Charlie sends θ to Eve.
4. Eve produces a guess $d \in \{0, 1\}$.
5. Charlie declares that Eve has won if and only (if $\theta = 0$ then $d = x$) *and* (if $\theta = 1$ then $\pi = x$).

With these simplifications in place, our goal is to show that no adversary can succeed in the game with probability that is too close to 1: the smaller a bound we can show the better. To do this we apply a similar proof strategy to our analysis of the BB'84 protocol in Chapter 8. Specifically, we start by considering a purified version of the game, as follows.

1. Eve is split in two parts, B and E. Eve prepares an arbitrary state ρ_{CBE}, where C is a single qubit, and sends C to Charlie.
2. Charlie selects a $\theta \in \{0, 1\}$ uniformly at random. He measures C in the basis indicated by θ to obtain an $x \in \{0, 1\}$.
3. B sends $\pi \in \{0, 1\}$ to Charlie.
4. Charlie sends θ to E, who responds with a $d \in \{0, 1\}$.
5. Charlie declares that Eve has won if and only (if $\theta = 0$ then $d = x$) *and* (if $\theta = 1$ then $\pi = x$).

Once again, this new, "purified" game gives more power to the adversary. To see why, observe that Eve could first prepare a state of the form $|\text{EPR}\rangle_{CE} \otimes |0\rangle_B$ and send C to Charlie; then, she could compute π from E and copy it to register B, and leave the post-measurement state in E until she receives θ. It's not hard to see that any adversary using a strategy of that form succeeds in the purified game with the same probability as in the nonpurified game. As usual, this is because measuring an EPR pair in any basis has the effect of collapsing both halves of the EPR pair to the same post-measurement state.

To conclude the analysis of the purified game we make use of an entropic uncertainty relation that is a generalization of the first inequality in (8.13). This relation can be stated as follows. For any state ρ_{ABE} where A is a single qubit, it holds that

$$\text{H}_{\max}(X_A | B) + \text{H}_{\min}(Z_A | E) \geq 1 \, , \tag{12.4}$$

where X_A is a random variable that denotes the outcome of a measurement of A in the Hadamard basis, and Z_A the outcome of a measurement of A in the standard basis.

The second entropy, $H_{min}(Z_A|E)$, we are already familiar with, and this is equal to $-\log(P_{guess}(Z_A|E))$, where $P_{guess}(Z_A|E)$ is exactly the maximum probability with which the adversary can succeed in the "(if $\theta = 0$ then $d = x$)" part of the security game. The quantity $H_{max}(X_A|B)$ is the *max-entropy*. This is defined a little bit differently from the min-entropy. For our purposes we only need to consider the case where X_A and B are both a single classical bit, since we may as well consider B to contain the proof $\pi \in \{0,1\}$. If we let $p(x,b)$ denote the joint distribution of two bits x and b, then

$$H_{max}(X|B) = \log\left(\sum_b \Pr(B = b)\left(\sum_x \sqrt{\Pr(X = x|B = b)}\right)^2\right).$$

While this expression may seem a little more complicated than we'd like, from a qualitative point of view we can observe that H_{max} is only close to 1 if the expression inside the log is close to 2, which requires $\sum_x \sqrt{\Pr(X = x|B = b)}$ to be close to $\sqrt{2}$ for both values of b. This, in turn, requires $\Pr(X = x|B = b)$ to be close to $1/2$ for both values of x. In other words, for H_{max} to be close to 1, x must be different from b with probability close to $1/2$.

To summarize our findings, qualitatively the entropic equation (12.4) implies the existence of a trade-off between the probability that $d = x$ (when $\theta = 0$) and that $\pi = x$ (when $\theta = 1$) in the purified version of the security game. This is because at least one of the two entropies must be larger than $1/2$, and the qualitative reasoning above suggests that this implies an upper bound on the adversary's probability of winning in the corresponding part of the security game. Concretely, this trade-off implies that there is a constant $0 \le p_s < 1$ such that no adversary can win in the game with probability larger than p_s. (It is possible to obtain precise estimates on p_s by carefully working through the definitions of the entropies and their relation to the guessing probabilities, but we satisfy ourselves with the qualitative statement.)

The constant p_s we have obtained bounds the success probability of an adversary in the certified deletion security game. However, this constant might not be very small! Instead we would like Eve to have a probability of cheating, i.e. providing a valid proof of deletion *and* being able to recover the plaintext, that is very small. This is why in the definition of the scheme we introduced a parameter λ that can be bigger than 1. From the point of view of the security game, considering higher parameters λ is equivalent to performing a repetition of the case $\lambda = 1$ in parallel, multiple times. To analyze the game for general λ we can proceed in two different ways. First of all, similar to our work in Chapter 8 we can consider the case of an adversary that behaves in an i.i.d. manner. In this case we directly obtain an upper bound of the form p_s^λ on the success probability in the λ-repeated game. This bound goes exponentially fast to zero with λ, and so by choosing λ sufficiently large we can make the success probability as small as we want. However, in general the adversary may not behave in an independent manner and can apply a global strategy in the security game. The analysis of such strategies is challenging technically, and lies beyond the scope of this book. Suffice it to say that, even in this more general setting, an exponentially decaying bound on the success probability can also be shown. This proves that the scheme introduced in the previous section is a good certified deletion encryption scheme. One more success for quantum information!

QUIZ 12.2.2 *Let's check the uncertainty relation* (12.4) *on a couple of examples. First, suppose that* ρ_{ABE} *consists of an EPR pair between A and E, and B is in state* $|0\rangle_B$. *In this case, what is the value of the pair* $(H_{max}(X_A|B), H_{min}(Z_A|E))$?

(a) $(0,0)$
(b) $(0,1)$
(c) $(1,1)$
(d) $(1,0)$

QUIZ 12.2.3 *Same question if AB is in an EPR pair, and E is in state* $|0\rangle_E$.

(a) $(0,0)$
(b) $(0,1)$
(c) $(1,1)$
(d) $(1,0)$

CHAPTER NOTES

For the general argument showing that perfect n-qubit quantum encryption schemes require keys of length $2n$, see the work of A. Ambainis, et al. (Private quantum channels. In *Proceedings 41st Annual Symposium on Foundations of Computer Science*, pp. 547–553. IEEE, 2000). The problem of approximate encryption is considered in a paper by P. Hayden, et al. (Randomizing quantum states: Constructions and applications. *Communications in Mathematical Physics*, **250**(2):371–391, 2004), where a randomized construction is given. For a deterministic construction along the lines mentioned in this chapter, see the paper by A. Ambainis and A. Smith in *Approximation, Randomization, and Combinatorial Optimization: Algorithms and Techniques* (Small pseudo-random families of matrices: Derandomizing approximate quantum encryption, pp. 249–260. Springer, 2004). The discussion in this chapter only scratches the surface of computational security, which is not a focus of this book. A comparison of many a priori different definitions of computational security for quantum encryption can be found in a paper by G. Alagic, et al. (Computational security of quantum encryption. In *International Conference on Information Theoretic Security*, pp. 47–71. Springer, 2016). For much much more on computational security in the classical setting, we refer to the classic introductory book *Introduction to Modern Cryptography* by J. Katz and Y. Lindell (CRC Press, 2020).

The notion of encryption with certified deletion is studied by A. Broadbent and R. Islam (Quantum encryption with certified deletion. In *Theory of Cryptography Conference*, pp. 92–122. Springer, 2020), from which the protocol given here is adapted. The uncertainty relation (12.4) is shown in the paper by M. Berta, et al. (The uncertainty principle in the presence of quantum memory. *Nature Physics*, **6**(9):659–662, 2010).

PROBLEMS

12.1 Approximate encryption from small-bias spaces

In this problem we show a property of a set of keys $\mathcal{K} \subseteq \{0,1\}^n \times \{0,1\}^n$ such that approximate encryption using the one-time pad restricted to keys $k \in \mathcal{K}$ is ε-secure.

First we need a definition. For a subset $S \subseteq \{0,1\}^n$, we say that S is δ-*biased* if for every $\alpha \in \{0,1\}^n$ such that $\alpha \neq 0^n$,

$$\left| \frac{1}{|S|} \sum_{s \in S} (-1)^{s \cdot \alpha} \right| \leq \delta .$$

Now let's fix a δ-biased subset $B \subseteq \{0,1\}^{2n}$. We can interpret each string $b \in B$ as a pair (k_1, k_2) of n-bit strings and define

$$\mathcal{E}(\rho) = \frac{1}{|B|} \sum_{b=(k_1,k_2) \in B} X^{k_1} Z^{k_2} \rho (X^{k_1} Z^{k_2})^\dagger .$$

Let's see how good an encryption scheme this is. As a warm-up, let's imagine that we try to distinguish $\mathcal{E}(\rho)$ from the totally mixed state $2^{-n}\mathbb{I}$ by making a Pauli measurement, i.e. using an observable of the form $i^{u \cdot v} X^u Z^v$ for $u, v \in \{0,1\}^n$.

1. Show that the expectation value $\text{tr}(X^u Z^v \mathcal{E}(\rho)) = \text{E}_{(k_1,k_2)\in B}[(-1)^{k_1\cdot u + k_2\cdot v}]\,\text{tr}(X^u Z^v \rho)$.
2. Using that B is a δ-biased set, deduce that $|\text{tr}(X^u Z^v \mathcal{E}(\rho))| \leq \delta\,|\text{tr}(X^u Z^v \rho)|$.
3. Show that for any matrix A, $\text{tr}(A^\dagger A) = \frac{1}{2^n}\sum_{u,v\in\{0,1\}^n}|\text{tr}(X^u Z^v A)|^2$. *[Hint: use that the Pauli matrices $\{X^u Z^v\}$ are orthonormal for the inner product $\langle A, B\rangle = \frac{1}{2^n}\text{tr}(A^\dagger B)$.]*
4. Deduce from the previous two questions that for any n-qubit density ρ,

$$\text{tr}(\mathcal{E}(\rho)^2) \leq \frac{1}{2^n} + \delta^2\,\text{tr}(\rho_0^2)\,.$$

For an n-qubit density matrix ρ, show that if $\text{tr}(\rho^2) \leq \frac{1}{2^n}(1+\varepsilon^2)$ for some $\varepsilon \geq 0$, then $D(\rho, 2^{-n}\mathbb{I}) \leq \varepsilon$.

5. Deduce a value of δ, as a function of ε and n, such that our approximate encryption scheme is ε-secure.
6. A δ-biased set of $2n$ bit strings can be constructed using $(2n)^2 \cdot (1/\delta)^2$ strings. How many keys does our ε-approximate encryption scheme use?

12.2 Uncertainty relation

In this problem we study some cases of equality in the entropic uncertainty relation, Eq. (12.4), which we restate for convenience: for any ρ_{ABE} such that A is a single qubit,

$$\text{H}_{\max}(X_A|B) + \text{H}_{\min}(Z_A|E) \geq 1\,.$$

Let's focus on the case where B is a single classical bit, E is a single qubit, and the quantum state $\rho_{AE} = |\psi\rangle\langle\psi|_{AE}$ is pure.

1. Suppose that we decompose $|\psi\rangle_{AE} = \alpha|0\rangle_A|u_0\rangle_E + \beta|1\rangle_A|u_1\rangle_E$, where $|u_0\rangle_A$ and $|u_1\rangle_A$ are abitrary (normalized) states of E. Show that $\text{H}_{\min}(Z_A|E) = 0$ if and only if $|u_0\rangle$ and $|u_1\rangle$ are orthogonal.
2. In this case, what is the value of $\text{H}_{\max}(X_A|B)$?
3. Show that $\text{H}_{\min}(Z_A|E) = 1$ if and only if $|u_0\rangle$ and $|u_1\rangle$ are parallel.
4. Give an example of a state of the previous kind, where $\text{H}_{\min}(Z_A|E) = 1$, such that in addition $\text{H}_{\max}(X_A|B) = 0$. Give another example where now $\text{H}_{\max}(X_A|B) = 1$.

13

Delegated Computation

Virtually all the cryptographic protocols that we studied in this book can be implemented using very simple equipment: essentially, a way to create and manipulate single qubits, to send them over a dedicated channel, and to measure (and sometimes store) them at the receiving end. And indeed, as discussed in Chapter 1, such equipment is already available today, making these protocols particularly appealing. Nevertheless, as experimental capabilities start to scale up to quantum computers of larger and larger sizes, cryptographic tasks that may involve more complex quantum computations start to become relevant. In this chapter we study the most fundamental such task, the problem of delegated computation. Delegated computation is a two-party task where there is a large asymmetry between the two parties: On the one hand, Alice would like to execute a quantum computation, but she does not have a powerful enough quantum computer to execute it. On the other hand, Bob has a quantum computer, but he is not trusted by Alice. Can Alice make sure that Bob executes her computation correctly for her? In the chapter we will see three very different approaches to this problem. Studying them gives us a good opportunity to introduce some basics of quantum computation, which you may find useful as you continue to study more and more complex tasks.

13.1 Definition of the Task

Let's formalize the problem of delegated computation and its security guarantees. Suppose a user Alice has a quantum circuit C in mind, which she would like to execute on some input x in order to learn the outcome $C(x)$. All of Alice's data, x and C, is classical: we will assume that x is a classical bit string, and C is specified by a sequence of two-qubit gates. We can also assume that the outcome is classical: let's say it is the outcome of a standard basis measurement performed on a specially designated output qubit of C. For simplicity we further assume that this circuit always returns deterministic outputs, and write $C(x)$ for the outcome.[1]

Now, unfortunately Alice herself does not have a universal quantum computer! Maybe she has a tiny desktop machine that lets her play around with BB'84-like operations: prepare or measure single qubits, possibly store a couple qubits at a time in

---

1 For example, C could be a circuit for factoring.

memory, but no more. Luckily, Alice has the possibility of buying computation time on a quantum server, appropriately named Bob, with which she could interact over the internet, or maybe even over a simple BB'84-type quantum communication channel that allows the exchange of one qubit at a time. So Alice could send x and the description of C to Bob, who would perform the computation and return the outcome – right?

Remember that this is a crypto book! Alice might not trust Bob. For one, she'd like to have a way to verify that the outcome provided to her is correct. What if Bob is lazy and systematically claims that the outcome of her computation, $C(x)$, equals "0"? Since Alice has no quantum computer herself she has no means of checking this! A second property Alice could require is that the computation remains private: while she certainly wants to learn $C(x)$, she'd rather not let Bob know that she is interested in circuit C, or in input x, as these might contain private data.

Let's restate these conditions as the requirements that the computation is *correct*, *verifiable*, and *blind*.

Definition 13.1.1 (Delegated computation). *In the task of delegated computation, a user, Alice, has an input (x, C), where x is a classical string and C the classical description of a quantum circuit. Alice has a multiple-round interaction with a quantum server, Bob. At the end of the interaction, Alice either returns a classical output y, or she aborts. A protocol for delegated computation is called:*

- Correct *if whenever both Alice and Bob follow the protocol, Alice accepts (she does not abort) and $y = C(x)$.*
- Verifiable *if for any server deviating from the protocol, Alice either aborts or returns $y = C(x)$.*
- Blind *if for any server deviating from the protocol, at the end of the protocol the reduced density matrix that describes the entire state of the server is independent of Alice's input (x, C).*

Each of these properties is stated informally. In particular, to make them formal we would have to introduce more quantitative versions, writing that "with high probability," and "almost independent," etc. We learned how to do this in many examples throughout the book. Delegated computation is the most complex task that we see so far, and to focus on the essentials we give ourselves a break and accept the informal definition for now.

Remark 13.1.1 In spite of being rather similar, neither the property of verifiability nor the property of blindness directly implies the other. In practice, verifiability often follows from blindness by arguing, using "traps," that if a protocol is already blind, the server's trustworthiness can be tested by making it run "dummy" computations for which Alice already knows the output, without the server being able to distinguish whether it is asked to do a real or dummy computation. We will see an example of this technique later on.

Are there good protocols for delegating quantum computations? It turns out that we don't have a fully satisfactory answer yet: this is an active area of research! In this chapter we'll outline three of the most prominent approaches. The first construction shows how arbitrary quantum circuits can be delegated, as long as the verifier has the ability to prepare certain specific single-qubit states and communicate them to the server. The second construction achieves a similar result, using a very different idea: *measurement-based* quantum computation. The third construction itself has a wholly different flavor. It achieves delegated computation by a purely classical Alice, with no quantum capabilities whatsoever. However, the downside is that Alice now has to interact with *two* isolated Bobs, which moreover need to share entanglement. This third method relies on similar techniques as we have seen in the analysis of device-independent quantum key distribution (QKD) in Chapter 9.

13.1.1 Preliminaries on Efficient Quantum Computation

Before we find out how to delegate quantum computations, let's first review briefly what a quantum computation *is*. As hinted earlier, there are many distinct models for quantum computation, each capable of universal computation (and thus of simulating each other).

The most straightforward model is called the quantum circuit model. Here a computation is represented by the action of a circuit C on n input qubits. The input qubits are initialized in an arbitrary quantum state that contains the input of the computation. Here we will only work with classical inputs of the form $|x\rangle$, where $x \in \{0,1\}^n$, but quantum inputs can be considered as well. The circuit C itself is specified by a list of m gates, which are one- or two-qubit unitaries that act on a subset of the qubits. For instance, a simple four-qubit circuit could be specified by the ordered list $((H,1),(\text{CNOT},(2,3)),(H,3),(Z,4))$. For simplicity we'll look only at circuits that return a single bit of output, which we'll take to be the result of a standard basis measurement of the first qubit. We will write $C(x)$ for the outcome of a measurement of the output qubit of C in the standard basis, when C is executed on the state $|x\rangle$. In general this would be a random value, but here we'll focus on circuits such that $C(x)$ takes a deterministic value for its output qubit.

In general we allow the circuit C to be based on any family of single- or two-qubit gates. An important theorem in quantum computing, the Solovay–Kitaev theorem, states that it is in fact possible to restrict the set of gates used to some simple sets of gates, called "universal gate sets": a gate set is universal if any circuit that can be implemented using *some* set of gates can also be implemented using that particular set of gates. Moreover, the number of gates required should be at most polynomially larger. An example of a universal gate set which we will use later on is the set

$$\mathcal{G} = \left\{ G = \begin{pmatrix} \cos(\pi/8) & -\sin(\pi/8) \\ \sin(\pi/8) & \cos(\pi/8) \end{pmatrix}, \text{CNOT} \right\},$$

where G implements a $\pi/4$ rotation around the y-axis of the Bloch sphere and CNOT is a controlled-X operation (on any two qubits of the circuit). Another example of a popular universal gate set is the set

$$\mathcal{G}' = \left\{ H, T = \begin{pmatrix} 1 & 0 \\ 0 & e^{i\pi/4} \end{pmatrix}, \text{CNOT} \right\},$$

and there are many others.

We will use one more useful feature of quantum (as well as classical) circuits, which is the notion of a "universal" circuit. Informally, a universal circuit is a circuit that can simulate the execution of any other circuit, provided that the other circuit's source code is provided to the universal circuit as input.

Theorem 13.1.2 (Universal circuit) *For any integer n and size parameter s there exists a fixed circuit $\mathcal{C}_U$ acting on $n + m$ qubits, where m is at most a polynomial in n and s, such that the following holds. For any circuit $\mathcal{C}$ of size at most s expressed using the gate set $\mathcal{G}$, and any input $x \in \{0,1\}^n$ to $\mathcal{C}$, there is a $z \in \{0,1\}^m$ which can be efficiently computed from $\mathcal{C}$ and is such that $\mathcal{C}_U(x,z)$ has the same distribution as $\mathcal{C}(x)$.*

13.1.2 The Pauli Group and Clifford Gates

We've already met the four single-qubit Pauli matrices, $\mathcal{P} = \{I, X, Y, Z\}$. These form a group, in the sense that for any two Pauli matrices P and Q, the product PQ is again a Pauli matrix (up to phase). A single-qubit *Clifford gate* U is any operation that preserves the Pauli group, in the sense that $UPU^\dagger$ is a Pauli matrix for any Pauli P. If U is a two-qubit gate, we similarly require that $UPU^\dagger \in \pm\mathcal{P}^{\times 2}$ for any P that is a tensor product of two Paulis.

Exercise 13.1.1   Verify that the Pauli matrices are Clifford gates. Show that the Hadamard, phase $P = T^2$, and CNOT gates are Clifford gates. Do you see other examples? Is the G gate also a Clifford gate? How about the T gate?

The defining property of Clifford gates is very useful, and plays an important role in delegated computation – we will see why. Unfortunately, it turns out that there is no universal gate set made only of Clifford gates: any universal set of gates for quantum computation must include at least one non-Clifford gate. This will be a source of many headaches when trying to implement delegated computation.

13.2 Verifiable Delegation of Quantum Circuits

Our first approach to delegated computation is based on the idea of *computing on encrypted data*. Recall the quantum one-time pad from Chapter 2. Suppose that we have an n-qubit density matrix ρ. To encrypt it using the one-time pad we select two n-bit strings $a, b \in \{0,1\}^n$ uniformly at random, and return $\tilde{\rho} = X^a Z^b \rho (X^a Z^b)^\dagger$, where $X^a = X^{a_1} \otimes \cdots \otimes X^{a_n}$ denotes applying a Pauli X operator on all qubits i such that $a_i = 1$, and the identity on all other qubits. The notation for Z^b is similar, with Pauli Z operators instead. If ρ represents the input x to the circuit, $\rho = |x\rangle\langle x|$, then the result of applying the quantum one-time pad is still classical, $\tilde{\rho} = |x \oplus a\rangle\langle x \oplus a|$.

So here is an idea: Alice can encrypt her input x into a quantum one-time padded state $\tilde{\rho}$, and send $\tilde{\rho}$ to Bob. If she keeps a copy of the strings a, b and does not communicate them to Bob then her input x remains perfectly private. This is a good start: we already understand how we can use the server, Bob, as a quantum memory and maintain privacy of our (classical) input.

Of course, there is much more that we would like to do: Alice wants to make Bob execute a circuit C. Can she somehow guide him through this by working directly on $\tilde{\rho}$? For this we need to find a quantum operation $\tilde{C}$ that the server could apply, such that $\tilde{C}(\tilde{\rho}) = \widetilde{C(\rho)}$, an encrypted version of $C(\rho)$ from which Alice can recover the real output $C(x)$. (And of course, we want even more: we want $\tilde{C}$ to hide C, to achieve blindness, and we haven't even discussed verifiability yet – one step at a time!)

The circuit C can always be expressed using gates from a universal gate set, for example the set $\mathcal{G}' = \{H, \mathrm{CNOT}, T\}$ introduced earlier. Even though it is not needed, to warm up let's assume that we also allow X gates, and that the first gate in C is an X applied on the first qubit. Now notice that

$$X\left(X^a Z^b \rho (X^a Z^b)^\dagger\right) X^\dagger = X^{a \oplus e_1} Z^b \rho (X^{a \oplus e_1} Z^b)^\dagger = \widetilde{X \rho X^\dagger},$$

where e_1 is the bit string with a single 1 in the first position. This equation shows that Bob can in fact directly apply the X gate on $\tilde{\rho}$, and the effect is as if it had been applied directly on the real ρ! So an X gate is easy, and you can check that any Pauli gate, single- or multi-qubit, will be similarly easy. The main property that is used here is that different Paulis either commute or anti-commute with each other. In other words, Pauli gates can be "commuted" past each other with at most a sign change, which disappears as a global phase.

Let's move one step further and consider a Clifford gate. Let's take the example of a Hadamard gate on the second qubit, H^{e_2}, with e_2 the n-bit string $e_2 = (0, 1, 0, \dots, 0)$, i.e. $H^{e_2} = \mathbb{I} \otimes H \otimes \mathbb{I} \otimes \cdots \mathbb{I}$. Using the equation $HXH = Z$, we get

$$(X^a Z^b)(H^{e_2} \rho (H^{e_2})^\dagger)(X^a Z^b)^\dagger = (-1)^{a_2 b_2} H^{e_2} X^{a'} Z^{b'} \rho (X^{a'} Z^{b'})^\dagger (H^{e_2})^\dagger,$$

where (a', b') is obtained from (a, b) by exchanging the bits a_2 and b_2. We see that if Alice instructs Bob to apply an H gate on the second *encrypted* qubit, the effect is the same as if the server had applied the H gate directly on the second *unencrypted* qubit – as long as she updates her one-time pad key (a, b) to (a', b') as described above. As long as Alice does this simple classical operation on her side, when Bob returns the encrypted qubits she will be able to undo the one-time pad to recover the correct outcome of the computation.

The following exercise asks you to show that a similar trick can be employed for any Clifford gate.

Exercise 13.2.1 Let U be any one- or two-qubit Clifford gate. Show that the effect of applying U to the encrypted state $\tilde{\rho}$ is equivalent to the application of U on ρ, up to an update rule on the one-time pad key (a, b). Work out the update rule in the case of the phase and CNOT gates.

Can Alice orchestrate the whole computation with Bob, while only having to keep track of simple updates on her one-time pad keys? Unfortunately, remember from Section 13.1.2 that no set of Clifford gates is universal – we need to show how to implement one more gate, for example the T gate considered in the universal set $\mathcal{G}'$. Because the T gate is non-Clifford, applying it to the encrypted state $\tilde{\rho}$ will have a more complicated effect, which we can't keep track of by a simple modification of the one-time pad keys. Instead, we'll show how Alice can make the server implement a T gate on the encrypted state by using the idea of *magic states*.

QUIZ 13.2.1 *Suppose that a server applies a T gate directly to an encrypted qubit $X^a Z^b |\psi\rangle$. Which of the states below represents the resulting state? Recall that the T gate is* $T = \begin{pmatrix} 1 & 0 \\ 0 & e^{i\frac{\pi}{4}} \end{pmatrix}$, *and* $P = \begin{pmatrix} 1 & 0 \\ 0 & i \end{pmatrix}$.

(a) $X^a Z^{a \oplus b} PT |\psi\rangle$
(b) $X^a Z^{a \oplus b} P^{a \oplus b} T |\psi\rangle$
(c) $X^a Z^{a \oplus b} P^a T |\psi\rangle$
(d) $X^a P^a T Z^b |\psi\rangle$

QUIZ 13.2.2 $P^c X^a Z^b |\psi\rangle$ *is equal to:*

(a) $X^a Z^a P^c |\psi\rangle$
(b) $X^a Z^{a \oplus b} P^c |\psi\rangle$
(c) $X^a Z^{a \oplus b \oplus c} P^c |\psi\rangle$
(d) $X^a Z^{a \cdot c \oplus b} P^c |\psi\rangle$

13.2.1 Computation with Magic States

The idea for magic states is that the computation of certain complicated gates on an arbitrary state can be replaced by a simple computation using certain auxiliary states called "magic states" as some form of catalyst. Let's see this for the T gate, the only gate that we still need to figure out how to implement. The magic state we need is

$$|\pi/4\rangle = T|+\rangle = \frac{1}{\sqrt{2}}|0\rangle + \frac{e^{i\pi/4}}{\sqrt{2}}|1\rangle . \tag{13.1}$$

Preparing this state itself requires applying a T gate. But the point is that we only need to apply the gate to a fixed, known input state, which is independent of the state $|\psi\rangle$ on which we really want to apply the T gate. So the preparation of single-qubit magic states is a relatively simple and computation-independent task which Alice should be able to perform herself, as long as she has access to a small single-qubit quantum computer.

Suppose we are given a single-qubit state $|\psi\rangle$ on a register A_1, and initialize a second qubit in register A_2 in the $|\pi/4\rangle$ state. Consider the following circuit: first, apply

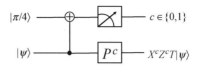

Fig. 13.1   Teleporting into a T gate. The state $|\pi/4\rangle$ is defined in (13.1).

a CNOT, controlled on A_2 and acting on A_1. Second, measure register A_1 in the computational basis, obtaining an outcome c. Third, if $c = 1$ we apply a P gate (recall that $P = T^2$) to register A_2, and if $c = 0$ we do nothing (see Figure 13.1). What is the post-measurement state in register A_2? It is a good exercise to verify that this is $X^c Z^c T |\psi\rangle$. That is, up to the possible apparition of a product (XZ) in front (depending on the value obtained for c), the effect of our small circuit is that the T gate got applied to $|\psi\rangle$. Moreover, the only resources we used to achieve this are a magic state, a CNOT gate, a measurement in the computational basis, and a controlled-P gate. Since the P gate is a Clifford gate, applying this transformation to all T gates in a circuit allows us to transform any quantum circuit into a circuit that only uses magic states and Clifford gates (with adaptive measurements).

> **Exercise 13.2.2**   Suppose that instead of being applied directly to the state $|\psi\rangle$, the circuit described above is applied to an encrypted version of $|\psi\rangle$, $X^a Z^b |\psi\rangle$. Show that the outcome of the circuit is then $P^a X^{d'} Z^{d'} T |\psi\rangle$, for some bits a' and b' depending on a, b, and c. (Convince yourself that the same calculation works out in the case when $|\psi\rangle$ is not pure, but a reduced density ρ on a single qubit.)

Note the P gate that we picked up in the exercise. This also needs to be corrected. But the gate is applied on the encrypted state. So Alice could instruct Bob to apply $(P^a)^\dagger$ directly, to remove the gate. Unfortunately, this would require revealing the bit a, which is part of Alice's secret one-time pad key. There is a way around this that involves adding a little bit of randomization in the choice of magic state we use (essentially, considering a one-time padded magic state). This will guarantee that the phase correction is always independent of the one-time pad key.

We now have the outline of a delegated computation protocol. Alice first prepares a one-time padded version of her classical input x, and sends it to Bob. She also prepares many (one-time padded) magic states, and sends them to Bob as well. Finally, Alice and Bob both go through the circuit $\mathcal{C}$ one gate at a time. For Clifford gates, Bob applies the gate directly on the encrypted qubit, and Alice updates her keys as in Exercise 13.2.1. For a T gate, Bob executes the circuit from Figure 13.1 and sends the outcome c back to Alice, who tells him to apply a gate $(P)^\dagger$ or not, depending on her one-time pad keys.[2]

---

[2] This last step was not described in detail; believe us that it can be done without leaking information about Alice's keys – or check the chapter notes for details!

13.2.2 Blindness

Is our delegation protocol blind? We were pretty careful to ensure that Alice's input x remains perfectly private. However, to implement the protocol she needs to completely reveal her circuit C! Luckily there is a simple way out: Alice can instruct Bob to execute a fixed "universal" circuit (as in Theorem 13.1.2), and instead encode the actual circuit C she is interested in as part of the input x.

13.2.3 Verifiability

Our protocol satisfies the blindness property (provided we use a universal circuit), but so far it is not verifiable: Alice has no guarantee that Bob performs the required computation! Indeed, no check is performed at all. For all we know, Bob never does anything and simply returns the original $\tilde{\rho}$ back to Alice at the end of the protocol. What kind of test could we make to check that Bob is not being lazy (or even malicious)?

The idea is to combine the protocol with some "test runs." The original protocol is now called a "computation run." In contrast, in a test run the computation is set up in such a way that Alice knows what the outcome should be, and she will check that Bob returns the correct value (after decoding). But Bob will not be able to distinguish test runs from computation runs, and as a consequence we'll have the guarantee that Bob is also being honest in a computation run.

There are two types of test runs, X-test and Z-test. In an X-test run, the computation is executed on an encryption of the all-0 input $|0\rangle^{\otimes n}$. In a Z-test run, the same computation is executed on an encryption of $|+\rangle^{\otimes n}$. The main trick to ensure that Alice can keep track of the computation is that all gates in a test run are replaced by *identity* gates, without the prover noticing! Note that we already know how to do this for Pauli gates, as these do not involve the server anyway (Alice only has to update her one-time pad keys). The H gate requires a bit more work, but the idea is simple: since an H exchanges the standard basis and the Hadamard basis, we can think of it as exchanging between an X-test run and a Z-test run. So in that case as well, Alice can perfectly keep track of the state that the encrypted quantum state maintained by Bob should be in. The T gate, of course, is the interesting one. The idea is to modify the implementation described in Section 13.2.1 by changing the magic state, as well as the update rule, in a way that is unnoticeable by Bob but will result in an application of the identity gate instead of the T. The following exercise asks you to work out how this can be done.

Exercise 13.2.3 Consider the following procedure for implementing a T gate on the single-qubit state $|\psi\rangle_{A_1}$ using a magic state in register A_2. Alice first selects two bits $d, y \in \{0, 1\}$ uniformly at random, and prepares the magic state $Z^d P^y T |+\rangle_{A_2}$, where $P = T^2$ is the phase gate. Alice sends register A_2 to Bob, who performs a CNOT controlled on A_1 and with target A_2. Bob measures A_1 in the computational basis, obtaining an outcome $c \in \{0, 1\}$ which he sends to Alice. Alice then sends back $x = y \oplus c$ to Bob, who applies a gate P^x to the remaining system A_2.

1. Show that the state of A_2 at the end of this procedure is $X^c Z^{c(y\oplus 1)\oplus d\oplus y}T\,|\psi\rangle$, i.e. it is an encryption (using a key known to Alice) of $T\,|\psi\rangle$.

Next let's suppose we're doing a computation run, so that $|\psi\rangle = X^a\,|0\rangle_{A_1}$ for some $a \in \{0,1\}$. Alice would like to perform the identity instead of a T gate, without Bob noticing. This can be done by executing precisely the same circuit, except that the magic state is replaced by $X^d\,|0\rangle_{A_2}$ (it does not depend on y).

2. Show that with the magic state replaced by $X^d\,|0\rangle_{A_2}$ the interaction results in a register A_2 in state $X^d\,|0\rangle_{A_2}$. Show that in this case the outcome c of Bob's measurement is deterministically related to a and d in a simple way.

3. Can you find a similar modification, with a different magic state, that will implement the identity for the case of a Z-test run, where $|\psi\rangle_{A_1} = Z^b\,|+\rangle_{A_1}$ for some $b \in \{0,1\}$?

The exercise shows that simply by changing the magic state used in the implementation of the T gate, Alice can force that gate to act as identity in an X- or Z-test run. Moreover, due to the random bits d, y used in the preparation of the magic state you can verify that, from the point of view of Bob, these magic states look uniformly distributed, and thus he has no way of telling which "gadget" – for a T gate or the identity – he is really implementing.

In a test run Alice knows exactly what the outcome of the circuit should be, so she can verify the answer provided by Bob. Is this enough to ensure that Bob cannot cheat in a computation run? After all, we can imagine that Bob may be able to perform certain "attacks" that do not affect simple computations, where the state is always a tensor product of single qubits encoded in the computational or Hadamard bases, but such that the attack would perturb the kind of highly entangled states that will show up at intermediate stages in Alice's more complex circuit C.

To show that this is not the case – that any significant attack will necessarily have a noticeable effect on either the X- or Z-test runs – the idea is to use an observation called the "Pauli twirl," which you are asked to work out in the next exercise.

 Exercise 13.2.4   Pauli twirl. Let ρ be a single-qubit density matrix, and $P, P' \in \mathcal{P}$, where $\mathcal{P} = \{I, X, Y, Z\}$ is the set of single-qubit Pauli operators. Show that $\frac{1}{4}\sum_{Q\in\mathcal{P}}(Q^\dagger P Q)\rho(Q^\dagger (P')^\dagger Q)$ equals $P\rho P^\dagger$ if $P = P'$, and is 0 otherwise. Show that the same result holds for n-qubit Pauli operators.

The Pauli twirl allows us to argue that, thanks to the use of the quantum one-time pad, any "attack" of Bob boils down to the application of a Pauli operator at the last step of the circuit. Indeed, suppose first that the interaction performed between Alice and Bob results in the correct circuit C being implemented, except at the last step Bob applies an arbitrary "deviating unitary" U. Thus the outcome is $U\tilde{C}\tilde{\rho}\tilde{C}^\dagger U^\dagger$, where $\tilde{C}$ is the unitary Alice instructed Bob to implement, and $\tilde{\rho}$ the initial one-time-padded state sent by Alice. Due to the one-time pad, $\tilde{\rho}$ has the form $\tilde{\rho} = \sum_{Q\in\mathcal{P}} Q\,|x\rangle\langle x|\,Q^\dagger$, where $|x\rangle$ denotes the real input state that Alice would like the computation to be performed on. Moreover, for any Q there is a correction $c(Q) \in \mathcal{P}$ applied by Alice, which is such that $c(Q)\tilde{C}Q|x\rangle\langle x|Q^\dagger\tilde{C}^\dagger(c(Q))^\dagger = C|x\rangle\langle x|C^\dagger$. Thus, after applying $c(Q)$ to the corrupted circuit,

$$\sum_{Q\in\mathcal{P}} c(Q)U\tilde{C}Q|x\rangle\langle x|Q^\dagger\tilde{C}^\dagger U^\dagger(c(Q))^\dagger$$

$$= \sum_{Q\in\mathcal{P}} c(Q)U(c(Q))^\dagger c(Q)\tilde{C}Q|x\rangle\langle x|Q^\dagger\tilde{C}^\dagger(c(Q))^\dagger c(Q)U^\dagger(c(Q))^\dagger$$

$$= \sum_{Q\in\mathcal{P}} c(Q)U(c(Q))^\dagger C|x\rangle\langle x|C^\dagger c(Q)U^\dagger(c(Q))^\dagger$$

$$= \sum_{P\in\mathcal{P}} |\alpha_P|^2 PC|x\rangle\langle x|C^\dagger P^\dagger \,,$$

where for the last step we decomposed $U = \sum_{P\in\mathcal{P}} \alpha_P P$ in the Pauli basis, and used the property of the Pauli twirl proved in Exercise 13.2.4.

This computation shows that any unitary applied by a malicious Bob at the end of the honest circuit is equivalent to a convex combination of Pauli operators. But any such non-trivial operator will be detected in either the X- or Z-test runs, as it will result in one of the outcomes being flipped in either the standard or the Hadamard basis.

To conclude, we need to deal with the case where Bob applies a deviating unitary, not at the end of the circuit, but at some intermediate step. This case can be reduced to the previous one! Indeed, we can always think of a "purified" version of the whole protocol, where all measurements are deferred until the end. Suppose the unitary $\tilde{C}$ that Bob is supposed to implement decomposes as $\tilde{C} = \tilde{C}_2\tilde{C}_1$, and that Bob applies a deviating unitary U in-between the two circuits. The result can be written as

$$\tilde{C}_2 U\tilde{C}_1 = (\tilde{C}_2 U\tilde{C}_2^\dagger)\tilde{C}_2\tilde{C}_1 = (\tilde{C}_2 U\tilde{C}_2^\dagger)\tilde{C} \,,$$

where we used that $\tilde{C}_2$ is unitary, and hence $\tilde{C}_2^\dagger\tilde{C}_2 = \mathbb{I}$. Thus the deviation U is equivalent to applying another deviating unitary $U' = \tilde{C}_2 U\tilde{C}_2^\dagger$ at the end of the circuit, and we are back to the analysis performed in the previous case: if the deviation has a nontrivial effect it will be detected by Alice in one of the test runs.

13.3 Delegation in the Measurement-Based Model

Our second scheme for delegated computation has a similar flavor to the previous one, but at its heart it is based on a completely different approach to universal quantum computation. So far we have encountered the circuit model for performing quantum computations. From the point of view of computer science this is the most straightforward model, as it is a direct analogue of the classical circuit model on which the architecture of our (classical) computers is based. However, quantum information allows for other, more exotic, models of computation. Many of these models were originally proposed with the idea that they might be more powerful than the circuit model, although ultimately they were proved equivalent. This includes the adiabatic model for computation and the measurement-based model that we will discuss in this section.

The highlight of measurement-based quantum computation (MBQC) is that it realizes an arbitrary quantum computation (specified by a circuit using some universal gate set) through an (adaptive) sequence of *single-qubit* measurements on a *fixed*, universal starting state. Seems impossible? Let's first give an overview of how this model works, and then we'll explain how MBQC can be used to achieve blind, verifiable delegated computation.

13.3.1 Measurement-Based Computation

Measurement-based computation is based on an idea very similar to *teleportation-based computation*, a model to which we return in the next section. This is the idea that a complete quantum computation, including the preparation of the initial state and the application of gates from a universal set, can be performed by making a sequence of adaptive measurements on a fixed universal state, successively "teleporting" the input state from one qubit to the next while at the same time applying unitary transformations on the state.

Let's do a simple example first. Suppose we have a qubit initialized in the state $|\psi\rangle_A = \alpha|0\rangle_A + \beta|1\rangle_A$. Suppose a second qubit is created in the state $|+\rangle_B$, and a CTL-Z operation is performed, controlling on the first qubit to perform a phase flip on the second. Then the joint state of the system becomes $|\psi\rangle_{AB} = \alpha|0\rangle_A|+\rangle_B + \beta|1\rangle_A|-\rangle_B$. Suppose now that we measure the first qubit in the Hadamard basis. What happens to the second qubit? Let's rewrite

$$|\psi\rangle_{AB} = \alpha|0\rangle_A|+\rangle_B + \beta|1\rangle_A|-\rangle_B$$
$$= \frac{1}{\sqrt{2}}|+\rangle_A(\alpha|+\rangle_B + \beta|-\rangle_B) + \frac{1}{\sqrt{2}}|-\rangle_A(\alpha|+\rangle_B - \beta|-\rangle_B).$$

The measurement rule states that if we get the outcome "+" the second qubit is projected to $|\psi'\rangle_B = \alpha|+\rangle_B + \beta|-\rangle_B$, and if we get a "−" it is projected to $\alpha|+\rangle_B - \beta|-\rangle_B$. In the first case, $|\psi'\rangle_B = H|\psi\rangle$, and in the second $|\psi'\rangle_B = XH|\psi\rangle$. More succinctly put, $|\psi'\rangle_B = X^m H|\psi\rangle$ where $m \in \{0,1\}$ denotes the outcome of the measurement: $m = 0$ in the case of "+" and $m = 1$ in the case of "−". Thus, up to a "Pauli correction" X^m, we managed to apply a Hadamard gate simply by making a single-qubit measurement on the appropriate state.

Exercise 13.3.1 Write the circuit described above in the same form as the T-gate gadget from Figure 13.1, where the state $|\pi/4\rangle$ is replaced by a different "magic" state, and the P correction by a different correction. (Be careful not to confuse CNOT, which is CTL-X, with CTL-Z!) How are these related?

For an arbitrary $\phi \in [0, \pi/2)$ let

$$|+_\phi\rangle = \frac{1}{\sqrt{2}}|0\rangle + e^{i\phi}\frac{1}{\sqrt{2}}|1\rangle \qquad \text{and} \qquad |-_\phi\rangle = \frac{1}{\sqrt{2}}|0\rangle - e^{i\phi}\frac{1}{\sqrt{2}}|1\rangle, \qquad (13.2)$$

and recall the rotations

$$R_z(\phi) = \begin{pmatrix} 1 & 0 \\ 0 & e^{i\phi} \end{pmatrix}, \qquad R_x(\phi) = HR_z(\phi)H.$$

The rotations $R_z(\phi)$ and $R_x(\phi)$ together generate a universal set of single-qubit gates. This is because any rotation on the Bloch sphere can be implemented as $R_z(\varphi_3)R_x(\varphi_2)R_z(\varphi_1)$ for an appropriate choice of φ_1, φ_2, and φ_3.

The following exercise asks you to generalize the example of the Hadamard gate to any single-qubit rotation that can be decomposed in this way.

Exercise 13.3.2 Modify the method we described to apply a Hadamard gate by instead performing a measurement of the first qubit in the basis $\{|+_\varphi\rangle, |-_\varphi\rangle\}$. Show that the second qubit is then projected on the state $X^m H R_z(\varphi)|\psi\rangle$, where $m \in \{0,1\}$ indicates the measurement outcome.

Now consider a sequence of three measurements with angles φ_1, φ_2, and φ_3. That is, suppose a first qubit is in state $|\psi\rangle_A$, and three additional qubits are created in the $|+\rangle$ state and organized on a line. Three CTL-Z operations are performed from left to right. Then the first qubit is measured in basis $\{|+_{\varphi_1}\rangle, |-_{\varphi_1}\rangle\}$, obtaining an outcome $m_1 \in \{0,1\}$, the second qubit is measured with angle φ_2, obtaining outcome m_2, and finally the third qubit is measured with angle φ_3, obtaining outcome m_3. Show that the state of the fourth qubit can then be written as

$$|\psi'\rangle_D = X^{m_3} Z^{m_2} X^{m_1} H R_z((-1)^{m_2}\varphi_3) R_x((-1)^{m_1}\varphi_2) R_z(\varphi_1)|\psi\rangle . \tag{13.3}$$

[Hint: you may use the identities $X R_z(\phi) = R_z(-\phi)X$ and $H R_z(\phi)H = R_x(\phi)$, valid for any real ϕ.]

The exercise *almost* lets us apply an arbitrary rotation $R_z(\varphi_3)R_x(\varphi_2)R_z(\varphi_1)$, except that there are these annoying "corrections," the X and Z operations and the Hadamard to the left, as well as extra $(-1)^{m_i}$ phases in the angles. But these are easy to handle! For the phases, note that we perform the measurements sequentially, and the phase flip that got applied to a certain angle only depends on the outcome of the measurement performed right before. For the case of the calculation performed in the exercise, if we *really* had wanted to end up with $U_x(\varphi_2)$, after having obtained outcome m_1 we could have updated our choice of angle in which to measure to $\varphi_2' = (-1)^{m_1}\varphi_2$. As for the X, Z, and H corrections at the end of the computation, we can handle those at the time of final measurement: they correspond to corrections that will need to be applied once we measure the final qubit (this is similar to how we handled the one-time pad in the previous section).

So we now know how to apply any sequence of single-qubit rotations to a qubit by using only measurements. To do this we start with a line of m qubits, each initialized in the $|+\rangle$ state. Then we apply CTL-Z operations on all pairs of neighboring qubits, from left to right. This corresponds to preparing a $1 \times m$-dimensional "brickwork state," a universal resource for single-qubit computation. Suppose for simplicity the initial qubit is meant to be initialized in the $|+\rangle$ state (if it is not you can modify the circuit so that the first gate applied prepares the correct qubit). Any rotation can be applied by decomposing it in the form $R_z(\varphi_3)R_x(\varphi_2)R_z(\varphi_1)$ and making the correct sequence of measurements on three qubits, keeping track of successive measurement outcomes to update the angles and the X, Z, and H "corrections" that tag along to the left of the description of the state of the qubit, as in (13.3) (note that you do not need to remember all measurement outcomes, but only their combined effect in terms of a power of X and a power of Z).

What if we have a multi-qubit computation? We won't give the details, but the general idea is the same. Since we already know how to implement arbitrary single-qubit gates, to get a universal gate set it suffices to implement a two-qubit CNOT gate. This can be done by using multiple lines of qubits, one for each qubit of the original computation. The lines are connected by vertical CTL-Z operations once every three qubits (in a slightly shifted manner). A two-qubit CNOT gate can then be applied using similar ideas as we described, but performing measurements on the two lines associated with the two qubits on which the gate acts. We'll leave the details as an exercise, and refer you to the chapter notes for detailed explanations.

13.3.2 Blind Delegation in the MBQC Model

Now that we have seen how to perform an arbitrary computation in the MBQC model, let's see how the computation can be delegated to an untrusted server Bob. Let's imagine that Alice has a sequence of single-qubit measurements, specified by angles $\{\varphi_{ij}\}_{1 \le i \le n, 1 \le j \le m}$ and update rules (depending on prior measurement outcomes), that she wishes to apply on an $n \times m$ brickwork state in order to implement an n-qubit quantum circuit that she is interested in. Let's also assume for simplicity that the outcome of the last measurement would (possibly after a Pauli correction if needed) give her the answer she is looking for.

Of course, Alice could tell the server to prepare the $n \times m$ brickwork state and then instruct it, through a classical interaction, to perform the measurements specified by the φ_{ij}. The server would report the outcomes, Alice would perform the updates, and tell the server the next angle to measure in. But clearly this would be neither blind nor verifiable.

The key idea is for Alice to (partially) prepare some kind of "one-time padded" version of the brickwork state, on which the server will implement the computation without ever having any information about the "real" angles φ_{ij}.

Consider the following protocol outline.

Protocol 13 Fix a set of "hiding angles" $D = \{0, \pi/4, 2\pi/4, \ldots, 7\pi/4\}$.

1. For each of the nm qubits of the brickwork state, Alice chooses a random $\theta_{i,j} \in D$, prepares the state $|+_{\theta_{i,j}}\rangle$, and sends it to Bob.

2. Bob arranges all the qubits he receives in the shape of an $n \times m$ brickwork state, and performs CTL-Z operations on neighboring qubits as required.

3. Alice and Bob have a classical interaction over nm rounds. In each round,

   1. Alice computes an angle δ_{ij} as a function of θ_{ij}, φ_{ij}, private randomness $r_{ij} \in \{0, 1\}$, and previous outcomes b_{ij} reported by Bob. She sends δ_{ij} to Bob.

2. Bob measures the (i, j)-th qubit of the brickwork state in the $\{|+\rangle_{\varphi_{ij}}, |-\rangle_{\varphi_{ij}}\}$ basis and reports the outcome $b_{ij} \in \{0, 1\}$ to Alice.

4. Alice infers the outcome of her circuit from her private data and Bob's last reported outcome.

There are many details missing to fully specify the protocol. The idea is to design rules for Alice to update the measurement angles δ_{ij} that she sends to Bob in such a way that, from the point of view of the server, δ_{ij} is always uniformly random in D (so it reveals no information about the computation being performed), yet Alice is able to keep track of the actual computation being performed under her one-time pad. To see how this can be done, first attempt the following exercise.

Exercise 13.3.3   Based on Exercise 13.3.2 we know that applying a Hadamard gate to a qubit A can be performed by measuring the qubit in the basis $\{|+\rangle, |-\rangle\}$ and adding an X^m correction, where m is the measurement outcome.

This is correct when the second qubit B has been initialized in a $|+\rangle$ state, as it would be for the unhidden brickwork state. Now suppose that the qubit has in fact been initialized in the state $|+_\theta\rangle$, for some real angle θ (and a CTL-Z operation has been performed on the two qubits). Show that the result of measuring the first qubit in the basis $\{|+_\delta\rangle, |-_\delta\rangle\}$ is to project the second qubit on $X^m HR_z(\theta + \delta)|\psi\rangle$, where m is the measurement outcome.

Suppose then that Alice would like to apply a rotation $R_z(\varphi)$, for some angle $\varphi \in D$. The exercise shows that communicating the angle $\delta = \varphi - \theta$ to Bob instead, where θ is the initial angle she used to prepare the corresponding qubit of the brickwork state, will have the desired effect of implementing $X^m HR_z(\varphi)$. However, this still poses a problem: if Bob is given both the quantum state $|+_\theta\rangle$ *and* the real angle $\varphi - \theta$, we can't argue that the computation is blind, as the joint distribution of these two pieces of information depends on φ.

Exercise 13.3.4   Fix φ, and suppose an adversary is given a classical value $\eta = \varphi - \theta$ and a single-qubit state $|\psi\rangle = |+_\theta\rangle$, where θ is chosen uniformly at random. Design a strategy for the adversary to recover φ, given $(\eta, |\psi\rangle)$. What is its success probability (averaged over the random choice of θ)?

The role of the additional values r_{ij} specified in the protocol is to hide φ_{ij} completely from Bob. Here r_{ij} is chosen uniformly at random in $\{0, 1\}$, and Alice communicates the angle $\varphi - \theta + r\pi$ to Bob. Based on Exercise 13.3.2, the effect of $r\pi$ on the computation is to add an extra Z^r correction, which Alice can easily keep track of. To see that it is sufficient to ensure blindness, imagine that instead $r\pi$ had been added to the initial angle θ. For any fixed θ, a random choice of $r \in \{0, 1\}$ suffices to make sure that Bob gains no information from receiving $|+_{\theta+r\pi}\rangle$, as $\frac{1}{2}|+_\theta\rangle\langle+_\theta| + \frac{1}{2}|+_{\theta+\pi}\rangle\langle+_{\theta+\pi}| = \frac{1}{2}\mathbb{I}$. But as θ varies in D the angle $\theta - \varphi$ itself is uniformly distributed in D. Therefore, from the point of view of Bob, the joint distribution of the pair $(|+_\theta + r\pi\rangle, \theta - \varphi)$ is indistinguishable from that of a uniformly random qubit and a uniformly random value from D. Bob receives completely random data, so the computation is perfectly blind.

13.3.3 Verifiability

In the previous section we showed that blind delegation could be implemented in the MBQC model. Can we make the protocol verifiable? Note that so far Alice does not perform any checks, so Bob could just as easily report random outcomes to her at each step. Already though, due to blindness there is no way that Bob can *force* a particular outcome on Alice; the best he can do is mislead her into thinking that the outcome of the computation is some random bit.

There are different techniques available to make the protocol verifiable. The main idea is to introduce *trap qubits*. Those are particular rows of the brickwork state that Alice randomly inserts into her circuit but on which the only operation performed is a sequence of identity gates: they are meant to remain in the $|0\rangle$ state (hidden, as usual, under the quantum one-time pad). By asking Bob to measure a qubit on such a line, Alice can verify the measurement outcome. Due to the blindness property, even the application of identity gates cannot be detected by Bob, so he does not know that he is being tested.

Implementing this idea requires a little care, as it is important to ensure that even the tiniest attack by Bob, such as reporting a single false measurement outcome, is detected with good probability: a single such deviation could suffice to ruin the whole computation. This can be achieved by introducing ideas from fault-tolerant computation, which we will not go into here.

13.4 Classically Delegating to Two Quantum Servers

Both schemes for delegated computation we've seen so far, in the circuit model or using measurement-based computation, require Alice to prepare single-qubit states taken from a small fixed set and send them to Bob. What if Alice has no quantum capability whatsoever? Intuitively, the goal of the qubits sent by Alice in the two previous schemes is to establish some kind of "trusted space" within Bob's quantum memory, in which he is constrained to perform the computation. The quantum one-time pad is used to guarantee that if Bob tries to cheat by not using these qubits then Alice interprets the results, at best, as garbage. In fact the verifiability property, enforced through the use of test runs or trap qubits, ensures that Bob's cheating will be detected with high probability.

How can we establish a "trusted computation space" without sending the qubits in the first place? You know the answer! In Chapter 9 we saw that simple tests based on the CHSH game could be used to guarantee that *two* arbitrary but *noncommunicating* players share a specific state, the EPR pair $|\text{EPR}\rangle$. Even if it is limited to a single qubit per player, this gives us a solid starting point: a test which ensures that a certain little corner of the servers' workspace behaves in a way that we can control.

Let's see how this idea can be leveraged to devise a scheme for delegated computation in which Alice is completely classical, but has access to *two* noncommunicating servers, both untrusted. To avoid confusion we'll call the servers Charlie and Dave – these are the Alice and Bob from Chapter 9, but we already have an Alice and a Bob here! This method is the most technical of the three we are presenting, and we'll remain at an intuitive level of presentation.

13.4.1 Establishing a Trusted Computation Space

In Chapter 9 we saw the CHSH rigidity theorem, which states that if Charlie and Dave successfully play the CHSH game then up to local isometries the operations they perform are equivalent to those specified in the ideal strategy for the CHSH game. Thus the CHSH game provides a simple test, to certify not only the presence of an EPR pair between the servers, but also the specific measurements that the servers perform on their respective half of the EPR pair when asked certain questions. The central idea for using this in delegated computation will be to alternate between playing the CHSH game with the servers and playing other games, some of which involve the actual computation Alice wants the servers to implement; this will be done in such a way that the servers individually can never tell whether they are being "CHSH-tested" or actually "used" to implement a useful part of the computation. Therefore the servers have to apply the honest CHSH strategy all the time, test or computation, and this gives us a way to control which operations they apply.

The first thing to deal with is that we're going to need many EPR pairs. One idea to certify n EPR pairs would be to play n CHSH games "in parallel": Alice could select n pairs of questions $(x_j, y_j)_{j=1,...,n}$ to send to Charlie and Dave, collect n pairs of answers (a_j, b_j), and check how many satisfy the CHSH condition $a_j \oplus b_j = x_j \cdot y_j$. If this estimate is close enough to the optimal $\cos^2(\pi/8) \cdot n$ she would accept the interaction. Although this is a sensible idea it is currently not known how well it works; in particular, the effect of small errors in Charlie and Dave's answers is not clear.

Instead of executing the games in parallel, Alice will perform them sequentially. That is, she sends the questions (x_j, y_j) to Charlie and Dave one pair at a time, waiting for their answer before sending the next pair of questions. After having repeated this procedure for n rounds, she counts the number of rounds in which the CHSH condition was satisfied, and accepts if and only if it is at least $(\cos^2(\pi/8) - \delta)n$, for some error threshold δ. The following sequential rigidity theorem states the consequences of this test in the idealized setting where $\delta = 0$.

Theorem 13.4.1 (Idealized) *Suppose the two servers, Charlie and Dave, successfully play n sequential CHSH games. Then, up to local isometries, their initial state is equivalent to $|EPR\rangle_{CD}^{\otimes n} \otimes |junk\rangle_{CD}$. Moreover, at each step $j \in \{1, \ldots, n\}$ the measurements performed by each server are equivalent to those of the ideal strategy for CHSH (Z and X for Charlie and H and $\tilde{H}$ for Dave) applied on the j-th EPR pair.*

You may notice that the protocol for the n sequential CHSH tests is similar to how the CHSH tests are performed in the protocol for device-independent QKD from

Chapter 9. The analysis uses similar tools: a first step uses a (martingale) concentration inequality to argue that, if a fraction about $\cos^2(\pi/8) - \delta$ of the games are won by the servers, then for most $j \in \{1, \ldots, n\}$ the a priori probability that the servers would have won in round j must be of the same order, say at least $\cos^2(\pi/8) - 2\delta$. For any such j the basic CHSH rigidity theorem can be applied to conclude that the measurements applied, and the state on which they were applied, are (up to local isometries) equivalent to the ideal CHSH strategy.

This reasoning by itself is not sufficient to imply that the servers' initial state is a tensor product of EPR pairs. Indeed, the different EPR pairs used in each round could partially "overlap," or even be the same pair! Intuitively we know this is not possible, as any measurement destroys the EPR pair, so it cannot be re-used. But this is delicate to establish rigorously, because the EPR pair need not be completely destroyed; could many "leftover EPR pairs" be combined together to make a fresh one? Nevertheless, the analysis can be done, and for the remainder of the section we will assume that a "robust" version of the "idealized" theorem above can be proven, dealing with the more realistic setting where the servers are not required to play the CHSH games strictly optimally, a far too stringent requirement for any practical application.

13.4.2 State Tomography

Now that we have a way to establish a "secure computation space," as a second step let's see how the client Alice can use that space, and additional CHSH tests, to certify that one of the servers has prepared certain single- or two-qubit states in that space.

Consider the following protocol. With Charlie, Alice behaves exactly as if she was executing the n sequential CHSH games described in the previous section. With Dave, however, she does something different: she instructs him to measure each half of the EPR pairs he is supposed to share with Charlie in a certain basis, say $\{|+_\theta\rangle, |-_\theta\rangle\}$ for some real θ (defined as in (13.2)), and to report the outcome.

Dave of course knows that something special is going on. So we have no guarantee as to what action he performs. In contrast, Charlie is told the exact same thing as in the n-sequential CHSH test. He must thus behave exactly as if this is the test Alice was performing, and Theorem 13.4.1 applies: in each round, Charlie applies the ideal CHSH measurements, in the standard or Hadamard bases, on his half of the j-th EPR pair, in a way that, if Dave had been measuring using his own CHSH measurements, they would have succeeded with near-optimal probability.

But now Dave is doing something different – we don't know what. But if Dave performs the measurement asked by Alice, and reports the right outcome, we know what should happen: Dave's half-EPR pair gets projected onto one of the basis states, $|+_\theta\rangle$ or $|-_\theta\rangle$, and by the special properties of EPR pairs so does Charlie's half. In particular, whenever Charlie performs a measurement in the Hadamard basis the average value of his outcome (considered as a value in $\{\pm 1\}$) should be $\langle +_\theta | X | +_\theta \rangle = \cos(\theta)$ or $\langle -_\theta | X | -_\theta \rangle = -\cos(\theta)$. Thus, by collecting all of Charlie's answers associated to measurements in the X basis, Alice can check whether the average outcome over the rounds in which Dave reported a $+$ is approximately $\cos(\theta)$, and $-\cos(\theta)$ over those rounds when Dave reported a $-$. Alice is using Charlie's answers to perform tomography on the state that Dave claims to have prepared, without Charlie being able to detect what is going on! (Even though Charlie knows that he *might* be currently tested, since he is aware of the structure of the protocol, there is nothing he can do about it – if he deviates he risks failing too many CHSH games, in case this is what Alice is doing.)

In the CHSH game the only measurements made by Charlie are in the computational or Hadamard bases. To perform tomography of arbitrary multi-qubit states we would also need him to sometimes apply a Pauli Y. It is possible to do this via a simple modification of the CHSH game. For our purposes the modification will not be necessary, as the set of states that are characterized by their expectation value with respect to Pauli X and Z observables (we call such states XZ-*determined*) is sufficient to implement the delegated computation protocol.

Exercise 13.4.1   Show that the family of all single-qubit states in the xz-plane of the Bloch sphere, i.e. all states of the form

$$\rho = \frac{1}{2}(\mathbb{I} + \cos(\theta)X + \sin(\theta)Z), \qquad \theta \in [0, 2\pi),$$

are XZ-determined.

Exercise 13.4.2   Show that the family of two-qubit states of the form

$$|\psi\rangle = U \otimes P |\text{EPR}\rangle,$$

for any single-qubit real unitary U and $P \in \{I, X, Y, Z\}$, are XZ-determined.

Exercise 13.4.3   Give an example of two distinct single-qubit states that have the same expectation values with respect to both X and Z observables, and are thus not XZ-determined.

13.4.3 Process Tomography

Beyond state tomography, our protocol for delegated computation will require us to implement some limited form of *process tomography*: we need to find a way to guarantee that at least one of the servers, Charlie or Dave, is performing the right computation! At first this task may appear overwhelming: While as described in the previous section it is possible to use one server to perform tomography against the other server's state, how can we test that a certain *gate* has been applied? For the case of state preparation we know what the correct states are, and as long as they are restricted to simple single- or two-qubit states we can do full state tomography. But our ultimate goal is to implement an arbitrary quantum circuit, which may generate highly entangled states of its n qubits; there is no hope of performing full tomography on such states, as it would require an exponential number of measurements.

We will sidestep the difficulty and use a model of computation that only requires the application of a very special type of gate – a measurement in the Bell basis, i.e. the simultaneous eigenbasis of $X \otimes X$ and $Z \otimes Z$, given by

$$|\psi_{00}\rangle_{AB} = \frac{1}{\sqrt{2}}(|00\rangle_{AB} + |11\rangle_{AB}), \qquad |\psi_{01}\rangle_{AB} = \frac{1}{\sqrt{2}}(|00\rangle_{AB} - |11\rangle_{AB}),$$

$$|\psi_{10}\rangle_{AB} = \frac{1}{\sqrt{2}}(|01\rangle_{AB} + |10\rangle_{AB}), \qquad |\psi_{11}\rangle_{AB} = \frac{1}{\sqrt{2}}(|01\rangle_{AB} - |10\rangle_{AB}),$$

where $|\psi_{00}\rangle = |\text{EPR}\rangle$ is the familiar EPR pair. This model of computation is called *teleportation-based computation* (recall that a measurement in the Bell basis is precisely the operation required of the sender in the teleportation protocol), and we'll review it in the next section. But let's already see how it can be used for delegated computation.

Similar to the previous section, suppose Alice instructs Dave to measure his n qubits in the Bell basis, where the qubits are paired in an arbitrary way chosen by Alice (so she tells Dave the whole set of measurements to be performed at the outset). Of course, as usual Dave does what he wants – he may not even have n qubits in the first place. But Alice also instructs Charlie to play sequential CHSH games, so that from his point of view the protocol is perfectly indistinguishable from the tests. Once Alice has collected all of Charlie and Dave's outcomes, she groups Charlie's outcomes when they are associated to the same state, and uses them to check that Dave did not lie. For example, if Dave reports $|\psi_{00}\rangle$ then whenever Charlie measured the two corresponding qubits using the same basis, computational or Hadamard, his two outcomes should be the same. (Note that not all Charlie's measurements are useful, as it will sometimes be the case that the qubits were measured in different bases, in which case there is no useful test Alice can perform – she simply discards those rounds.)

The following exercise asks you to make this argument more formal.

Exercise 13.4.4 Suppose that Charlie and Dave share two EPR pairs, $|\text{EPR}\rangle_{C_1D_1} \otimes |\text{EPR}\rangle_{C_2D_2}$. Dave measures his two halves, D_1D_2, using an arbitrary four-outcome POVM, obtaining a result $(d_1, d_2) \in \{0,1\}^2$. Charlie measures each of C_1 and C_2 using observables $O_1, O_2 \in \{X, Z\}$ chosen uniformly at random.

Suppose that if $(O_1, O_2) = (X, X)$ then Charlie's outcomes (as values in $\{\pm 1\}$) satisfy $c_1c_2 = a$, and if $(O_1, O_2) = (Z, Z)$ they satisfy $c_1c_2 = d$, for some fixed values $a, d \in \{\pm 1\}$ (i.e. imagine the same experiment is repeated many times, and Charlie's outcomes consistently satisfy these equations, for the same values of a and d). Show that Dave must have been implementing a measurement in the Bell basis. Which Bell state is associated to each of the four possible values for (a, d)?

The exercise shows that, provided we can trust that Charlie and Dave indeed share EPR pairs, and Charlie's measurements are made in the computational or the Hadamard basis, then Alice has a way to verify that Dave has been implementing a Bell basis measurement on certain pre-specified pairs of qubits. Just as for the case of state tomography, these assumptions are guaranteed by the fact that Charlie cannot tell the difference between when Alice is executing the process tomography protocol described here and when she is executing sequential CHSH games.

13.4.4 Teleportation-Based Computation

The final ingredient needed for our delegation protocol is a method of computation adapted to the kinds of operations we are able to certify: preparation of EPR pairs and single- or two-qubit XZ-determined states (Exercise 13.4.2), and measurements of pairs of qubits in the Bell basis (Exercise 13.4.4).

Computation by teleportation is a model of computation which does just that. The main idea is that a gate can be applied to a qubit by "teleporting the qubit into the gate." The following exercise fleshes out the main gadget used in computation by teleportation.

Exercise 13.4.5 Let $|\psi\rangle_A$ be an arbitrary single-qubit state and let $|\phi\rangle_{BC} = (I \otimes UP|\text{EPR}\rangle)$, where U is an arbitrary single-qubit unitary and $P \in \{I, X, Y, Z\}$. Suppose a measurement of qubits A and B is performed in the Bell basis, yielding a pair of outcomes $(b_1, b_2) \in \{0, 1\}^2$. Show that there exists a Pauli operator Q (depending only on (b_1, b_2)) such that the post-measurement state of the qubit in C is $(UQPU^\dagger)U|\psi\rangle$.

The idea is then the following. Suppose that Alice wishes to implement an arbitrary computation on n qubits, specified by a circuit $\mathcal{C}$ using the universal gate set $\mathcal{G} = \{\text{CNOT}, G\}$ introduced in Section 13.1.1. Assume for simplicity the input to the circuit is $|0\rangle^{\otimes n}$; this is without loss of generality since the input can always be hardcoded into the circuit by using X gates where appropriate. Alice initializes her workspace with a large number of "magic states" from the set

$$\left\{ |0\rangle, (I \otimes H)|\text{EPR}\rangle, (I \otimes G)|\text{EPR}\rangle, \text{CNOT}_{B_1 B_2}(|\text{EPR}\rangle_{A_1 B_1}|\text{EPR}\rangle_{A_2 B_2}) \right\}. \tag{13.4}$$

At each stage of the computation Alice keeps track of a special set of n qubits that represent the current state of the circuit. We can label these as $A_1 \cdots A_n$, even though they will change over time. Initially $A_1 \cdots A_n$ point to any n of the "magic" $|0\rangle$ qubits she has prepared in her workspace.

Now suppose Alice would like to apply a gate to one of her qubits A_j, for example a G gate. Then she can perform the circuit described in Exercise 13.4.5, where the role of A is played by A_j, and the roles of B and C by one of her "magic" $(I \otimes G)|\text{EPR}\rangle$. As a result the state of C is projected to $(GQG^\dagger)G|\psi\rangle_C$, where initially A_j is in state $|\psi\rangle$ (the same computation would work for mixed states as well). This is the operation Alice wanted to perform, except for the correction $GQG^\dagger$. How do we deal with this?

Depending on Q, $GQG^\dagger$ will amount to a Pauli correction, possibly multiplied by a Hadamard: $GXG^\dagger = iHY$, $GYG^\dagger = Y$, and $GZG^\dagger = H$. By now we are used to Pauli corrections: Alice can keep track of these as a form of one-time pad that is tagged along the whole computation. The Hadamard gate is a little more annoying, but in fact it can be easily corrected using one more step of "teleportation," this time using a "magic" $(I \otimes H)|\text{EPR}\rangle$. This will induce yet another correction $HQ'H^\dagger$, but this time whatever Q' is the result is a Pauli correction that Alice can again tag along as part of the one-time pad.

Thus, aside from the preparation of the magic states, the whole computation boils down to a simple sequence of Bell basis measurements. Note however that, due to the necessity of performing Hadamard corrections in an unpredictable way (as it depends on measurement outcomes obtained when teleporting into a G gate), this sequence is adaptive. This is similar to the scenario of MBQC, but it will require us to proceed with a little extra care in the final delegation protocol.

QUIZ 13.4.1 *True or false? Consider the following circuit on two qubits. Qubit 1 is in state* $|\psi\rangle$. *Qubits 2 and 3 are prepared in the state* $(I \otimes T)|EPR\rangle$. *Perform a Bell measurement on Qubits 1 and 2. If the two-bit outcome is* a, b *corresponding to eigenvector* $|\psi_{a,b}\rangle = I \otimes X^a Z^b |EPR\rangle$, *apply the unitary* $X^b Z^a$ *to the third qubit. Then the state of the third qubit after the measurement is* $T|\psi\rangle$.

QUIZ 13.4.2 *True or false? Same setup as in Quiz 13.4.1, with the only difference that qubits 2 and 3 are initially prepared in the state* $(I \otimes H)|EPR\rangle$. *Then the state of the third qubit after the measurement is* $H|\psi\rangle$.

QUIZ 13.4.3 *Suppose Alice and Bob share two EPR pairs. Bob then applies an H gate on the second of his two qubits. He then performs a Bell measurement on his two qubits. What is the resulting post-measurement state on Alice's side, up to Pauli corrections?*

(a) $(\mathbb{I} \otimes H)|EPR\rangle$
(b) $|+\rangle \otimes H|+\rangle$
(c) $(CNOT_{1 \to 2})|EPR\rangle$
(d) $|EPR\rangle$

QUIZ 13.4.4 *True or false? Suppose Alice and Bob share two EPR pairs. Bob then applies a G gate on the second of his two qubits. Recall that* $G = \begin{pmatrix} \cos(\frac{\pi}{8}) & -\sin(\frac{\pi}{8}) \\ \sin(\frac{\pi}{8}) & \cos(\frac{\pi}{8}) \end{pmatrix}$. *He then performs a Bell measurement on his two qubits. The resulting post-measurement state on Alice's side, up to Pauli corrections, is* $(I \otimes G)|EPR\rangle$.

13.4.5 Blind and Verifiable Delegated Computation

With all the ingredients in place we are ready to describe a delegation protocol for the classical client Alice to delegate the computation of an n-qubit quantum circuit $\mathcal{C}$, specified using the universal gate set $\{CNOT, G\}$, to two isolated quantum servers, Charlie and Dave, sharing entanglement.

Protocol 14 Alice compiles a description of $\mathcal{C}$ in the teleportation-based model. Let $t = \text{poly}(|\mathcal{C}|)$ be chosen large enough, and $\delta = \text{poly}^{-1}(|\mathcal{C}|)$ small enough. Alice performs one of the following four tests (the first three with equal probability, and the last with smaller probability depending on the size of $\mathcal{C}$):

1. She plays t sequential CHSH games with the servers. If the fraction of successes is below $\cos^2(\pi/8) - \delta$ she rejects.
2. Alice executes a state tomography protocol in which Dave is asked to perform measurements that correspond to $t/9$ copies of each of the four magic states in (13.4). Charlie is instructed to play t CHSH games. Charlie's measurement outcomes are used to check the results reported by Dave, as described in Section 13.4.2.
3. Alice executes a process tomography protocol in which Charlie is instructed to perform Bell basis measurements according to the pattern that corresponds to the teleportation-based circuit associated with $\mathcal{C}$. Charlie is instructed to play t CHSH games. Dave's measurement outcomes are used to verify that Charlie is reporting the correct outcomes for his Bell measurements, as described in Section 13.4.3.
4. Alice orchestrates the computation of $\mathcal{C}$: she instructs Dave to prepare the magic states as in step 2, and Charlie to perform Bell basis measurements as in step 3. She uses the measurement outcome reported by Charlie for the output qubit as the output of the computation (after having applied any required Pauli corrections).

The main reason this protocol works is that Charlie cannot distinguish a computation run from a process tomography run, or a CHSH run from a state tomography run; similarly, Dave cannot distinguish between a computation run and a state tomography run, or between a CHSH run and a process tomography run. The protocol can be made blind: blind to Charlie, who without knowledge of which magic states his Bell basis measurements are performed on gains no useful information from the pattern of Bell measurements Alice instructs him to perform (the pattern can be made independent of the circuit $\mathcal{C}$, aside from its size); blind to Dave, who prepares magic states in a way that is completely independent of the computation. Verifiability follows directly from the tests performed in cases 1, 2, and 3 of the protocol.

There is one difficulty we hinted at earlier and we have glossed over so far. This is the fact that, after application of a Bell basis measurement corresponding to teleportation into a G gate, Alice needs to make an adaptive choice: either apply an H correction, or not. However, Charlie should be ignorant of this choice, as otherwise the protocol would no longer be blind. The solution is to switch the focus over to Dave. Charlie will always be asked to perform the same pattern of Bell basis measurements, but Dave will be (adaptively) asked by Alice to create certain magic states as $|\text{EPR}\rangle$, and others as $(I \otimes H)|\text{EPR}\rangle$, as a function of the outcomes reported by Charlie. Since these outcomes are uniformly distributed, the pattern of state preparation requests Dave sees is still random, so he does not gain any information about the computation either. (Note, however, that a third observer able to eavesdrop on the messages exchanged with both Bob and Dave would learn valuable information about the computation; however, such an attack falls outside of the scope of the security definition of delegated computation.)

Only one task remains: performing a soundness analysis of the protocol! Given that it is not possible to require that the servers *exactly* pass all the tests, some error should be tolerated. How does this error affect the quality and trustworthiness of the computation? This is quite delicate. The best analysis known to-date makes this protocol, compared to the ones we saw in the previous two sections, highly inefficient, as it requires T to be a very large power of n before even relatively weak security guarantees can be obtained. Nevertheless, it is the only protocol known for purely classical delegated computation, and improving it is an important research problem.

CHAPTER NOTES

In our definition we stated the requirements of verifiability and blindness rather informally. A precise definition satisfying all the desired properties (universal composability in particular) would take many pages. Such a definition was given using the framework of *abstract cryptography* in a paper by V. Dunjko, et al. (Composable security of delegated quantum computation. In *International Conference on the Theory and Application of Cryptology and Information Security*, pp. 406–425. Springer, 2014).

There are many works on delegated quantum computation. Already in 2001, A. M. Childs (Secure assisted quantum computation. *Quantum Information and Computation*, **456**(5), 2005) provided a protocol that allowed the blind delegation of quantum computations to a server, provided the client has a quantum memory and the ability to implement Pauli gates. Futher protocols in the circuit model, which substantially weakened the requirements on the client and introduced verifiability, were given by D. Aharonov, et al. (Interactive proofs for quantum computations. arXiv:0810.5375, 2008; Interactive proofs for quantum computations. arXiv:1704.04487, 2017). The protocol that we present here is due to A. Broadbent (How to verify a quantum computation. *Theory of Computing*, **14**(1):1–37, 2018). Delegated quantum computation in the measurement-based model was put forward by A. Broadbent, J. Fitzsimons, and E. Kashefi (Universal blind quantum computation. In *50th Annual IEEE Symposium on Foundations of Computer Science*, pp. 517–526. IEEE, 2009) and J. Fitzsimons and E. Kashefi (Unconditionally verifiable blind quantum computation. *Physical Review A*, **96**(1):012303, 2017). For some lecture notes on measurement-based quantum computation itself, we refer to D. E. Browne and H. J. Briegel (One-way quantum computation: A tutorial introduction. quant-ph/0603226, 2006). The delegated computation protocol in the two-server model presented in this chapter is adapted from the protocol by B. W. Reichardt, et al. (Classical command of quantum systems. *Nature*, **496**(7446):456–460, 2013). A different class of delegation protocols, which we did not describe here, operate in the so-called "receive-and-measure" model. In this type of protocol it is the server, Bob, who sends single qubits to Alice, and Alice measures them. See, for example, the work of J. F. Fitzsimons, M. Hajdušek, and T. Morimae (Post hoc verification of quantum computation. *Physical Review Letters*, **120**(4):040501, 2018).

A survey on protocols may be found in the paper by J. F. Fitzsimons (Private quantum computation: An intoduction to blind quantum computing and related protocols. *NPJ Quantum Information*, **3**(23), 2017). For work focusing on optimizing delegated quantum computation protocols for real-world implementations, we refer to the work of T. Kapourniotis, et al. (Unifying quantum verification and error-detection: Theory and tools for optimisations. arXiv:2206.00631, 2022).

PROBLEMS

13.1 Concentration bounds

Consider the following toy setup. Alice and Bob perform a certain physical experiment with their quantum devices. They repeat the same experiment N times in a row, and each time they observe an outcome $Z_i \in \{0, 1\}$. Think of Z_i as representing "success" of the experiment. For example, they play the CHSH game with their devices and note $Z_i = 1$ whenever the CHSH test is passed, $Z_i = 0$ when it fails.

Let ω^* be the maximum probability with which any two quantum devices may produce outcomes that result in a setting $Z_i = 1$. For example, in the case of CHSH $\omega^* = \cos^2 \frac{\pi}{8}$.

Suppose that Alice and Bob perform the experiment N times, but only check the result for half of the experiments they perform, chosen at random. That is, they choose a random $S \subseteq \{1, \ldots, N\}$ uniformly at random such that $|S| = N/2$, and evaluate $\omega_{\text{est}} = \frac{2}{N} \sum_{i \in S} Z_i$. The goal of this problem is to determine when useful consequences of this estimate can be derived on the outcomes of rounds that were not explicitly tested, $i \notin S$.

Our main tool is the following concentration inequality, which we already introduced in Chapter 8. Let $N = n + k$ and consider binary random variables $Z_1, \ldots, Z_N$ (the Z_i may be arbitrarily correlated). Let S be a uniformly random subset of $\{1, \ldots, N\}$ of size k. Then for any $\delta, \nu > 0$,

$$\Pr\left(\sum_{j \in S} Z_j \geq \delta k \wedge \sum_{j \in \{1, \ldots N\} \setminus S} Z_j \leq (\delta - \nu)n \right) \leq e^{-2\nu^2 \frac{nk^2}{(n+k)(k+1)}}.$$

1. Suppose Alice and Bob observe that $\omega_{\text{est}} = \omega^* - \varepsilon$, for some quantity $\varepsilon > 0$ (which may depend on n). Suppose they would like to conclude that, if they randomly choose an $i \in \{1, \ldots, N\} \setminus S$, then the probability that the corresponding Z_i is such that $\Pr(Z_i = 1) \leq \omega^* - 2\varepsilon$ is very small – at most $2^{-N/100}$ (as long as N is large enough). Among the following possibilities, which ones let them reach this conclusion, based on the above concentration inequality?

   (a) $\varepsilon = 2^{-N}$
   (b) $\varepsilon = \frac{1}{\sqrt{N}}$
   (c) $\varepsilon = 0.01$

2. Let $\mathcal{E}$ be the event that Alice and Bob observe $\omega_{\text{est}} = \omega^* - \varepsilon$. Suppose they would like to conclude that, whenever this event happens, then also $\frac{2}{N} \sum_{i \notin S} Z_i \geq \omega^* - 2\varepsilon$, at least with probability $1 - e^{-N/200}$. Among the following possibilities, select those that will let them reach this conclusion, based on the above concentration inequality.

   (a) $\varepsilon = 2^{-N}$
   (b) $\varepsilon = \frac{1}{\sqrt{N}}$
   (c) $\varepsilon = 0.01$

3. Now assume also that $\Pr(\mathcal{E}) \geq e^{-N/500}$. In the same scenario as described in the previous question, but under this additional assumption, select the valid value(s) for ε. Recall that the size of the set S that Alice inspects is $\frac{N}{2}$.

   (a) $\varepsilon = 2^{-N}$
   (b) $\varepsilon = \frac{1}{\sqrt{N}}$
   (c) $\varepsilon = 0.02$

Index